Foundations of Generic Optimization

MATHEMATICAL MODELLING:
Theory and Applications

VOLUME 20

This series is aimed at publishing work dealing with the definition, development and application of fundamental theory and methodology, computational and algorithmic implementations and comprehensive empirical studies in mathematical modelling. Work on new mathematics inspired by the construction of mathematical models, combining theory and experiment and furthering the understanding of the systems being modelled are particularly welcomed.

Manuscripts to be considered for publication lie within the following, non-exhaustive list of areas: mathematical modelling in engineering, industrial mathematics, control theory, operations research, decision theory, economic modelling, mathematical programmering, mathematical system theory, geophysical sciences, climate modelling, environmental processes, mathematical modelling in psychology, political science, sociology and behavioural sciences, mathematical biology, mathematical ecology, image processing, computer vision, artificial intelligence, fuzzy systems, and approximate reasoning, genetic algorithms, neural networks, expert systems, pattern recognition, clustering, chaos and fractals.

Original monographs, comprehensive surveys as well as edited collections will be considered for publication.

The titles published in this series are listed at the end of this volume.

Foundations of Generic Optimization

Volume 1: A Combinatorial Approach to Epistasis

by

M. Iglesias
Universidade da Coruña,
A Coruña, Spain

B. Naudts
Universiteit Antwerpen,
Antwerpen, Belgium

A. Verschoren
Universiteit Antwerpen,
Antwerpen, Belgium

and

C. Vidal
Universidade da Coruña,
A Coruña, Spain

edited by

R. Lowen and A. Verschoren
Universiteit Antwerpen,
Antwerpen, Belgium

Springer

A C.I.P. Catalogue record for this book is available from the Library of Congress.

ISBN-13 978-90-481-6922-1 (PB)
ISBN-13 978-1-4020-3665-1 (e-book)

Published by Springer,
P.O. Box 17, 3300 AA Dordrecht, The Netherlands.

www.springeronline.com

Printed on acid-free paper

Printed in the Netherlands.

Do or do not – there is no try

(Yoda, The Empire Strikes Back)

Preface

This book deals with combinatorial aspects of epistasis, a notion that existed for years in genetics and appeared in the field of evolutionary algorithms in the early 1990s. Even though the first chapter puts epistasis in the perspective of evolutionary algorithms and artificial intelligence, and applications occasionally pop up in other chapters, this book is essentially about mathematics, about combinatorial techniques to compute in an efficient and mathematically elegant way what will be defined as normalized epistasis. Some of the material in this book finds its origin in the PhD theses of Hugo Van Hove [97] and Dominique Suys [95]. The sixth chapter also contains material that appeared in the dissertation of Luk Schoofs [84]. Together with that of M. Teresa Iglesias [36], these dissertations form the backbone of a decade of mathematical ventures in the world of epistasis.

The authors wish to acknowledge support from the Flemish Fund of Scientific research (FWO-Vlaanderen) and of the Xunta de Galicia. They also wish to explicitly mention the intellectual and moral support they received throughout the preparation of this work from their family and their colleagues Emilio Villanueva, Jose María Barja and Arnold Beckelheimer, as well as our local TeXpert Jan Adriaenssens.

Contents

Chapter O

Genetic algorithms: a guide for absolute beginners

In this preliminary chapter, we will describe in an intuitive way what genetic algorithms are about, referring to the literature (and the rest of this book) for details. This chapter (which, at some point, we intended to call "Genetic algorithms for dummies") is written and included for readers for whom the term "genetic algorithm" is completely new. Readers with some basic background may skip it and start immediately with Chapter I.

Every day, one is almost continuously confronted with questions of the type: "What is the best way to ...?", "What is the shortest way to go to ...?", "What is the cheapest ...?". All of these questions are examples of so-called *optimization problems*, i.e., one is given a set of data, of possible solutions of a given problem, and one is asked to find the "best" solution within this "search space". In order to make sense, there should, of course, be some way to measure this idea of "best": every item in the search space, every possible solution to the given problem should be given a value, and one should be looking for elements in the search space for which this value is maximal (or minimal, depending on the problem).

Formally, one may thus think of an optimization problem to be represented as follows. First, one is given a set Ω of possible solutions, of data to be optimized. This set Ω may be very general, finite or infinite, but in general it consists of numbers, of vectors, of paths from one city to another, graphs, or whatever type of object one wants to study. Next, there is given some function f which associates with every object in Ω a value, which expresses its quality with respect to the problem one wishes to solve. This value could be the price of some product, the distance covered by traveling from one city to another, ... Although any set of values could do, in practice one prefers to work with real values, i.e., we work with a function $f : \Omega \to \mathbb{R}$. The function f is usually referred to as *fitness function* or *objective function.*

The associated optimization problem may then be formulated as follows: find the element(s) $s \in \Omega$, for which $f(s)$ is minimal (or maximal).

Let us already point out here that there is no real restriction in limiting ourselves to maxima: if we define $g : \Omega \to \mathbb{R}$ by letting $g(s) = -f(s)$ for every $s \in \Omega$, then, clearly, f reaches its minimal values exactly where g reaches its maximal values. Finding the minimum for f is thus just the same as finding the maximum for g. Moreover, for practical reasons, one usually assumes the fitness function to only have positive values – if necessary, one may always add a constant to realize this.

So, how does one proceed to find a maximum (or minimum) for $f : \Omega \to \mathbb{R}$? If Ω is a subset of n-dimensional real space $\mathbb{R}^n$, high school mathematics is very clear about this: just try and find $s \in \Omega$ such that

$$\frac{\partial f}{\partial x_1}(s) = \ldots = \frac{\partial f}{\partial x_n}(s) = 0 \ (*)$$

Well, of course, one has to impose some conditions on Ω, e.g., Ω has to be an open subset of the space $\mathbb{R}^n$. But that is not the real point – just do not believe everything your maths teacher taught you:

1. calculating partial derivatives looks fine, but has it ever occurred to you that most functions one may want to optimize in real life do not have derivatives? That they are usually even not continuous? And that the search space is almost always discrete or finite?

2. and even if the partial derivatives exist, the resulting equations (*) will probably look ugly, if not horrible, i.e., they may be extremely hard to solve, even

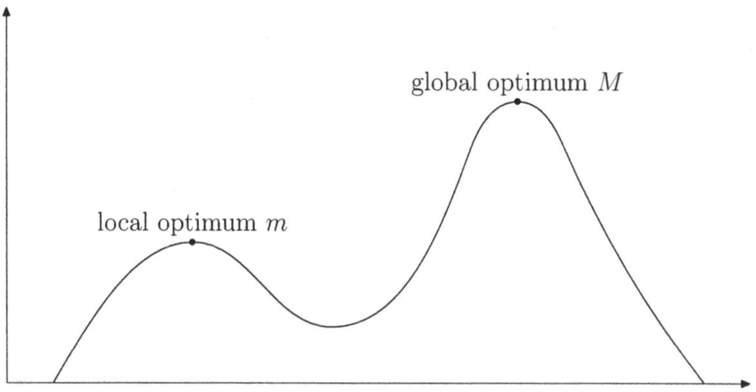

Figure O.1: A local and a global optimum.

numerically. (For example, in one variable, just look at the function $f(x) = x^2 + \exp(\cos x) + x$, which leads to the equation $2x - \sin x \exp(\cos x) + 1 = 0$).

3. and even when (if!) we find a solution, are we certain that we are not stuck in a local optimum? (See figure O.1.)

Fortunately, there are alternative search methods to find the optimum, e.g., so-called *gradient methods.* One of these is what one uses to refer to as *hill-climbing.* Roughly speaking, this method starts from a random point in search space and iteratively moves to points with a higher fitness value, or in the steepest direction in the neighborhood of this point, until one reaches an optimal value. But again, how can we be certain that we do not get stuck in a local optimum? (See figure O.2.)

Of course, in real life, our search space is, indeed, always finite (albeit sometimes very big), so that we definitely need other methods.

For small search spaces, we might try an exhaustive search. For small, really small spaces, this clearly works, but again, unless we do restrict to "toy problems", this approach definitely does not work.

Let us give an example. The so-called Traveling Salesman Problem is a classic in optimization theory and may be formulated as follows. Given a set of N cities and their mutual distances, starting from a fixed city, try and find a way to visit each of

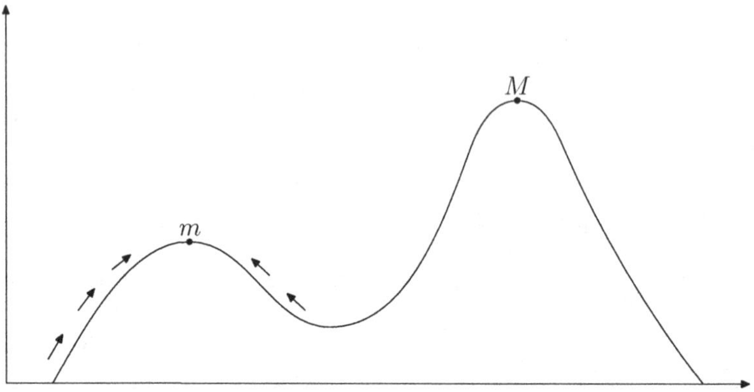

Figure O.2: Stuck in a local optimum.

these cities once and such that the total distance covered is minimal.

If we do this for $N = 5$ cities, we have to compare $4! = 24$ different circuits, and it is easy to find the shortest one amongst these. But who is interested in only 5 cities? If we look at a somewhat less silly situation, say $N = 100$ cities to be visited, exhaustive search leads to comparing $99! \approx 10^{156}$ travel lengths. Taking into account that the total number of atoms in our universe is of the order of 10^{78}, it should be obvious that no computer will ever be able to solve the problem of finding the shortest circuit in this case.

Fortunately, for this type of problem, excellent heuristics have been developed, leading to excellent sub-optimal solutions in a reasonable time.

Random search then? If people keep buying lottery tickets (and even win, now and then), why not try this approach in optimization? Why not pick sample points and check whether one is lucky? Again, it is obvious that this will not work, unless one works with very small search spaces or if one is willing to wait for a long, long time before finding a reasonable solution. And even then, how can one be certain to be even close to an optimum or local optimum?

What appears to be the case is that for specific, individual problems one may be given an algorithm, frequently highly problem-dependent, that leads to reasonable solutions. Moreover, for really hard problems, it appears that a probabilistic ap-

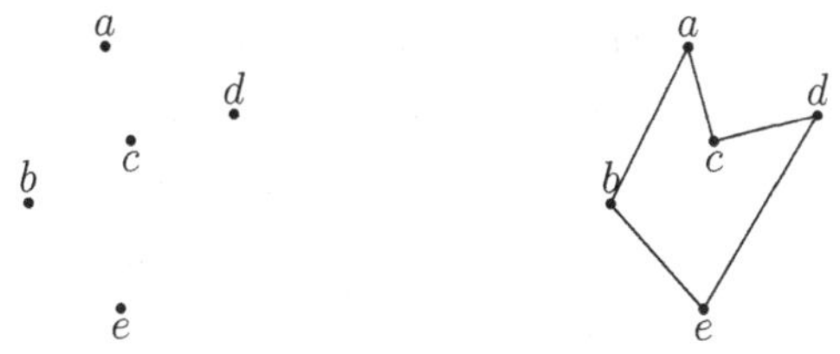

Figure O.3: Five cities, and one possible way of visiting each city once.

proach is sometimes very useful. By "probabilistic", we, of course, do not mean "random search" (as indicated above), but rather a guided random search, i.e., a deterministic search algorithm, that uses some aspects of randomization in its initialization or in directing its search path.

Several "universal" algorithms of this type have been developed and studied during the last decades, amongst them simulated annealing (which, by the way, yields nice results for the traveling salesman problem) and the so-called genetic algorithm(s), which will be studied extensively below.

Genetic algorithms are inspired by nature, by evolution and Mendel's ideas about this.

The underlying idea is extremely simple. Let us consider a population P of prey, with characteristics making them more or less likely to be eaten by predators surrounding them. These characteristics may involve speed, mimicry or even intelligence. Let us suppose that we can describe these features that permits an individual p to survive by some "fitness function" $f : P \to \mathbb{R}$, i.e., the higher the value of $f(p)$, the higher the probability of survival of $p \in P$. The population P is, of course, not static, it evolves in time: some prey is eaten, there is some breeding, ... For obvious reasons, one expects the prey with high fitness $f(p)$ to eventually dominate the population P: individually they have more chances of surviving (and thus of breeding!) and one may expect that strong, fit parents (with high f!) produce strong offspring. Of course, this is just theory: some weak animals (with low f!) may survive by chance and offspring of strong parents could still be relatively weak. Moreover, there is also the dynamics of mutation: if no new genetic material is thrown into the pool, the flock will tend to stabilize and not improve anymore. Also, maybe the characteristics that made the individuals fit for survival is superseded by another, new feature, which has even stronger effect (intelligence over speed, e.g.).

On the average, it appears that either the prey becomes extinct (if hardly any individuals were strong, quick or smart enough) or tends to increase its overall fitness.

Well, this is exactly how a basic genetic algorithm works. Of course, we will not consider a herd of prey, but population P within a search space Ω and instead of measuring the fitness, the aptness to survive of individual prey, we will work with some fitness function $f : \Omega \to \mathbb{R}$, which may be applied to any member of Ω, hence of P. Note that repetitions will be allowed in P, which makes it different from an ordinary subset of Ω (whence the terminology "population" or "multiset", instead of just "subset"). This is related to the fact that we are not just interested in finding optimal or, at least, good solutions with respect to f, but rather also the structure of these solutions, the reason why they produce high values for f – see also below, when we talk about schemas. In fact, instead of working with elements in an arbitrary search space Ω, we will usually codify these elements as binary strings[1] $s = s_{\ell-1} \ldots s_0$ of fixed length ℓ, say, in order to be able to manipulate them in a uniform way. Moreover, encoding data by binary strings is not so unnatural: in real life, many kinds of data are encoded this way – just think of the number of enquiries which ask you to answer a variety of questions just by "yes" or "no".

So, let us assume we are given a function

$$f : \Omega_\ell = \{0,1\}^\ell \to \mathbb{R},$$

which we want to optimize. Maybe we should stress that "being given" this function f means that we are able to calculate the value $f(s)$ for each string $s \in \Omega$. This is not the same thing as being given, initially, all of the values $f(s)$, for every $s \in \Omega$. As an example, in the Traveling Salesman Problem, we are perfectly able to calculate

[1]Note that if we consider strings of length ℓ, we identify Ω with the set $\{0,1\}^\ell$ and, hence, we silently assume that our search space has cardinality 2^ℓ; of course, in practice, the number of elements of Ω is not necessarily a power of 2; several methods have been developed to remedy this – we refer to the literature for details.

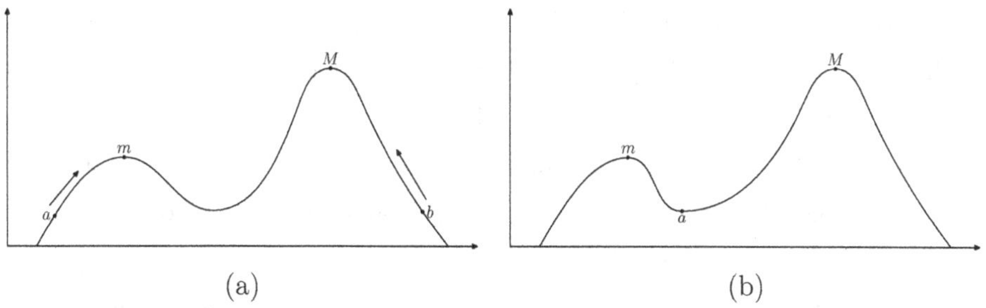

Figure O.4: (a) If one starts in a, one reaches the local maximum m, starting from b one reaches the real maximum M. (b) Starting in a, moving in the direction of the local maximum m is much steeper – the real maximum M will not be reached.

the length of each circuit (just adding distances), but we are, of course, not given the whole set of lengths (otherwise, there would be no optimization "problem"; finding the optimum would just amount to looking at a (large!) list of values and just picking the best one!).

As already indicated, one usually tries to tackle the problem of optimizing $f : \Omega \to \mathbb{R}$ by gradient methods or, somewhat easier and more straightforward, by hill-climbing. This method essentially reduces to starting from an arbitrary point and always moving in the direction of the "best" neighboring point. Of course, sometimes this works, sometimes it does not – much depends on the starting point and the geography of f, as shown in figure O.4.

To remedy this, one might try and consider a whole group of hill-climbers, starting at different points. But then again, unless the hill-climbers interact in some way, exchanging information, one cannot be certain whether, on the average, they will move in the right direction: if some of them are pertinently moving in the wrong way, they should be retrieved and switch to "another hill", in order to help the others.

In order to attain this information exchange, one proceeds as follows. First, we start from a random population $P(0)$ of fixed size $N < 2^\ell$ of possible candidate solutions (repetitions are allowed!). These strings are chosen randomly – their number N is a fixed quantity controlled by the user. For each $s \in P(0)$, we may calculate the value

$f(s)$ and the idea is to use the different values $f(s)$ for $s \in P(0)$ to help increase the overall or average value of the population. Note again that we only have to calculate $f(s)$ for at most N values (recall that P admits repetitions).

We view the strings in the population as prey or, better, as genetic material or chromosomes describing them, measuring their overall fitness with respect to survival resp. the problem we wish to solve.

Just as in genetics, reproduction involves combining chromosomes, exchanging genetic material and applying changes, e.g., through crossover or mutation.

So, let us mimic these operations in our present context.

The first operator, *selection* or *reproduction*, essentially just picks two parents to produce offspring. Of course, if we wish strong offspring, we should better pick good parents, i.e., strings with high fitness. One might thus be tempted to restrict choices to strings in $P(0)$ with maximal fitness. This is a rather bad idea, however. Just like what happens in nature, an individual may be very fast, but just too stupid to run. Combining this with an intelligent, but unfortunately very slow partner, may still lead to offspring which is both fast and bright (or, of course, slow and stupid – but these tend to disappear anyway, remember the hunter/prey model). To allow these "accidental" good strings to be produced, we will include some probabilistic dynamics.

For each string $s_i \in P(0)$, the probability p_i of being selected as a parent will be put, e.g., proportional to its fitness (several variants are possible!). Hence, if $f_i = f(s_i)$, this probability is $f_i / \sum_{s_i \in P(0)} f_i$. One may simulate this through the so-called *roulette* or *casino* model. We assign to each s_i a sector of the roulette wheel, with size proportional to its fitness. (See figure O.5.)

We then spin the wheel around and pick the string corresponding to the place where the ball stops. Since the "good" strings correspond to large sectors of the wheel, these have a higher probability of being picked than their "bad" counterparts, corresponding to smaller sectors. But then again, one never knows – and this is good for the dynamics of the system.

So, once we picked two parents, what do we do? Well, reproduce, of course. This works as follows.

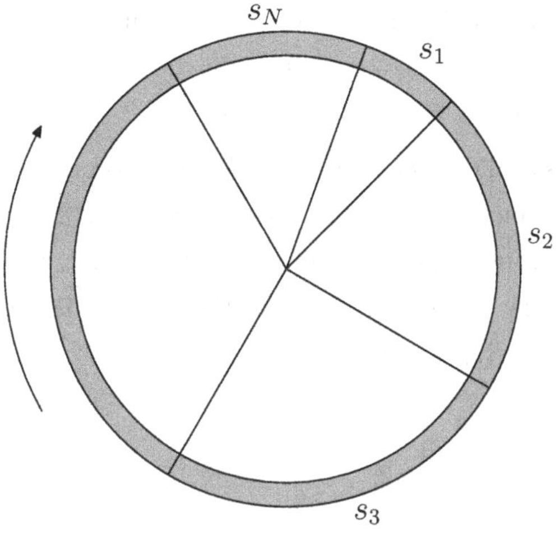

Figure O.5: Roulette wheel selection.

First, we apply *crossover*. Assume we picked two strings

$$s = s_{\ell-1} \dots s_0$$
$$t = t_{\ell-1} \dots t_0,$$

then we first randomly pick a crossover site $0 < i < \ell - 1$ and we exchange heads and tails at this site, to obtain

$$s' = s_{\ell-1} \dots s_i t_{i+1} \dots t_0$$
$$t' = t_{\ell-1} \dots t_i s_{i+1} \dots s_0.$$

We then replace the original parents s, t by the offspring s', t'. Repeating this for each selected pair of parents, we thus replace the original population $P(0)$ by a new one (of the same size).

However, just as in genetics, crossover does not always occur when we pick two strings. In practice, we will thus only apply crossover with a fixed probability, say $p_c = 0.3$, for example.

The second operator one usually applies is *mutation*. Exactly as happens in real life, where mutation occasionally changes the genetic contents of individual genes, we will apply something similar to the bits of which our strings are composed. What we essentially will do is to change bits with value 0 to 1 and with value 1 to 0, with a low probablity p_m, say $p_m = 0.01$, for example.

And that's it! The so-called *simple genetic algorithm* proceeds exactly the way we just described: start from a random generation of strings $P(0)$, select pairs of strings through the roulette principle, apply crossover and mutation, and repeat this process until one obtains a new populaton $P(1)$; we then iterate this procedure to obtain successive populations $P(t)$, $t \geq 0$.

Somewhat more formally:

```
procedure: genetic algorithm
   begin
   t <-- 0
   initialize P(t)
   evaluate P(t)
   while (not-termination condition) do
      t <-- t+1
      select P(t-1)' from P(t-1)
      apply crossover on P(t-1)'
      apply mutation on P(t-1)'
      P(t) <-- P'(t-1)
   end
end
```

What one hopes to obtain through this process are populations $P(t)$ whose average fitness increases in time, i.e., we want each generation to contain more and more strings with fitness converging to the optimum of the function we are studying.

Before including an easy example, let us point out that this is just a particular instance of a genetic algorithm – there is a huge folklore in the field of GAs, as they are usually referred to, involving several types of sometimes rather exotic operators. In this introduction, we will restrict ourselves to the "simple GA" described above.

So, let us give an example to show how the GA acts in practice. Consider the space Ω_5 of strings of length 5, which we identify with the set $\{0, 1, \ldots, 31\}$ in the obvious way, i.e., $0 \leftrightarrow 00000$, $1 \leftrightarrow 00001$, ..., $31 \leftrightarrow 11111$. We wish to optimize the function

$$f : \Omega_5 \to \mathbb{R} : x \mapsto x^2.$$

(Yes, the authors are aware of the fact that the maximum of f is reached at $f(31) = 961$!)

We start with the following initial random population $P(0)$, where for each $s \in P(0)$ we also include the corresponding fitness value $f(s)$:

$P(0)$	f	$P(0)$	f
01011	121	10100	400
00001	1	00100	16
00111	49	11100	784
11110	900	11010	676
10101	441	10011	361

Note that the maximum value obtained within $P(0)$ is 900 and that the average is 375.

We construct the next generation $P(1)$ by first selecting an intermediate population $P'(0)$ through the roulette model and then applying successively crossover for each pair of selected parents and mutation to their offspring. In the table below we indicate by | the randomly selected crossover site and by underscore the bits where mutation has been applied:

$P'(0)$	$P(1)$	f
111|10	11101	841
101|01	10110	484
11|100	11111	961
00|111	00100	16
10|011	10110	484
11|110	11011	761
111|00	11110	900
110|10	11001	625
0|1011	01110	296
1|1110	11011	761

Note that the population now has maximal value 961 (the maximum of f!) and average fitness 613.

Iterating this procedure, we obtain:

$P'(1)$	$P(2)$	f	$P'(2)$	$P(3)$	f	$P'(3)$	$P(4)$	f
111|11	11111	961	111|11	11110	900	1111|1	11110	900
110|11	11011	761	111|10	11111	961	1111|0	11111	961
1110|1	11100	784	110|11	11001	625	11|111	11111	961
1111|0	11111	961	111|01	11111	961	11|111	11111	961
111|11	11110	900	1|1110	11000	576	1111|0	11111	961
101|10	10111	529	1|1100	11110	900	1100|1	11000	576
1|1001	11011	761	1111|1	11111	961	11|100	11111	961
1|1011	11101	841	1101|1	11011	761	11|111	11000	576
1110|1	11100	784	11|100	11111	961	1|1111	11110	900
1111|0	11111	961	11|111	11100	784	1|1110	11111	961

If we look at the evolution of the maximum and the average through these successive generations, we find the following values for the maximum value in the population, its multiplicity and the average:

population	max	mult	av
$P(0)$	900	1	375
$P(1)$	961	1	613
$P(2)$	961	3	814
$P(3)$	961	4	840
$P(4)$	961	6	872

It appears that the string 11111 which corresponds to the maximum of f quickly starts to dominate the population – actually, we could use the fact that a certain string accounts for more than half of the population as a stopping criterion. Moreover, we may also observe that the average fitness of the population gradually increases through successive generations.

In view of this example, it thus seems that the GA does indeed function well, i.e., that each successive generation tends to be better, in the sense that its average fitness increases, that it contains an increasing number of strings which are close to realizing the optimum.

Of course, an obvious but fundamental question is: why does this (seem to) work? In order to try and give an answer to this, let us take another look at the previous example. It appears that strings starting with 1 definitely have a higher fitness than those starting with 0. The reason for this behavior is, of course, trivial, since the first bit accounts for an extra value of 16 if it is set to 1, in the identification $\Omega_5 \leftrightarrow \{0, \ldots, 31\}$. In general, however, we are only able to calculate $f(s)$ for individual strings $s \in \Omega_\ell$ (where ℓ may be large); in particular, it is thus initially unclear whether certain bits or combinations of them are "more important" than others.

Nevertheless, since it appears that the structure of certain strings somehow makes them of higher fitness, let us introduce the notion of schema, in order to be able to describe the kind of structure we are interested in. By definition, a *schema* is an element of the space $\{0, 1, \#\}^\ell$, i.e., a string of length ℓ involving the bits 0 and 1 and the "don't care" symbol #. For example, $H = 01\#1\#$ is a schema of length 5. We say that a string s *belongs to* the schema H if s and H coincide at all places where H is different from #. For example, $s = 01110$ belongs to the schema $H = 01\#1\#$. We will frequently identify a schema with the set of strings belonging to it, so, for example

$$H = 01\#1\# \longleftrightarrow \{01010, 01011, 01110, 01111\}.$$

Formally: if $H = h_{\ell-1} \ldots h_0$ and $s = s_{\ell-1} \ldots s_0$, then $s \in H$, exactly when $s_i = h_i$, whenever $h_i \neq \#$. Note that $H = \# \ldots \#$ may be identified with the whole search space Ω_ℓ. If we define the order $o(H)$ of H to be the number of 0 and 1 positions, i.e., the *fixed positions*, as opposed to the *don't care positions*, then it is clear that H corresponds to a subset of Ω_ℓ of cardinality $|H| = 2^{\ell - o(H)}$.

Returning to the above example, it should now be clear that the schema $H_1 = 1\#\#\#\#$ is better than the schema $H_0 = 0\#\#\#\#$, in the sense that the strings in H_1 have higher fitness than those in H_0. In fact, if we denote by $f(H)$ the average fitness of H, i.e.,

$$f(H) = \sum_{s \in H} f(s)/|H| = \sum_{s \in H} f(s)/2^{\ell - o(H)},$$

then an easy calculation shows that $f(H_1) = 573.5$ and $f(H_0) = 77.5$.

If we consider the runs of the GA in the above example, then we may observe that the number of strings belonging to H_1 is increasing gradually, whereas the number of strings belonging to H_0 is decreasing. This is exactly what we want: we would like the GA to mainly produce strings performing well, belonging to schemas whose structure is related to high fitness.

Let us briefly sketch some of the maths behind this.

For any schema H, denote by $P(H,t)$ the set of strings in the t-th population $P(t)$, which also belong to H and by $m(H,t)$ the cardinality of $P(H,t)$. In particular, $P(\Omega, t) = P(t)$ and $m(\Omega, t) = N$.

During the selection procedure (recall the roulette wheel model!), an intermediate population of N strings is created, where each string $s \in P(t)$ has a probability $p_s = f(s)/\sum_{r \in P(t)} f(r)$ of being selected.

For each $s \in P(t)$, one thus expects $N.p_s$ copies of s in this intermediate population. Restricting to H, one may expect

$$m(H, t+1) = N \sum_{s \in P(H,t)} p_s$$

strings in $P(H, t+1)$. Denote by $f(H, t)$ the average of f on $P(H, t)$, i.e.,

$$f(H, t) = \sum_{r \in P(H,t)} f(r)/m(H, t),$$

and note that

$$f(\Omega, t) = \sum_{r \in P(t)} f(r)/N$$

is the average of f on the whole population $P(t)$.

Then the above identity yields

$$\begin{aligned} m(H, t+1) &= N \sum_{s \in P(H,t)} p_s \\ &= N \sum_{s \in P(H,t)} \frac{f(s)}{\sum_{r \in P(t)} f(r)} \\ &= N \frac{\sum_{s \in P(H,t)} f(s)}{\sum_{r \in P(t)} f(r)} \\ &= N \frac{m(H,t) f(H,t)}{\sum_{r \in P(t)} f(r)} \\ &= m(H, t) \frac{f(H,t)}{f(\Omega,t)}. \end{aligned}$$

Let us stress that this identity just says that the *expected* value of $m(H, t+1)$ is equal to $m(H, t)\frac{f(H,t)}{f(\Omega,t)}$ and that we did not yet apply genetic operators like crossover or mutation.

As an immediate, first corollary, it is clear that if $f(H, t) > f(\Omega, t)$, i.e., if, on the average, the strings in H score better than those in the whole population, then $m(H, t+1)$ is higher than $m(H, t)$, whereas $m(H, t+1)$ is lower than $m(H, t)$ in the other case. Otherwise put: if H is a "good" schema, its presence will be higher in the next population, if it is "bad', then its presence will be lower.

If H remains a fixed percentage $a > 0$ above average, i.e., if $f(H, t) = (1+a)f(\Omega, t)$ throughout, then the previous formula yields

$$m(H, t) = m(H, 0)(1 + a)^t.$$

This means that one expects the number of strings in $P(H,t)$ to increase exponentially! In a similar way, it is easy to see that the number of strings in $P(H,t)$ will decrease exponentially, if H constantly remains below average.

Let us stress that this statement only has a theoretic value: (1) we already mentioned that we are talking about "expected behavior" and (2) the assumption that $f(H,t)$ constantly remains a factor $1+a$ above average cannot hold permanently, as this would imply $P(H,t)$ to continue growing which is, of course, impossible within the (finite!) search space Ω.

The previous results seem very promising, but unless we include some extra dynamics, we are essentially just looking for the best solution within an arbitrary but fixed population (all of whose strings might well be of very low fitness!). As we pointed out before, we need genetic operators like crossover and mutation to remedy this.

Let us start with crossover. Consider the following two schemas of length 8:

$$H_1 = \#\#\#\#1\#1\#$$
$$H_2 = \#1\#\#\#\#1\#$$

and the string $s = 11111111$, which belongs to both H_1 and H_2.

If we combine s with the string $t = 00000000$, say, and if we choose the crossover site between the fourth and the fifth bit, for example, then we obtain new strings

$$s' = 00001111$$
$$t' = 11110000.$$

It appears that s' belongs to H_1, whereas neither s not t' belongs to H_2. In other words, the schema H_1 survives in the offspring, whereas H_2 does not.

It is easy to see that there are 5 crossover sites for which H_1 always survives, whereas there are only 2 (between the first and the second bit and between the seventh and the last bit) where this holds for H_2. Since the crossover site is chosen randomly (and uniformly) amongst the $\ell - 1$ possible ones, it appears that the probability of survival of H_1 is thus equal to $5/7$ and that of H_2 to $2/7$.

This is tightly linked to the notion of *defining length* of a schema H, denoted by $\delta(H)$ and defined to be the distance between the first and the last fixed string position.

For example,

$$\delta(H_1) = 7 - 5 = 2 \text{ resp. } \delta(H_2) = 7 - 2 = 5.$$

For an arbitrary schema H, it is easy to see that the probability of destruction through crossover is equal to $\delta(H)/(\ell - 1)$ and the probability of survival is thus

$$1 - \frac{\delta(H)}{\ell - 1}.$$

As we pointed out, crossover is, in general, just applied with some probability p_c, which implies that the probability of survival for H is thus

$$p_c(H) = 1 - p_c \frac{\delta(H)}{\ell - 1}.$$

Again we should not view this as an exact statement: even if a "bad" crossover site is selected, the schema H could still survive "accidentally" though the choice of the partner of the string we consider. To given an example, suppose that

$$s = 11111111$$
$$t = 01000001,$$

where, again, $s \in H_1$ Applying crossover between the seventh and last bit yields

$$s' = 01000011$$
$$t' = 11111101,$$

and we see that, although the crossover site was "bad", the schema H_1 still survives (through t'). So, to be more precise, we should put

$$p_c(H) \geq 1 - p_c \frac{\delta(H)}{\ell - 1}.$$

Combining this with the general formula, this leads to

$$m(H, t+1) \geq m(H, t) \frac{f(H, t)}{f(\Omega, t)} (1 - p_c \frac{\delta(H)}{\ell - 1}).$$

Finally, let us include the effect of mutation. As we mentioned before, mutation randomly changes bits from 0 to 1 and vice-versa, with some fixed, small probability p_m.

As an example, let us consider the string $s = 01110$, which belongs to the schema $H = 01\#1\#$. Flipping bits at the third position yields the string $s' = 01010$, which still belongs to H, whereas flipping bits at the second position yields $s'' = 00110$, which does not.

In general. it should be clear that the schema will only survive, if we apply mutation at the non-fixed positions of the schema. Since $1 - p_m$ is the probability of *not* changing a certain, single bit and since we do not want to touch the fixed positions (whose number is $o(H)$, the order of H), the probability of survival of a schema H is

$$p_m(H) = (1 - p_m)^{o(H)}.$$

Note also that we assumed p_m to be small (usually of the order of 0.01, for example), hence we may approximate this by

$$p_m(H) \approx 1 - o(H)p_m.$$

If we thus, finally, combine the effects of selection, crossover and mutation, we get

$$\begin{aligned} m(H, t+1) &\geq m(H,t)\frac{f(H,t)}{f(\Omega,t)}(1 - p_c\frac{\delta(H)}{\ell - 1})(1 - o(H)p_m) \\ &\approx m(H,t)\frac{f(H,t)}{f(\Omega,t)}(1 - p_c\frac{\delta(H)}{\ell - 1} - o(H)p_m). \end{aligned}$$

Let us call a schema a *building block* (with respect to f) if it is short ($\delta(H)$ is mall), of low order ($o(H)$ is small) and above average (throughout $f(H,t) > f(\Omega, t)$). Since for a building block the factor $1 - p_c\frac{\delta(H)}{\ell-1} - o(H)p_m$ is close to 1, it thus follows that building blocks still tend to dominate the population, as $m(H, t+1) > m(H,t)$. More precisely, we obtain the following

Theorem O.1 (Schema Theorem). *By applying a genetic algorithm, building blocks receive an exponentially increasing number of trials through the successive generations.*

Intuitively, this result says that if the structure of "good" solutions of our optimization problem may be described by "simple" schemas (building blocks), then these good solutions will tend to dominate the population in an exponentially increasing way.

Of course, as we already stressed before, this result is mainly of theoretic value as, in practice, several phenomena may complicate search by a genetic algorithm, including deception or epistasis. For the former, we refer to existing literature, the latter is exactly the subject of this book.

Chapter I

Evolutionary algorithms and their theory

We do not aim at completeness in this review of genetic algorithms and their theory. Our wish is to provide a brief overview that is broad enough to show the richness of the field, yet focused enough to mention a number of important results that help put the rest of the book in good perspective. We only occasionally fill in details but refer amply to existing literature.

1 Basic concepts

Generate-and-test is an important paradigm in search and optimization. It groups algorithms which iteratively generate a *candidate solution* and then test whether this candidate solution satisfies the goal of the search problem. *Random search* is the simplest of all generate-and-test methods: according to predefined probabilistic rules, it generates an arbitrary candidate solution, tests it, stops when it is successful, and does another iteration when it is not. While usually not very competitive, this algorithm is sometimes used as a basis for measuring the performance of other search algorithms.

Search is easily cast to optimization by providing a binary function with value 1 if the candidate solution is indeed a solution, and 0 otherwise. In most cases, however, the function is more fine-grained, and interpreted as a measure of how far away the candidate solution is thought to be from a solution. In this book, it is called the *fitness function*; other names apply in other contexts (heuristic function, objective function, penalty function, cost function, ...), but they always refer to the same concept. Similarly, "candidate solution" can be replaced by *configuration*, or, common in genetic algorithms literature, *individual*. In our setup, finding the optimal individual will always accord to maximizing (rather than minimizing) the fitness function.

At the basis of the generator is a *representation* of the candidate solutions. In the case of random search, the form of this representation is irrelevant as long as all candidate solutions can be represented in a unique way. More typically the outcome of the tester is used to guide the generation of the next candidate solution. The algorithm then explicitly exploits the relation between the representation and the fitness function by making modifications in the representation based on fitness information.

The *stochastic hill-climber* is a simple form of such an algorithm. Initially, an individual is selected at random, and tested. Then a small modification is made in the representation of this individual, yielding a new individual which is also tested. If the fitness of the new individual is better than that of the original, the modification is accepted: the new individual replaces the old one. If the fitness of the new individual is worse, the modification is rejected and a new modification is tried. A *neutral* modification, one which does not change the fitness value, may be accepted or rejected. When all possible modifications to a (suboptimal) individual yield a strictly inferior individual, this individual is called a *local optimum*.

An example optimization problem of high tutorial value in genetic algorithm research is the *onemax* problem. Its individuals are defined by their representation, in casu bit strings of length ℓ (which allows us to use the words *string* and *individual* interchangeably). The optimum is the string of all 1s, and the fitness function maps a string $s \in \Omega = \Sigma^\ell = \{0,1\}^\ell$ to the number of 1s in this string. Note that there are $\binom{\ell}{n}$ strings with fitness value n; the distribution of fitness values is a Binomial with ℓ draws and probability $\frac{1}{2}$ of drawing a 1. As a consequence, a *randomly generated string*, i.e., a string where the value of each position is independently drawn from a uniform distribution on $\{0, 1\}$, is likely to have a fitness value around $\ell/2$, where the mass of the distribution is.

The stochastic hill-climber described above can be applied to the onemax problem if we specify what a modification in the representation of an individual means. Usually, one chooses to *flip* the value of one of the bits: 0 becomes 1 and 1 becomes 0. In genetic algorithms language, this modification is called a *(single point) mutation*. Equipped with this modification operator, the stochastic hill-climber has no difficulties optimizing the onemax problem. This might seem strange at first sight, for when ℓ is sufficiently large, the number of high fitness strings in the onemax problem is extremely small compared to the number of average fitness strings. The relation between the fitness function and the representation, however, is benign and responsible for the immediate success of the algorithm.

The *Hamming distance* is a metric on Ω, the space of binary strings of length ℓ, which reflects the minimal number of single point mutations required to change one string into another. Said differently, it counts the number of bit positions where two strings differ from each other. One easily observes that the onemax function evaluated in a string t is nothing but the Hamming distance between the string of all 0s and this string t. Mutations which increase the fitness value automatically decrease the distance to the optimum. No local optima exist. Somewhat heuristically, one can say that there is an *easy path* toward the optimum.

It is easy to modify the onemax problem to obtain a search problem which shows no benign correlation whatsoever between representation and fitness. Suppose, for example, that we select a symmetric encryption scheme (DES, IDEA, Rijndael, ...), and encrypt each string before applying the fitness function to it. Regardless of whether neutral modifications are allowed or not, the single-point mutation stochastic hill-climber described above will not even reach fitness value $\frac{2}{3}\ell$, because there is no information to guide it toward the increasingly rare high fitness individuals.

This setup allows us to introduce the terms *genotype* and *phenotype*, adapted from genetics to distinguish between the individual (the genotype) and its representation (the phenotype). In almost all of this book, we will identify the individual with its representation, ignoring the distinction. But when we apply the encryption before the fitness function, we can speak of a genotype (unencrypted string, identified with individual) and a phenotype (the encrypted string). Using this terminology, we observe that a small modification to a genotype yields a phenotype which differs in on average $\frac{\ell}{2}$ bits from the original phenotype. In this way, the correlation between a change in fitness and a change in Hamming distance toward the optimum is completely lost. The probability of hitting the optimum with the stochastic hill-climber has effectively become that of finding a *needle-in-a-haystack* (i.e., 1 out of 2^ℓ).

The well-known *Metropolis* algorithm [58] differs from the basic stochastic hill-climber in only one rule: instead of always rejecting an inferior individual, the algorithm accepts an inferior individual with a probability which is a function of the difference in fitness and the *temperature* at which the algorithm is operating. Concretely, a modification is accepted with probability $\min(1, \exp(-\beta \Delta f))$, where $\beta = 1/T$ denotes the inverse temperature and Δf the fitness difference. This Markov Chain Monte Carlo algorithm can be used to draw independent samples from the fitness distribution at a fitness level corresponding to the temperature. The lower the temperature, the higher the fitness, the more the samples come from the interesting tail of the fitness distribution, and the slower the process becomes.

Simulated annealing [101] is the process of repeatedly applying the Metropolis algorithm at well-chosen, ever decreasing temperatures. It turns Metropolis into a "hands free" optimization algorithm which is guaranteed to find the optimum of any reasonably behaving search problem, if, and there is the catch, the temperature is decreased *slowly enough*.

Three closely interacting features distinguish a *genetic algorithm* (from now on abbreviated to GA) from a hill-climber: a GA maintains a *population* of individuals, and applies a second modification operator, called *crossover*, which creates two new individuals (called *children*) by exchanging parts of the representation of two individuals in the population (called *parents*). A *selection scheme* keeps the population at a fixed size by removing the least fit individuals to make place for (the offspring of) the fitter ones.

The traditional references to GAs are Holland [34] and Goldberg [26]. More recent overviews include [62] and [3, 4].

2 The GA in detail

A GA, in its canonical form, maintains a population $\mathcal{P}$ of individuals $I_1, \ldots, I_n$, identified with their representation as a bit string of length ℓ. Although we will rarely use it, we mention that the terminology borrowed from genetics is as follows: each bit in the string is called a *gene*, its value is called the *allele* (here 0 or 1), and the position of the bit in the string is termed the *locus* (here between 0 and $\ell - 1$; the plural of "locus" is "loci").

Initially, the population is filled with randomly generated strings. Each of these strings is evaluated, and their fitness values are stored. Then the following steps are iterated until some stopping criterion is fulfilled, e.g., the number of iterations has reached a bound or an optimum of sufficient quality is found:

1. *selection.* Fill a temporary population by independently drawing individuals, with replacement, from the current population according to some probability distribution based on their fitness. If the probability of selecting individual I equals $f(I)/f_{\mathcal{P}}$, where the denominator represents the average fitness of the population, we speak of *fitness proportional selection.*

2. *crossover.* Arbitrarily partition the temporary population into pairs of strings called *parents.* Perform crossover on each pair to obtain new pairs called *children,* which replace their parents in the temporary population. *One-point crossover* is defined as follows. With probability χ, called the *crossover rate,* we draw a *crossover point* p from a uniform distribution on $\{1, 2, \ldots, \ell - 1\}$ (where string positions are labeled 0 up to $\ell - 1$). We then swap the tails of the strings of the parents starting from bit position p to obtain the children. For example,

$$\begin{array}{c|c} 101 & 0011 \\ 011 & 1001 \end{array} \quad \Longrightarrow \quad \begin{array}{c} 1011001 \\ 0110011 \end{array}$$

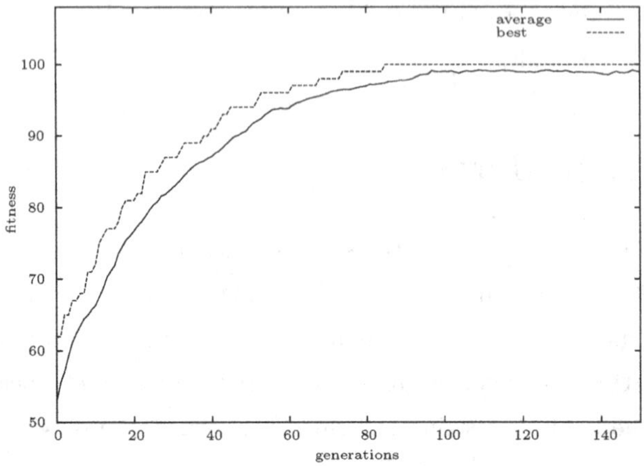

Figure I.1: Average and best fitness in the population of one run of a generational GA on a 100-bit onemax problem. The population size is 100, the selection strategy is binary tournament selection, one-point crossover is applied at a rate of 0.8, and the mutation rate is 0.1.

With probability $1 - \chi$, the children are exact clones of their parents (no crossover is performed).

3. *mutation*. Perform mutation on each string in the temporary population. The common mutation operator flips each bit independently with a (small) probability μ, called the *mutation rate*. For example, 0111001 becomes 0110001.

4. Accept the temporary population as the new generation, and reevaluate the strings in this new population. In a time-static environment, where the fitness function does not change each generation (which is what we have assumed so far) a *cache* may be used to avoid reevaluation of the same string over and over again.

Figure I.1 shows a typical statistics of a GA run on the onemax problem: the fitness of the best individual in the population, and the average fitness of the population. The wealth of parameters (population size, selection method, crossover method and crossover rate, mutation rate are the most common) and the large number of algorithmic variants which we do not even begin to describe, make it hard to speak of *the* GA. Rather, one should always try to specify its main components. The GA variant described above is sometimes referred to as *the simple GA*. Let us discuss some of the main components and parameters in more detail, referring to *the GA* when the details of its components do not matter that much or can be inferred from the context.

The GA described above is a *generational GA*. In each generation, the whole population is updated. This is in contrast to *steady state* GAs where in one generation, only a small portion of the population is changed. The generation gap quantifies the position of the algorithmic variant between the extremes of a generational GA and a steady state GA where in one generation, only one individual is changed.

Many different selection schemes have shown up over the years. The one we will use throughout this work is called *binary tournament selection*, a scheme which fills the temporary population by repeatedly drawing, with replacement, two individuals from the population and selecting the fittest of this "tournament" of size two. Each scheme, usually equipped with some parameters, can be classified according to the amount of *selective pressure* it exerts on the individuals. Fitness proportional selection is classified as weak compared to *ranking selection* and *truncation selection*, to name only two other schemes. For our purposes, it suffices to know that the selection pressure can be chosen by selecting an appropriate selection scheme. An additional feature, implicitly present in only a few schemes and often explicitly forced, is *elitism*. An elitist GA never throws away its best individual.

The mutation operator is not considered to be the driving force of the optimization process. Rather, it is seen as a background process which adds *diversity* to the population, supplementary to crossover. Many variations of the canonical GA decrease the mutation rate as the search process advances. Note that an extremely high mutation rate of $\frac{1}{2}$ turns each new generation into a population of random strings.

Crossover is the feature that distinguishes a GA from other generic optimization techniques. It generates individuals that are composed of parts of the genotype of their parents. The implicit assumption here is, of course, that swapping tails or exchanging substrings is beneficial for the search process. In many real-world applications this appears to be the case. Theoretically, however, proving that crossover actually improves performance is non-trivial and only recently example problems have shown up that are next to impossible to solve for mutation-based algorithms and easy to solve for any variant of a GA that uses crossover.

We throw in a brief note on the role of mutation and crossover. The preceding paragraphs contain phrases like "is seen as" and "is considered to be". The point of view expressed above is the traditional one. GAs have not been developed by theoretical computer scientists specialized in algorithms, nor by mathematicians or physicists. Many people with this background view the GA as a mathematical object or an efficient heuristic rather than an algorithm mimicking a natural process. To them, the question of importance is replaced by the question of how these operators interact with each other.

The term *fitness landscape* refers to a fitness function in combination with a metric on the search space. The metric defines which points in the search space are near to a point or a set, and which are far away. If one visualizes a landscape as a three-dimensional (real world) landscape, where the height is a function of longitude and latitude, then the terms *local optimum, rugged landscape, smooth landscape* and *flat landscape* obtain intuitive meanings.

We have already encountered one important metric on the space of binary strings: the Hamming distance. It is the natural metric to describe the action of the common mutation operator of GAs since the Hamming distance between two strings is nothing but the minimal number of bits that need to be flipped to move from one string to the other. In terms of this metric, the common mutation operator is a *local search operator*, since on average it flips one bit. The mutation landscape of the onemax problem is ideal to optimize: it is unimodal (it has one broad peak) and regular (in the sense that a single mutation either increases or decreases the fitness by one; no neutral steps are possible). A question that is more difficult to answer is what the landscape of a crossover operator looks like, since it is a binary operator that is applied to individuals drawn from a population that changes over time. We will elaborate more on this theme in section 7, and finish here by mentioning that the amount of local optima and the sizes of their basins of attraction can be estimated using different statistical techniques [21, 78].

3 Describing the GA dynamics

The GA can be modeled exactly as a homogeneous Markov chain whose states are all possible populations of finite size n. Because the mutation operator can theoretically transform one string into any other string with a non-zero probability, this chain is ergodic, and yields a limiting distribution where all populations have a non-zero probability of being visited. When elitism is added to the algorithm, only optimal populations will have a non-zero probability in the limiting distribution, and the GA can be said to converge [82]. The vast number of different states makes the Markov matrix difficult to analyze in all but the most trivial cases. A general convergence result is of limited practical value since any finite search space can be enumerated and hence optimized in a finite time.

The approach presented in [105] also contains the previous result but is a much more extended and richer body of theory. Let $p \in \Lambda$ be a population vector, describing for each string its proportion in the population. We call $\Lambda = \{(p_i); \sum_i p_i = 1\}$ the *simplex.* The *heuristic function* $\mathcal{G}$ embodies all GA operations and maps p to a population vector $\mathcal{G}(p)$ whose entries determine the probability that each individual will be contained in the next generation. The next generation is then constructed by sampling n times from this distribution with replacement. This approach allows the separation of the GA dynamics into a signal, given by $\mathcal{G}$, and stochastic noise introduced by the sampling of a finite population. When the population size is taken to infinity, the stochastic noise disappears and the dynamics of the (now deterministic) system is governed by $\mathcal{G}$ only.

The latter situation is commonly referred to as *Vose's infinite population model.* It is an approximation of the GA dynamics in the sense that the trajectory of the GA in the simplex does not necessarily follow that of $\mathcal{G}$. For large enough population sizes, the presence of *punctuated equilibria* [105, chapter 13], periods of relative stability near a fixed point of $\mathcal{G}$ interrupted by the occurrence of a low probability event moving the population away into a basin of attraction of another fixed point, reconcile the ergodicity of the GA with the potentially converging infinite population dynamics.

The heuristic $\mathcal{G}$ can be written in an elegant and relatively compact form but typically results in a set of minimally 2^ℓ coupled difference equations to be solved if one wants a closed form for the dynamics. For this reason, aggregation of the variables (*coarse graining*) is necessary to reduce the number of equations. Keeping track of all details is unnecessary or at best intractable. The question is, however, which macroscopic variables to choose? In a number of concrete cases, ad-hoc choices are possible, but it remains unclear how to proceed in a systematic way.

From the introduction of GAs onward, *schemata* have played a prominent role in the efforts of understanding how the GA works. (A schema is a hyperplane in Ω and is typically denoted by a string over the alphabet $\{0, 1, \#\}$, where the # is used as a "don't care" symbol. We elaborate further on schemata and their properties in the next section.) In his seminal work, Holland introduced the *schema theorem* which gives a lower bound on the expected presence of a schema in the next generation given the information about schemata in the current one. Over the years, this relation has received much criticism and many exact versions (sets of difference equations that can be iterated) have shown up (e.g., [105, chapter 19], [91]).

However, in [105, chapter 19] it is also shown that as natural macroscopic variables, schemata are *incompatible* with the heuristic $\mathcal{G}$ for most common situations; i.e., for non-trivial fitness functions, coarse graining a population vector and then applying $\mathcal{G}$ is not identical to coarse graining after the application of $\mathcal{G}$. This implies that no exact coarse grained version of $\mathcal{G}$ can exist, although approximations are possible. The fitness component of $\mathcal{G}$ is responsible for the incompatibility, for the mixing component (mutation and crossover) can be shown to be compatible. The latter result helps understanding the interest in schemata as a tool for understanding the working of crossover. It also explains why computing the dynamics of a GA under non-trivial fitness is (up to now) very difficult within the "exact schema theorem" framework.

An altogether different approach to describing the dynamics has come from physicists studying disordered systems. Starting from only two macroscopic variables, the mean and variance of the fitness distribution of the population, they have succeeded in approximating the dynamics of a GA operating on a number of search problems whose complexity is beyond that of *onemax* [74, 85, 86]. The approximations have been improved by computing higher moments of the fitness distribution, and adding finite population corrections.

We finally mention the approach followed in [63, 64] for modeling GA dynamics. Here a probability distribution over the state space is modeled by *factorizing* the distribution to obtain a limited number of parameters whose evolution can be computed. The first order approximation, where each of the ℓ variables is considered independent, corresponds to a population which is permanently in *linkage equilibrium* (this notion is defined in the next section). More accurate models introduce dependencies between the variables. An alternative family of search algorithms is based on this theory: they start from an initial distribution, sample the search space according to this distribution, and then adjust the parameters of the distribution according to the observations. They can even adapt the structure of the factorization dynamically.

4 Tools for GA design

The engineer's approach to genetic algorithms is concerned with finding reliable "rules of thumb" that help design an algorithm appropriate for a search problem at hand. Since the current theory of GA dynamics is insufficiently rich to provide such guidelines, the engineer will need to use more approximate and heuristic arguments. Focusing, for the sake of brevity, on population sizing guidelines, we will discuss the basics of a model based on *building block* properties of the search problem [30]. Before defining the term building block, however, we need some more terminology about schemata.

As defined before, a *schema* is a hyperplane of the search space $\Omega = \Sigma^\ell$. It is usually written as a string over the augmented alphabet $\Sigma' = \Sigma \cup \{\#\} = \{0, 1, \#\}$, where the # plays the role of a "don't care" symbol. Technically, an individual[1]

[1]Unconventionally, we write strings with the highest index first, ending with index 0. This

$s_{\ell-1}\dots s_0$ belongs to a schema $h_{\ell-1}\dots h_0$ if $s_i = h_i$ for all i such that $h_i \neq \#$. We use the term *schema fitness distribution* to denote the distribution of fitness values of all individuals belonging to a schema. The fitness of a schema is defined as the expected value of this distribution. The *order* of a schema is given by the number of non-# symbols in the schema, the *length* of a schema is given by the largest distance between two non-# symbols. The term building block, finally, refers to a fit, short, low-order schema.

The concepts of a schema and its fitness distribution are central to a simple and intuitive hypothesis about the dynamics of a GA, which we present in the form of the *static building block hypothesis* [29]. Knowing that a *hyperplane partition* consists of all schemata with #s on the same positions, and a *schema competition* is defined as the comparison of the average fitness values of all schemata in a hyperplane partition, the hypothesis sounds:

> Given any short, low order hyperplane partition, a GA is expected to converge to the winner of the corresponding schema competition.

(Note that the word *static* is used to stress that no actual GA dynamics is involved.) Using this hypothesis as a starting point, Goldberg [27] decompose the problem of understanding GA behavior into seven points:

1. know what the GA is processing: building blocks
2. solve problems that are tractable by building blocks
3. supply enough building blocks in the initial population
4. ensure the growth of necessary building blocks
5. know the building block takeover and convergence times
6. decide well among competing building blocks
7. mix the building blocks properly.

notation is more natural when we consider the correspondence between natural numbers and their binary expansion.

Elaborating on items 3 and 6, these authors apply the Gambler's ruin model to estimate the probability of mistakingly choosing an inferior building block over a better one for a given population size. Their result is a population sizing equation

$$n = -2^{k-1} \ln(\alpha) \frac{\sigma_{BB}\sqrt{\pi m'}}{d}.$$

The equation indicates that the population size n gets larger as the average variance σ_{BB} of the building blocks increases, and smaller as the signal difference d (the difference between the average fitnesses in the competition) increases. The parameter $m' = m - 1$ determines the number of competing building blocks; α is 0.1 or 0.05, the probability with which an error is allowed.

A schema competition is said to be *deceptive* when no solution of the problem is contained in the fittest schema of the partition. The parameter k indicates the size of the largest such deceptive competition. The equation shows that the population size is exponential in k; keeping k small is therefore essential, and this is expressed in item 2 of the decomposition.

We end this section with a little more terminology. A schema is called deceptive when (a) it is the winner of its schema competition and (b) no solution is contained in it. A search problem is called deceptive when it contains deceptive low order schemata. We refer to [108] for a discussion of deception. According to the building block hypothesis, deceptive problems should be hard for a GA because they mislead its search for a solution.

5 On the role of toy problems...

Both for empirical and theoretical GA research, simple and extreme fitness functions provide a first starting point. Different forms of GA behavior can be cast to a number of archetypal forms induced by these extreme problems. Understanding the GA on them is a prerequisite for understanding it on daily-life search problems.

5.1 Flat fitness

The simplest of all fitness functions is the constant function. Yet, the dynamics of the GA on a flat fitness landscape are non-trivial, and understanding them helps to predict how the algorithm will behave in a flat or uninformative area of a much more complex fitness landscape. We briefly discuss two issues here, genetic drift and Geiringer's theorem. Both have extensively been studied in population genetics.

Genetic drift is a phenomenon that is inherently related to sampling and finite populations. Suppose an initial population contains n different individuals and selection (without any selective pressure) is repeatedly applied. All individuals have an equal probability of being selected, but chances are that some will be lost, and others duplicated. In this way, the population looses individuals, and therefore diversity, and ends up with one individual having spread n copies of itself. In general, selective pressure and genetic drift are the two factors that reduce the diversity in the population.

Geiringer's theorem [22] shows that for a fairly general set of crossover operators, the limit of repeatedly applying such an operator (without mutation and selection) results in a population that is in *linkage equilibrium*. The probability of a string $p = p_{\ell-1}\dots p_0$ appearing in a population $\mathcal{P}$ in linkage equilibrium is given by Robbins' proportions

$$P_{\mathcal{P}}(p) = \prod_{i=0}^{\ell-1} P_{\mathcal{P}}(p_i),$$

where $P_{\mathcal{P}}(p_i)$ is the marginal frequency of value p_i at bit position i. The ultimate question that arises in the GA setting is to determine the dynamics of this process under the influence of finite sampling (genetic drift) and mutation (see e.g., [90]).

5.2 One needle, two needles

One step away from the constant fitness function is the *needle-in-a-haystack* problem. Here, exactly one out of the 2^ℓ individuals is assigned a non-zero fitness value. It is straightforward to show that when the location of the needle is unknown, exhaustive enumeration is the most efficient algorithm to find it; clearly, the expected time to reach the solution is exponential in the string length. Many optimization problems contain needle-in-a-haystack problems, although they do not always present themselves so clearly. When the global optimum is hidden in an exponential set of local optima of almost the same fitness, for example, we also speak of a needle-in-a-haystack.

GAs were originally conceived as algorithms to mimic natural evolution, rather than as optimization algorithms. In this context, the needle-in-a-haystack problem transforms from an extreme optimization problem to a simple model problem for evolution in a changing environment, very similar to Eigen's quasi-species model. The setting is as follows. Every τ time steps, the needle is replaced by a new one located in the near Hamming neighborhood of the old one. Thanks to its population and the constant source of diversity in the population provided by the mutation operator, the GA stands a chance of tracking the movement of the needle. Under proper conditions on the mutation rate and the cycle length τ, the population contains a number of copies of the needle (the species) and a number of mutants of this needle (the quasi-species).

This situation touches the notion of *effective fitness* [89]. Although the mutants have the same, low fitness as any other string in the search space except for the needle, it is opportune to have the mutants in the population since one of them may become the next needle. In a way their effective fitness is higher than that of the strings far away.

Occurring further in this book (chapter II, section 4) is a fitness function with two peaks at maximal Hamming distance from each other. We have dubbed this function the *Camel function*, though it is probably better known as a two-peak problem. With a population sufficiently large, it is possible to maintain copies of both peaks in the population in a stable way. In this situation, most of the so called "interspecies" crossovers, crossovers where each of the parents belong to the (quasi)species of different peaks, are wasted: they yield offspring that are far away in Hamming distance from both peaks, and they stand no chance to survive more than a few generations.

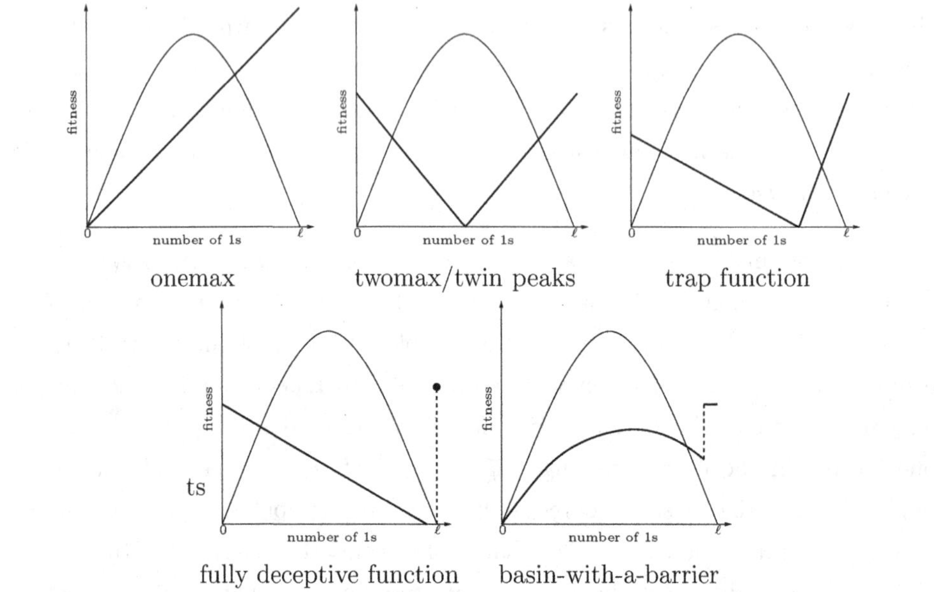

Figure I.2: A number of unitation functions. The parabola in each figure gives an indication of the number of individuals that are mapped to a fitness value, in logarithmic scale.

5.3 Unitation functions

A *unitation function* is a function on $\Omega = \{0,1\}^\ell$ that is defined in terms of the number of 1s in a string. Formally,

$$f : \Omega \to \mathbb{R} : s \mapsto h(u(s)),$$

where $u(s)$ denotes the number of 1s in s and h is a function which maps $\{0,\ldots,\ell\}$ to the real numbers. A few examples are shown in figure I.2; actual fitness values have been omitted.

We have already encountered the *onemax* problem in section 1. It is about the most important toy problem, and the dynamics of many variants of GAs (quite often without crossover) running on onemax have been studied in great detail (e.g., [65], [73]). To give an impression about the complexity of the GA dynamics on this simple problem, we note that using the dynamical systems approach, Wright and co-workers only recently obtained exact equations for the infinite population model and a GA with a crossover that permanently maintains linkage equilibrium [110].

The *twomax* or *twin peaks* [16] problem is a typical example of a problem where more than one area of the search space is worth investigating. Initially, the population contains individuals with low fitness and an average number of 1s of $\ell/2$. Some individuals will be fitter, i.e., have either an excess of 0s or 1s, but due to the symmetry of the problem, neither of the directions will be significantly overrepresented. After only a few generations, however, stochastic errors break this balance, and the problem becomes similar to a onemax problem [100]. One optimum is reached quickly, the other optimum is ignored.

Breaking the symmetry of the twomax problem may result in a deceptive problem. The particular form shown is called a *trap function* [1]. Depending on the relative heights of the optimal and sub-optimal peaks and the location of the fitness 0 strings (closer to all 1s, or closer to all 0s), one can control the fraction of times the GA is deceived and led in the direction of the sub-optimal peak.

The extreme case of deception is shown by the *fully deceptive* function [28, 108], where the optimum is a lonely individual and the GA is led to its complement. The probability of the GA ever hitting this optimum is next to zero; a black-box algorithm will not be able to distinguish the problem from a zeromax problem where the optimum is missing.

Finally we list the *basin-with-a-barrier* problem [81], for which approximate GA dynamics have been obtained using statistical mechanics techniques. It contains one local optimum that attracts the population, which will then stay there (due to entropy, the average number of 1s in the population will be slightly less that that of the local optimum) until a mutation causes a jump out of the basin of attraction to the global optimum.

$H_1 =$	11111111	########	########	########	########	########	########	########
$H_2 =$	########	11111111	########	########	########	########	########	########
$H_3 =$	########	########	11111111	########	########	########	########	########
$H_4 =$	########	########	########	11111111	########	########	########	########
$H_5 =$	########	########	########	########	11111111	########	########	########
$H_6 =$	########	########	########	########	########	11111111	########	########
$H_7 =$	########	########	########	########	########	########	11111111	########
$H_8 =$	########	########	########	########	########	########	########	11111111
$Opt =$	11111111	11111111	11111111	11111111	11111111	11111111	11111111	11111111

Figure I.3: The schemata defining Royal Road function R_1.

$H_9 =$	11111111	11111111	########	########	########	########	########	########
$H_{10} =$	########	########	11111111	11111111	########	########	########	########
$H_{11} =$	########	########	########	########	11111111	11111111	########	########
$H_{12} =$	########	########	########	########	########	########	11111111	11111111
$H_{13} =$	11111111	11111111	11111111	11111111	########	########	########	########
$H_{14} =$	########	########	########	########	11111111	11111111	11111111	11111111
$Opt =$	11111111	11111111	11111111	11111111	11111111	11111111	11111111	11111111

Figure I.4: The extra schemata defining Royal Road function R_2.

5.4 Crossover-friendly functions

Royal Road functions

In this book the Royal Road functions developed by Mitchell, Forrest and Holland [60] are studied. Initially developed to yield a "royal road" for the crossover operator, it was shown about a year later by the same authors [18] that a mutation based hill climber actually outperforms the GA by a factor of ten if one counts the number of fitness evaluations as a measure of complexity.

This is in contrast to the Real Royal Road functions that appeared only recently [44]. They are real in the sense that they do what they promise: any mutation based optimization algorithm requires exponential time to solve them, and a very broad class of GAs with crossover (uniform or one-point) can get to the solution in polynomial time. They have been carefully designed to achieve this goal; for this reason, they are less likely to give rise to equal insights in the working of the crossover operator.

The original Royal Road functions R_1 and R_2 are defined as follows. Consider the schemata

$$H_1 = 1^{(8)}\#^{(56)}, H_2 = \#^{(8)}1^{(8)}\#^{(48)}, H_3 = \#^{(16)}1^{(8)}\#^{(40)}, \ldots, H_8 = 1^{(56)}\#^{(8)},$$
$$H_9 = 1^{(16)}\#^{(48)}, H_{10} = \#^{(16)}1^{(16)}\#^{(32)}, \ldots, H_{12} = \#^{(48)}1^{(16)},$$
$$H_{13} = 1^{(32)}\#^{(32)}, H_{14} = \#^{(32)}1^{(32)},$$

where we denote by $a^{(n)}$ the string $aa\ldots a$ consisting of n copies of a. Figures I.3 and I.4 show the schemata fully written out. Then

$$R_1(s) = 8\sum_{i=1}^{8} [s \in H_i],$$
$$R_2(s) = 8\sum_{i=1}^{8} [s \in H_i] + 16\sum_{i=9}^{12} [s \in H_i] + 32\sum_{i=13}^{14} [s \in H_i],$$

with the brackets denoting the indicator function. The functions were designed as the simplest search problems on which the GA performs "as expected" if one assumes correctness of the building block hypothesis. The different schemata are the intermediate stepping stones or building blocks which need to be combined by crossover to reach the optimum. The average fitness of R_1 is

$$\frac{1}{2^{64}}\sum_{s\in\Omega} R_1(s) = \frac{1}{2^{64}}\sum_{s\in\Omega} 8\sum_{i=1}^{8}[s \in H_i] = \frac{8}{2^{64}}\sum_{i=1}^{8}|H_i| = \frac{1}{4},$$

with $|H_i|$ denoting the number of elements of H_i. We compute the the fitness of the schema H_i ($i = 1,\ldots,8$) under R_1 as 8 plus the average fitness of a 56-bit Royal

Road function $R'_1 : s \in \Omega' = \{0,1\}^{56} \mapsto R_1(11111111s)$:

$$\begin{aligned}
f(H_i) &= \frac{1}{|H_i|}\sum_{s\in H_i} f(s)\\
&= 8 + \frac{1}{2^{56}}\sum_{s\in\Omega'} R'_1(s)\\
&= 8 + \frac{1}{2^{56}}\sum_{s\in\Omega'} 8\sum_{i=2}^{8}[s \in H'_i]\\
&= 8 + \frac{8}{2^{56}}\sum_{i=2}^{8}\sum_{s\in\Omega'}[s \in H'_i]\\
&= 8 + \frac{8}{2^{56}}\sum_{i=2}^{8}|H'_i| = 8 + \frac{8}{2^{56}}\cdot 7\cdot 2^{48} = \frac{263}{32} = 8.218\ldots
\end{aligned}$$

This is a lot higher than the function average. In the function R_2, "reinforcements" are built in that provide additional stepping stones. It was therefore expected that a GA would perform better on R_2 than on R_1.

A phenomenon called *hitch-hiking* prevents this from happening; it fact, the GA consistently performs better on R_1. An individual A contained in H_{12} but not in H_6 dominates an individual B not contained in H_7 or H_8 (and therefore not in H_{12}) but contained in H_6, assuming that A and B are similar on the other schemata. Even with relatively weak selection, A will suppress B after a few generations — and make building block H_6 disappear. This situation is visible in figure I.5, where H_6 is almost omnipresent until, around generation 20, H_{12} appears. The few 0s that cause A not to belong to H_6 are said to hitch-hike along with the strong building block H_{12}. The GA then has to wait for a combination of mutation and crossover events that recreate this building block. Due to lower relative differences in the fitness of building blocks, R_1 provides better opportunities for the GA to combine the blocks.

H-IFF and Ising

The *hierarchical if-and-only-if* (H-IFF) problem [107] has been designed by Watson and co-workers as a problem which is extremely hard to solve for mutation-based

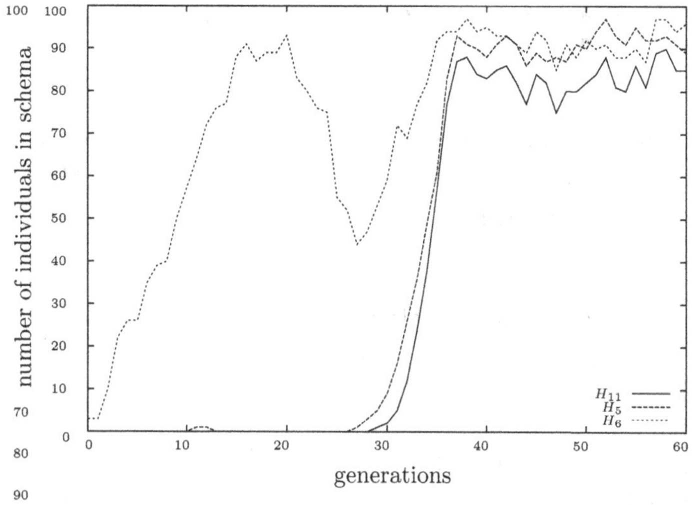

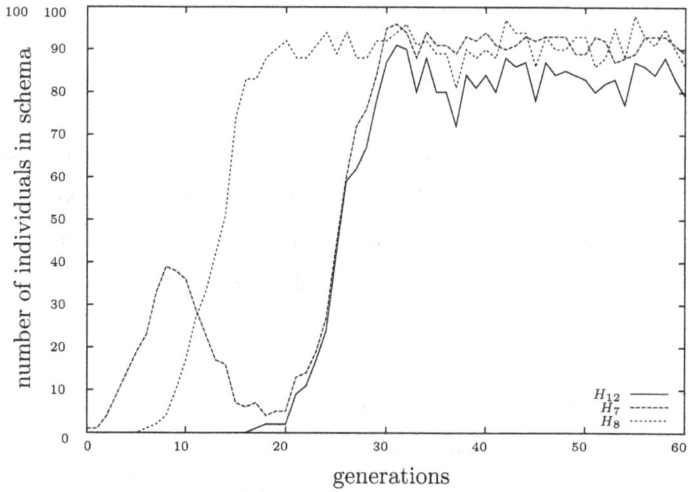

Figure I.5: Number of individuals in schemata H_5, H_6 and H_{11} (top plot) and H_7, H_8 and H_{12} (bottom plot) for one GA run on R_2 with population size 100, binary tournament selection, one-point crossover at rate 0.8, and ordinary mutation at rate $1/64$. This figure illustrates the hitch-hiking effect detailed on page 39.

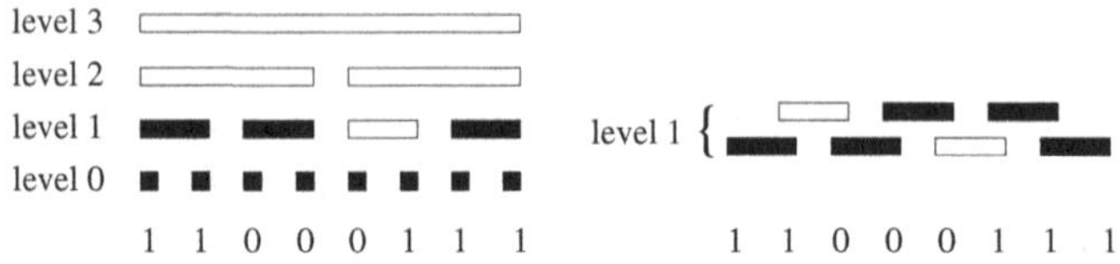

Figure I.6: The building block representation of a H-IFF problem and a one-dimensional, non-circular Ising problem, both with string length 8. All black building blocks are satisfied; all white building blocks are not satisfied.

algorithms, and as easy as possible for algorithms that employ crossover. It is defined by the recursive function

$$f(A) = \begin{cases} 1, & \text{if } |A| = 1, \\ |A| + f(A_L) + f(A_R), & \text{if } |A| > 1 \text{ and } (\forall i : a_i = 0 \text{ or } \forall i : a_i = 1), \\ f(A_L) + f(A_R), & \text{otherwise,} \end{cases}$$

with A a block of bits $\{a_1, \ldots, a_n\}$, $|A|$ the size of the block and A_L and A_R respectively the left and the right halves of A (i.e., $A_L = \{a_1, \ldots, a_{n/2}\}$, $A_R = \{a_{(n/2)+1} \ldots, a_n\}$). The string length ℓ must equal 2^k, with k an integer indicating the number of hierarchical levels. Figure I.6 gives a building block representation of a H-IFF problem with string length 8. A building block is satisfied if all corresponding string positions have the same value. The fitness contribution of a satisfied building block is equal to its size. In the figure, all black building blocks are satisfied, all white ones are not.

Note that the H-IFF problem has two optima, the string containing only 1s and the string containing only 0s. Moreover, it contains *spin-flip* or *bit flip symmetry* (see [70] in the context of GAs), a symmetry which is characterized by fitness invariant permutations on the alphabet. These properties are shared by the one-dimensional Ising problem, whose roots lie in statistical physics [43]. Its most convenient form for this introduction is

$$f(s) = \sum_{i=0}^{\ell-1} \delta_{s_i, s_{i+1}} \quad \text{with} \quad \delta_{s_i, s_j} = \begin{cases} 1 \text{ if } s_i = s_j; \\ 0 \text{ otherwise.} \end{cases}$$

Here, the problem is seen as a circle, with $s_\ell \equiv s_0$. The problem has the same optima as H-IFF: the strings of all 1s and all 0s.

The two problems induce, to a large extent, similar behavior on both mutation-based and crossover-based algorithms. The problems of mutation are related to the fact that the operator has no "global view" of the problem: improvements can be made in different parts of the individual, but in an unsynchronized way: sometimes to 0, sometimes to 1, but rarely consistently to one of the two. Consider for example the 16-bit string

$$0001111100001111 \tag{I.1}$$

In the circular Ising problem, the fitness of this individual is only 4 away from the optimum. No mutations exist that improve the fitness, but a number of neutral steps are possible: every bit whose left or right neighbor has a different value can be mutated without changing the fitness. The following are examples of neutral mutations:

$$000\mathbf{0}111100001111 \qquad 00\mathbf{1}1111100001111 \tag{I.2}$$

The first mutation leads to an individual which is in Hamming distance even further away from one of the two optima; the second mutation leads to a string with more 1s than 0s and can be considered as a step in the right direction. But how does the mutation operation know?

In the H-IFF problem, individual (I.1) has a fitness of $16 \times 1 + 7 \times 2 + 3 \times 4 = 42$. Two mutations can improve this value; they lead to the first individual of (I.2), with fitness $16 \times 1 + 8 \times 2 + 4 \times 4 = 48$, and the second of (I.2), with fitness $16 \times 1 + 8 \times 2 + 3 \times 4 = 44$. Both are local optima. To reach the individual

$$1111111100001111$$

the fitness has to drop temporarily to 42; to then reach the string of all 1s, the third block of 4 bits needs to be broken up! But because the mutation operator has no global overview, it could also choose to break up the fourth block...

The crossover operator can combine large blocks of bits together, and succeed in jumping the large fitness barriers for mutation. The main condition, however, is sufficient genetic diversity: these blocks actually need to exist in the population. Without a system that forces this diversity, the GA evolves to populations which at certain positions only contain blocks of 1s or 0s. It then has to rely on the mutation operator to introduce these blocks – which is an extremely slow process. An in-depth discussion can be found in [100].

6 ... and more serious search problems

Of course, the ultimate goal is to know how a GA behaves on problems that you actually care about optimizing. A large part of the evolutionary algorithms literature contains papers whose main message is: "I am working on such and such search problem, and by fiddling a bit with the representation and the parameters, and adding some gizmos, I have constructed a GA that is pretty good at solving my problem. In fact, it performs even better than the three other algorithms I was using before." This type of literature is very useful if the problem at hand is very similar to one for which a successful GA has been found. Even if the problem is dissimilar, reading this type of papers is important in the sense that they teach the tricks that you need to get a GA working efficiently.

One of the main goals of GA theory is to give hard evidence of why and when certain 'tricks' increase the performance of a GA. The first step towards this goal, however, is to understand the structure of these realistic problems. Initially, this need not be done in the context of the very complex GA; actually, it need not even be done in the context of any search algorithm.

The theory of computational complexity (e.g., [71]) classifies search problems on a mathematical basis, without being much concerned about the pecularities of specific algorithms. If a problem is in the class *P*, to name a first, important class, it has been proved that an algorithm exists that can solve all instances of this problem in a time which is bounded by a polynomial in the size of the instance. Many real world search problems have been shown to be in *NP*, a class with two important properties: once an algorithm is shown to solve all instances of one *NP*-complete problem in polynomial time, *all* problems in *NP* are solvable in polynomial time and as a result, *P* would equal *NP*. The second property is that this equality seems very unlikely. Of course, when a problem is in *P*, there is no implication that a GA will be able to solve all instances in an expected time that is bounded by a polynomial of the size of the instance. Similarly, a fairly large set of instances of a problem that is shown to be in *NP* may be solved efficiently by a GA. To conclude: the complexity classification of search problems is important, but it rarely affects the GA practitioner. If a problem is not (yet) shown to be in *P*, a heuristic algorithm needs to be devised, and it might well turn out that this algorithm will efficiently solve most of the real world instances that show up. Some instances, however, will present difficulties for the heuristic, resulting in a running time that is exponential in the size of the instance.

The instances of an *NP*-complete problem share important structural properties. The word 'structure' here is used in the context of a topology on the search space; for simplicity, the metric and hence search landscape of the most obvious mutation operations are considered first. An important result of landscape theory shows that many such obvious choices of operators are indeed very natural choices, in the sense that the landscape turns out to be an eigenspace of the Laplacian matrix [88] of the operator graph. This is the case for the Travelling Salesman Problem and transposition or inversion as mutation operator, for the graph coloring problem with ordinary mutation (color-replacement), and so on.

Even though they share a number of structural properties, different instances may present different degrees of problem difficulty for an optimizer such as a GA. This can be exploited to construct a diverse set of problem generators, where both structural properties and problem difficulty can be predicted beforehand. Consider, for example, the *NP*-complete class of *binary constraint satisfaction problems* [96]. They have the general form

$$f : \Sigma^\ell \to \mathbb{R} : s \mapsto \binom{\ell}{2} - \sum_{0 \leq i < j < \ell} g_{ij}(s_i, s_j),$$

were Σ is a multary alphabet ($|\Sigma| > 2$), and $g_{ij} : \Sigma \times \Sigma \to \{0, 1\}$ for all $i < j$. A string for which the fitness is maximal, i.e. $\binom{\ell}{2}$, is called a solution. Instances can be randomly generated by, for example, performing a Bernoulli trial for each of the $g_{ij}(a, b)$, with a fixed probability p of success. When p is close to zero, instances are likely to have many solutions; when p is close to one, the opposite is true. When going from zero to one, a region called the *mushy region* is observed where the average number of solutions of the instances is around one. These instances are typically hard to solve: it is hard to find the solution, or equally hard to disprove the existence of a solution. With the problem size increasing, the mushy region becomes smaller; in the limit of infinite problem size, a *phase transition* from solvable to unsolvable can be shown to occur [109].

7 A priori problem difficulty prediction

The aim of *a priori* problem difficulty prediction is to make an educated guess about the performance of a certain type of GA on a particular search problem *without actually running the GA*. The prediction should be based on a number of properties of the search problem and its representation. Preferrably, the amount of fitness evaluations carried out by a statistic to assess a property is lower than the amount the GA is expected to do, but this is not really the point. If information about the whole search space is needed, so be it. The more fundamental question is: what are the properties that capture the landscape as the GA sees it?

7.1 Fitness–distance correlation

A well-known example of a property of a search problem and its representation is *fitness–distance correlation* [46]. The statistical correlation between the fitness values of individuals and their Hamming distance to the optimum is computed and plotted. When there is a negative correlation, i.e., decreasing the distance to the optimum increases the fitness, then the search problem is predicted to be easy. The onemax problem is the archetypal problem where this negative correlation is perfect. When there is no correlation, the search problem is predicted to be difficult. The case of a positive correlation (the closer one gets to the optimum, the worse the fitness becomes) relates to deception: the algorithm is predicted to be led away from the optimum.

Of course the fitness–distance correlation as described here is actually a property of the Hamming landscape, and not of the "GA landscape": it does not include any information about crossover, selection or population effects. To obtain an idea about the performance or usefulness of a crossover operator, the *correlation coefficient* [55] of this operator can be computed. The procedure is similar to that of fitness–distance correlation: repeatedly sample parent individuals, cross them over, and compute the correlation between the average fitness of the children and that of the parents. If the sample is well chosen, i.e., it contains individuals that are likely to be crossed over by a GA, this coefficient will be informative about the action of the crossover operator in a GA.

Especially for fitness–distance correlation, a fair number of publications (e.g., [2, 48]) have demonstrated that a good correlation in the Hamming landscape does not necessarily lead to a problem that is easy for a GA, and the other way round. Note that this does not invalidate the concept of computing a correlation between fitness and distance as a predictor; it merely says that this property of the Hamming landscape is not representative for GA behavior. Ideally, one should compute whether populations that are fitter, for example in the sense that the average fitness of their individuals is higher, are also closer to a population containing optima, where closer is defined in terms of the number of generations that the GA requires to change the suboptimal population into the optimal [45]. Problems for which this *true* fitness–distance correlation is negative can be predicted to be easy with higher confidence. Sadly enough it seems unrealistic to expect such a statistic to exist.

7.2 Interactions

The term *epistasis* in a biological context refers to a linkage between genes. In the context of search landscapes, we use the term *interaction* and distinguish between the *range* of the interactions, denoting the amount of variables involved in the interactions, and their *structure*, which determines which interactions are present and which are not. Functions of the form

$$f : \{0,1\}^\ell \to \mathbb{R} : s \mapsto \sum_{0 \leq i < \ell} g_i(s_i) \tag{I.3}$$

are clearly free from interactions between the variables. We call them *first order functions* and note that they are not the only functions that are epistatically free.

Replacing the sum by a product, for example, yields another epistatically free class. The class of *second order functions* is defined as

$$f : \{0,1\}^\ell \to \mathbb{R} : s \mapsto \sum_{0 \le i < \ell} g_i(s_i) + \sum_{0 \le i < j < \ell} g_{ij}(s_i, s_j). \tag{I.4}$$

Next to the first order contributions to the fitness sum, contributions are possible whose values depend on two variables simultaneously. Typical members of this class are the (*NP*-complete) graph coloring [20] and binary constraint satisfaction problems as shown in section 6. A search problem consisting of interactions of order 3 or less is the famous (*NP*-complete) 3-SAT problem [20] ; the generalisation called K-SAT extends this to interactions of order K or less.

The longer the interactions, the higher is the potential effect of a small move in Hamming distance on the fitness. The *correlation length* [55] measures the predictability of fitness values as one goes along a random walk in the Hamming landscape. Problems with short range interactions have a high correlation length, whereas long range interaction problems may show much lower correlation lengths. In general, problems with short range interactions are easier to solve than problems with long range interactions: the extremes are the first order functions (no interactions, easy to solve) and random functions (full interactions, extremely hard to solve).

Next to the maximal order of the interactions, the *interaction structure* may also be important. It defines which interactions are present in a problem, and which are not. An example to illustrate this is the Ising model: the variables interact only with their nearest neighbor if ordered on a line. A two-dimensional Ising model also exists: the variables are put on a grid and again each variable interacts with its four nearest neighbors. Whatever values the interactions take, the one-dimensional Ising problem is optimizable in polynomial time by a divide-and-conquer algorithm. If the values of the interactions of the two-dimensional model are randomly drawn from $[0, 1]$ (this is the SK-model [87]), finding an optimal configuration becomes an *NP*-hard[2] search problem.

We end this section by mentioning the famous *NK-landscapes*, developed by Kauffman [49]. They are defined as the sum of N random subfunctions; each subfunction

[2] *NP*-hard is the optimization variant; *NP*-complete stands for decision problems.

f_i is defined on position i and K other, arbitrarily chosen bit positions. Clearly, an NK-landscape with $K = 1$ belongs to the class of second order functions. Only N pairs of variables are allowed to interact.

7.3 The epistasis measure

The epistasis measure that we study in this book computes the least squares distance of a fitness function or search problem to the class of first order functions defined above. A mathematical definition and in-depth treatment is given in chapter II, but we do not want to delay answering the question of its practical relevance. In the early 1990s, epistasis measures were "hot" and actively investigated. The results of research in the last decade, however, have shed light on many of its shortcomings. Even though a number of them can be easily be overcome, some even in a fairly trivial way, enough evidence remains to mistrust a classification of problem difficulty for GAs based on this measure. We now present some of its limitations as an a priori measure, and possible remedies.

First of all, the epistasis measure as we study it in this book is a static measure, i.e., it incorporates no information about the dynamics of a GA. An example in chapter III (page 99) makes it very clear that different GA settings may result in very different behavior on exactly the same function. One way to add a touch of GA to it is to use samples of the search space generated by successive populations of one or more GA runs to compute it [67].

Secondly, the measure can separate first order functions from higher order functions, but by randomly generating instances, one can easily show that it cannot reliably differentiate between different higher orders of interaction. In favor of the measure, however, we note that experiments show that on average, and in the case of all interactions randomly drawn from the same set, n-th order functions have a lower epistasis value than $n+1$-th order functions.

Thirdly, Reeves [77] showed that it is invariant to changes in sign of certain interactions, which means that it cannot distinguish between an interaction which merely reinforces and one which counteracts. An example of the former are the schemata $H_9, \ldots, H_{14}$ in the Royal Road function R_2. A counteracting interaction would for example penalize the presence of H_9, such that the presence of H_1 and H_2 together is inferior to the presence of either of them.

In general, it seems overly ambitious to try and capture all the richness of the vast amount of very different search problems into this one number representing the distance to one particular class of simple search problems. If the only easy problems for a GA looked like first order functions, the idea of measuring the distance from one set seems reasonable. As the case of the reinforcing and counteracting interactions shows, however, this is certainly not the case.

As a result of these weaknesses, no significantly large class of relevant search problems has shown up where the epistasis measure can be used as a reliable problem difficulty predictor. We will show in this book, however, that it can accurately classify instances belonging to the class of generalized Royal Road functions, for example. It performs equally well within the class of template functions, but fails for some simple unitation functions.

The remainder of this book deals with the mathematical aspects of one epistasis measure. Starting from the original definition by Davidor, it introduces, in chapter II, a different formulation based on linear algebra. In chapter III, the epistasis of a number of example function classes is computed analytically. Chapter IV shows that it is easier to compute epistasis in Walsh space. Finally, in chapters V and VI, the results of chapters II and IV are generalized to fitness functions over non-binary alphabets.

Chapter II

Epistasis

The goal of chapter I was to put epistasis measures in the context of GA research. This chapter deals with epistasis variance in full technical detail. We start with Rawlins' definition and Davidor's formalization, and transform this formalization into a convenient matrix formulation which is used throughout the book. We then deal with the rank and eigenspaces of the epistasis matrix $\boldsymbol{G}$, and present a couple of simple examples to illustrate the techniques. Next, we show in two different ways that the class of functions with minimal epistasis is exactly the class of first order functions defined by (I.3) on page 46. Finally, we show which class of functions yield a maximal epistasis value.

Note that Appendix B gives a brief account of the linear algebra used in this and the following chapters.

1 Introduction

Rawlins [75] was the first author to study epistasis in the context of GAs. He points out that the objective function may intuitively behave in two extreme ways:

- *There is zero epistasis.* In this case, every gene is independent of every other gene. That is, there is a fixed ordering of fitness (contribution to the overall objective value) of the alleles of each gene. It is clear that this situation can only occur if the objective function is expressible as a linear combination of the individual genes.

- *There is maximum epistasis.* In this case, no proper subset of genes is independent of any other gene. Each gene is dependent on every other gene for its fitness. That is, there is no possible fixed ordering of fitness of the alleles of any gene (if there where, then at least one gene would be independent of all other genes). This situation is equivalent to the objective function being a random function.

The first formal definition of the notion of epistasis was given by Davidor [11, 12] — we refer to the next section for details. Let us just point out here that Davidor's definition is based on the idea that, if a representation has very low epistasis, it should probably be processed more efficiently by a GA. If it contains high epistasis, there is too little structure in the solution space and the search process will probably settle on a local optimum. Starting from these principles, Davidor linearly decomposes the representation in order to develop a method for the prediction of the amount of nonlinearity embedded in a given representation. Quantifying this amount of nonlinearity should provide an estimate of the suitability of the problem to be processed efficiently by a GA.

2 Various definitions

In this section, we will define the notion of epistasis variance introduced in [12], as well as other epistasis measures derived from the original definition.

2.1 Epistasis variance

Let us fix a fitness function f on the search space $\Omega = \{0,1\}^\ell$. Given a population P in the search space Ω, the *average fitness value* of the population is given by

$$f(P) = \frac{\sum_{s \in P} f(s)}{|P|},$$

where $|P|$ denotes the cardinality of P. The *excess fitness value* of a string $s \in \Omega$ with respect to P is then given by $f(s) - f(P)$.

If we denote by $P_{i,a}$ the subset of P consisting of all strings with allele a on locus i, the i-th *average allele value* of s is defined to be

$$A_{i,P}(s) = \frac{\sum_{t \in P_{i,s_i}} f(t)}{|P_{i,s_i}|},$$

the *excess allele value* as $E_{i,P}(s) = A_{i,P}(s) - f(P)$ and the *excess genic value* as

$$E_P(s) = \sum_{i=0}^{\ell-1} E_{i,P}(s).$$

A linear "prediction" of the fitness value of a string may then be given by $A_P(s) = f(P) + E_P(s)$. The difference $\varepsilon_P(s) = f(s) - A_P(s)$ may thus be viewed as a measure of the epistasis of a string. If we expand the previous expression, we obtain the closed formula

$$\varepsilon_P(s) = f(s) - \sum_{i=0}^{\ell-1} \frac{1}{|P_{i,s_i}|} \sum_{t \in P_{i,s_i}} f(t) + \frac{\ell - 1}{|P|} \sum_{t \in P} f(t).$$

As a special case, if we consider $P = \Omega$, we deal with "global" epistasis:

$$\varepsilon(s) \equiv \varepsilon_\Omega(s) = f(s) - \sum_{i=0}^{\ell-1} \frac{1}{2^{\ell-1}} \sum_{t \in \Omega_{i,s_i}} f(t) + \frac{\ell - 1}{2^\ell} \sum_{t \in \Omega} f(t). \qquad \text{(II.1)}$$

Let us also introduce the following notions:

- the *epistasis variance* of a function is the variance of the fitness values of each string in the search space with respect to the predicted string value $A(s)$:

$$\sigma_\varepsilon^2 = \frac{1}{|\Omega|} \sum_{s \in \Omega} \varepsilon^2(s),$$

- the *fitness variance* is defined as

$$\sigma^2 = \frac{1}{|\Omega|} \sum_{s \in \Omega} (f(s) - f(\Omega))^2,$$

- the *genic variance* is defined as

$$\sigma_a^2 = \frac{1}{|\Omega|} \sum_{s \in \Omega} E(s)^2.$$

These notions are linked through the following result due to Manela and Campbell [56]:

Theorem II.1. *Let σ^2 be the fitness variance, σ_a^2 the genic variance and σ_ε^2 the epistasis variance. Then*

$$\sigma^2 = \sigma_a^2 + \sigma_\varepsilon^2.$$

Let us stress the fact that whereas epistasis is a property concerned with interactions between bits, epistasis variance essentially measures the amount of epistasis in a given function representation.

2.2 Normalized epistasis variance

Epistasis variance is meant to measure interactions between genes. These interactions will, of course, not change if we multiply the fitness function by a constant. Since, however, the epistasis variance does change, we are led to remedy this by considering normalized versions of epistasis variance.

A first way to realize this was proposed by Manela [56], who used the *proportion of epistasis*, defined as $P_\varepsilon = \sigma_\varepsilon^2/\sigma^2$.

Alternatively, one may also first normalize the fitness function and then calculate the associated epistasis variance (we introduce a dependency on f in our notations):

$$\sigma^*(f) = \sigma_\varepsilon^2 \left(\frac{f}{||\boldsymbol{f}||} \right) = \frac{\sigma_\varepsilon^2(f)}{||\boldsymbol{f}||^2}.$$

Here we use

$$\boldsymbol{f} = \begin{pmatrix} f(00\ldots0) \\ f(00\ldots1) \\ \vdots \\ f(11\ldots1) \end{pmatrix}.$$

We call $\sigma^*(f)$ the *normalized epistasis variance* of f. Omitting the factor $\frac{1}{|\Omega|}$ in the definition of epistasis variance, we obtain the notion of *epistasis value*, given by

$$\varepsilon_P^2(f) = \sum_{s \in P} \varepsilon_P^2(s)$$

and

$$\varepsilon^2(f) \equiv \varepsilon_\Omega^2(f) = \sum_{s \in \Omega} \varepsilon^2(s),$$

and that of *normalized epistasis value*, defined as

$$\varepsilon^*(f) = \varepsilon^2 \left(\frac{f}{||\boldsymbol{f}||} \right) = \frac{\varepsilon^2(f)}{||\boldsymbol{f}||^2}.$$

As we will see later, this normalized epistasis value (or *normalized epistasis*, as we will usually refer to it) only takes values between 0 and 1. If we restrict to positive functions, the actual maximum bound will be lower.

2.3 Epistasis correlation

Let us conclude this section with a last quantitative approach to epistasis, proposed by Rochet et al [80]. The notion we refer to is the so-called *epistasis correlation*, defined to be the correlation between the fitness function and the approximation A, essentially defined in section 2.1, over a given population. More precisely,

$$corr_\varepsilon(f) = \frac{\sum_{s \in P}(f(s) - f(P))(A(s) - A(P))}{\sqrt{\sum_{s \in P}(f(s) - f(P))^2} \sqrt{\sum_{s \in P}(A(s) - A(P))^2}},$$

where $A(P) = \frac{1}{|P|} \sum_{s \in P} A(s)$ is the average over the population P of the approximation A. Epistasis correlation takes values between 0 and 1. Note also that maximal epistasis occurs for an epistasis correlation value of 0 and minimal epistasis for the value 1.

3 Matrix formulation

3.1 The matrices $\boldsymbol{G}_\ell$ and $\boldsymbol{E}_\ell$

The main purpose of this section is to reformulate the definition of normalized epistasis, using elementary linear algebra, in order to simplify both its calculation and the study of its main properties. Let us start by introducing the vector

$$\boldsymbol{\varepsilon} = \begin{pmatrix} \varepsilon\,(00\ldots0) \\ \varepsilon\,(00\ldots1) \\ \vdots \\ \varepsilon\,(11\ldots1) \end{pmatrix}.$$

We will also write $f_0, \ldots, f_{2^\ell-1}$ for $f(00\ldots0), \ldots, f(11\ldots1)$, so

$$\boldsymbol{f} = \begin{pmatrix} f_0 \\ \vdots \\ f_{2^\ell-1} \end{pmatrix}.$$

For any $0 \leq i, j < 2^\ell$, put

$$e_{ij} = \frac{1}{2^\ell}\left(\ell + 1 - 2\mathrm{d}_{ij}\right),$$

where d_{ij} is the Hamming distance between i and j (the number of bits in which the binary representations on i and j differ; see section 1 of chapter I). We will rewrite the identity (II.1) as follows. First, note that for all $t \in \Omega$, the value $f(t)$ occurs $\ell - \mathrm{d}_{st}$ times in the expression $\sum_{i=0}^{\ell-1} \frac{1}{2^{\ell-1}} \sum_{t\in\Omega_{i,s_i}} f(t)$, hence

$$\sum_{i=0}^{\ell-1} \frac{1}{2^{\ell-1}} \sum_{t\in\Omega_{i,s_i}} f(t) = \frac{1}{2^\ell} \sum_{t\in\Omega} [2(\ell - \mathrm{d}_{st}) f(t)].$$

The entries e_{ij} define a matrix $\boldsymbol{E}_\ell = (e_{ij}) \in M_{2^\ell}(\mathbb{Q})$ (the set of all rational-valued square matrices of size 2^ℓ). From the previous remark, it now easily follows that

$$\begin{aligned} \varepsilon(s) &= f(s) - \frac{1}{2^\ell} \sum_{t\in\Omega} \left(2(\ell - \mathrm{d}_{st}) - (\ell - 1)\right) f(t) \\ &= f(s) - \sum_{t\in\Omega} e_{st} f(t), \end{aligned}$$

and therefore

$$\boldsymbol{\varepsilon} = \boldsymbol{f} - \boldsymbol{E}_\ell \boldsymbol{f}.$$

It thus finally follows that the epistasis value of f is given by

$$\varepsilon^2(f) = \sum_{s\in\Omega} \varepsilon^2(s) = \|\boldsymbol{\varepsilon}\|^2.$$

Define $\boldsymbol{G}_\ell = 2^\ell \boldsymbol{E}_\ell \in M_{2^\ell}(\mathbb{Z})$, i.e., $\boldsymbol{G}_\ell = (g_{ij})$, with $g_{ij} = \ell + 1 - 2\mathrm{d}_{ij}$ for all $0 \le i, j < 2^\ell$. For small values of ℓ, the corresponding matrices are

$$\boldsymbol{G}_0 = (1), \quad \boldsymbol{G}_1 = \begin{pmatrix} 2 & 0 \\ 0 & 2 \end{pmatrix}, \quad \boldsymbol{G}_2 = \begin{pmatrix} 3 & 1 & 1 & -1 \\ 1 & 3 & -1 & 1 \\ 1 & -1 & 3 & 1 \\ -1 & 1 & 1 & 3 \end{pmatrix}.$$

The matrices $\boldsymbol{G}_\ell$ may also be defined recursively by

$$\boldsymbol{G}_0 = (1), \quad \boldsymbol{G}_\ell = \begin{pmatrix} \boldsymbol{G}_{\ell-1} + \boldsymbol{U}_{\ell-1} & \boldsymbol{G}_{\ell-1} - \boldsymbol{U}_{\ell-1} \\ \boldsymbol{G}_{\ell-1} - \boldsymbol{U}_{\ell-1} & \boldsymbol{G}_{\ell-1} + \boldsymbol{U}_{\ell-1} \end{pmatrix},$$

where

$$\boldsymbol{U}_\ell = \begin{pmatrix} 1 & \cdots & 1 \\ \vdots & \ddots & \vdots \\ 1 & \cdots & 1 \end{pmatrix} \in M_{2^\ell}(\mathbb{Z}),$$

for every positive integer ℓ.

An interesting and useful property of $\boldsymbol{G}_\ell$ is given by the following result:

Lemma II.2. *With the previous notations, we have:*

1. *The sum of the elements of any row or column of $\boldsymbol{G}_\ell$ is given by*

$$\sum_{j=0}^{2^\ell-1} g_{ij} = \sum_{i=0}^{2^\ell-1} g_{ij} = 2^\ell.$$

2. *The sum of all the elements of $\boldsymbol{G}_\ell$ is given by*

$$\sum_{i,j=0}^{2^\ell-1} g_{ij} = 2^{2\ell}.$$

Proof. Take $i \in \{0, \ldots, 2^\ell - 1\}$. For any $0 \le j < 2^\ell$, clearly $\hat{\jmath} = 2^\ell - 1 - j \in \{0, \ldots, 2^\ell - 1\}$ has the property that $\mathrm{d}_{ij} + \mathrm{d}_{i\hat{\jmath}} = \ell$. It thus follows that

$$\sum_{j=0}^{2^\ell-1} g_{ij} = \sum_{j=0}^{2^{\ell-1}-1} (g_{ij} + g_{i\hat{\jmath}}) = \sum_{j=0}^{2^{\ell-1}-1} 2(\ell + 1 - (\mathrm{d}_{ij} + \mathrm{d}_{i\hat{\jmath}})) = 2^{\ell-1}2 = 2^\ell,$$

hence also that

$$\sum_{i,j=0}^{2^\ell-1} g_{ij} = \sum_{i=0}^{2^\ell-1}\sum_{j=0}^{2^\ell-1} g_{ij} = \sum_{i=0}^{2^\ell-1} 2^\ell = 2^{2\ell}.$$

□

Another recursion formula for $\boldsymbol{G}_\ell$ involves the Kronecker product of matrices (see section 1.1 of appendix B). Indeed, we may prove:

Lemma II.3. *For any pair of positive integers $q \le p$, we have:*

$$\boldsymbol{G}_p = \boldsymbol{U}_{p-q} \otimes \boldsymbol{G}_q + (\boldsymbol{G}_{p-q} - \boldsymbol{U}_{p-q}) \otimes \boldsymbol{U}_q.$$

Proof. As $\boldsymbol{G}_0 = \boldsymbol{U}_0 = (1)$, the statement is obvious for $q = 0$ and $q = p$. Take $q = 1$ and note that for any $0 \le i < 2^{\ell-1}$ and $0 \le j < 2^{\ell-1}$ the difference between $g_{2i,j}$ and $g_{2i+1,j}$ and between $g_{i,2j}$ and $g_{i,2j+1}$ is always 2. So,

$$\begin{aligned}\boldsymbol{G}_p &= \boldsymbol{G}_{p-1} \otimes \boldsymbol{U}_1 + \boldsymbol{U}_{p-1} \otimes (\boldsymbol{G}_1 - \boldsymbol{U}_1)\\ &= \boldsymbol{U}_{p-1} \otimes \boldsymbol{G}_1 + (\boldsymbol{G}_{p-1} - \boldsymbol{U}_{p-1}) \otimes \boldsymbol{U}_1.\end{aligned}$$

For $q = p - 1$, note that

$$\begin{aligned}\boldsymbol{U}_1 \otimes \boldsymbol{G}_{p-1} + (\boldsymbol{G}_1 - \boldsymbol{U}_1) \otimes \boldsymbol{U}_{p-1} &= \begin{pmatrix} \boldsymbol{G}_{p-1} & \boldsymbol{G}_{p-1} \\ \boldsymbol{G}_{p-1} & \boldsymbol{G}_{p-1} \end{pmatrix} + \begin{pmatrix} 1 & -1 \\ -1 & 1 \end{pmatrix} \otimes \boldsymbol{U}_{p-1}\\ &= \begin{pmatrix} \boldsymbol{G}_{p-1} + \boldsymbol{U}_{p-1} & \boldsymbol{G}_{p-1} - \boldsymbol{U}_{p-1} \\ \boldsymbol{G}_{p-1} - \boldsymbol{U}_{p-1} & \boldsymbol{G}_{p-1} + \boldsymbol{U}_{p-1} \end{pmatrix}\\ &= \boldsymbol{G}_p.\end{aligned}$$

Let us now argue by induction on q. Pick $0 < q \le p$, then

$$\begin{aligned}&\boldsymbol{U}_{p-q} \otimes \boldsymbol{G}_q + (\boldsymbol{G}_{p-q} - \boldsymbol{U}_{p-q}) \otimes \boldsymbol{U}_q\\ &= \boldsymbol{U}_{p-q} \otimes (\boldsymbol{U}_1 \otimes \boldsymbol{G}_{q-1} + (\boldsymbol{G}_1 - \boldsymbol{U}_1) \otimes \boldsymbol{U}_{q-1}) + (\boldsymbol{G}_{p-q} - \boldsymbol{U}_{p-q}) \otimes \boldsymbol{U}_q.\end{aligned}$$

As this is equal to

$$\boldsymbol{U}_{p-q+1} \otimes \boldsymbol{G}_{q-1} + (\boldsymbol{U}_{p-q} \otimes \boldsymbol{G}_1 + (\boldsymbol{G}_{p-q} - \boldsymbol{U}_{p-q}) \otimes \boldsymbol{U}_1 - \boldsymbol{U}_{p-q+1}) \otimes \boldsymbol{U}_{q-1},$$

we obtain

$$\boldsymbol{U}_{p-q+1} \otimes \boldsymbol{G}_{q-1} + (\boldsymbol{G}_{p-q+1} - \boldsymbol{U}_{p-q+1}) \otimes \boldsymbol{U}_{q-1} = \boldsymbol{G}_p.$$

This proves the assertion. □

We may also prove:

Proposition II.4. *For any positive integer ℓ, we have $\boldsymbol{G}_\ell^2 = 2^\ell \boldsymbol{G}_\ell$.*

Proof. For $\ell = 0$, clearly $\boldsymbol{G}_0^2 = (1)$, and the statement is true. Let us assume the statement to hold true for length $0, \ldots, \ell - 1$ and let us prove it for length ℓ. Since $\boldsymbol{U}_\ell^2 = 2^\ell \boldsymbol{U}_\ell$, it follows from the induction hypothesis that

$$\begin{aligned}
\boldsymbol{G}_\ell^2 &= \begin{pmatrix} \boldsymbol{G}_{\ell-1} + \boldsymbol{U}_{\ell-1} & \boldsymbol{G}_{\ell-1} - \boldsymbol{U}_{\ell-1} \\ \boldsymbol{G}_{\ell-1} - \boldsymbol{U}_{\ell-1} & \boldsymbol{G}_{\ell-1} + \boldsymbol{U}_{\ell-1} \end{pmatrix}^2 \\
&= \begin{pmatrix} 2\boldsymbol{G}_{\ell-1}^2 + 2\boldsymbol{U}_{\ell-1}^2 & 2\boldsymbol{G}_{\ell-1}^2 - 2\boldsymbol{U}_{\ell-1}^2 \\ 2\boldsymbol{G}_{\ell-1}^2 - 2\boldsymbol{U}_{\ell-1}^2 & 2\boldsymbol{G}_{\ell-1}^2 + 2\boldsymbol{U}_{\ell-1}^2 \end{pmatrix} \\
&= 2^\ell \begin{pmatrix} \boldsymbol{G}_{\ell-1} + \boldsymbol{U}_{\ell-1} & \boldsymbol{G}_{\ell-1} - \boldsymbol{U}_{\ell-1} \\ \boldsymbol{G}_{\ell-1} - \boldsymbol{U}_{\ell-1} & \boldsymbol{G}_{\ell-1} + \boldsymbol{U}_{\ell-1} \end{pmatrix} \\
&= 2^\ell \boldsymbol{G}_\ell.
\end{aligned}$$

□

Note that the previous result implies that the matrix $\boldsymbol{G}_\ell$ has only eigenvalues 0 and 2^ℓ. Indeed, if $\boldsymbol{x} \in \mathbb{R}^{2^\ell}$ is an eigenvector with eigenvalue λ, it follows from $\boldsymbol{G}_\ell \boldsymbol{x} = \lambda \boldsymbol{x}$ that

$$2^\ell \lambda \boldsymbol{x} = 2^\ell \boldsymbol{G}_\ell \boldsymbol{x} = \boldsymbol{G}_\ell^2 \boldsymbol{x} = \lambda^2 \boldsymbol{x},$$

whence $\lambda = 0$ or $\lambda = 2^\ell$, as claimed.

Corollary II.5. *For any positive integer ℓ, the matrix $\boldsymbol{E}_\ell$ is idempotent.*

Proof. This follows immediately from the fact that $\boldsymbol{E}_\ell = 2^{-\ell} \boldsymbol{G}_\ell$. □

As a consequence, $\boldsymbol{E}_\ell$ has eigenvalues 0 and 1.

The previous results allow for an elegant description of the epistasis value of f in terms of matrices:

Theorem II.6. *Let f be a fitness function over the search space Ω. The epistasis value of f is given by*

$$\varepsilon^2(f) = {}^t\boldsymbol{f}\boldsymbol{f} - {}^t\boldsymbol{f}\boldsymbol{E}_\ell \boldsymbol{f},$$

and the normalized epistasis value of f by

$$\varepsilon^*(f) = \frac{\varepsilon^2(f)}{\|\boldsymbol{f}\|^2} = 1 - \frac{{}^t\boldsymbol{f}\boldsymbol{E}_\ell\boldsymbol{f}}{{}^t\boldsymbol{f}\boldsymbol{f}}.$$

Proof. This follows immediately from

$$\begin{aligned}\varepsilon^2(f) &= \|\boldsymbol{\varepsilon}\|^2\\ &= {}^t(\boldsymbol{f} - \boldsymbol{E}_\ell\boldsymbol{f})(\boldsymbol{f} - \boldsymbol{E}_\ell\boldsymbol{f})\\ &= {}^t\boldsymbol{f}\boldsymbol{f} - {}^t\boldsymbol{f}\boldsymbol{E}_\ell\boldsymbol{f}.\end{aligned}$$

□

As the matrix $\boldsymbol{E}_\ell$ is idempotent and symmetric, it is an orthogonal projection (theorem B.29), hence

$$0 \leq \varepsilon^*(f) \leq 1.$$

3.2 The rank of the matrix G_ℓ

In order to describe the eigenspaces of the matrix $\boldsymbol{G}_\ell$ (and of $\boldsymbol{E}_\ell$), we first calculate its rank:

Proposition II.7. *For any positive integer ℓ, we have*

$$\operatorname{rk}(\boldsymbol{G}_\ell) = \ell + 1.$$

Proof. Let us again argue by induction, the statement being obvious for $\ell = 0$. Clearly,

$$\boldsymbol{G}_\ell = \begin{pmatrix} \boldsymbol{G}_{\ell-1} + \boldsymbol{U}_{\ell-1} & \boldsymbol{G}_{\ell-1} - \boldsymbol{U}_{\ell-1} \\ \boldsymbol{G}_{\ell-1} - \boldsymbol{U}_{\ell-1} & \boldsymbol{G}_{\ell-1} + \boldsymbol{U}_{\ell-1} \end{pmatrix}$$

may be transformed to

$$\begin{pmatrix} \boldsymbol{G}_{\ell-1} & 0 \\ 0 & \boldsymbol{U}_{\ell-1} \end{pmatrix}$$

by elementary row and column operations, showing that

$$\operatorname{rk}(\boldsymbol{G}_\ell) = \operatorname{rk}\begin{pmatrix} \boldsymbol{G}_{\ell-1} & 0 \\ 0 & \boldsymbol{U}_{\ell-1} \end{pmatrix} = \operatorname{rk}(\boldsymbol{G}_{\ell-1}) + 1 = \ell + 1,$$

indeed. □

Denote by V_0^ℓ and V_1^ℓ the eigenspaces in $\mathbb{R}^{2^\ell}$ corresponding to the eigenvalue 0 and 2^ℓ, respectively, of $\boldsymbol{G}_\ell$ (or, equivalently, to 0 and 1 as eigenvalues of $\boldsymbol{E}_\ell$). Then $\mathbb{R}^{2^\ell} = V_0^\ell \oplus V_1^\ell$ and as $V_0^\ell = \mathrm{Ker}\,(\boldsymbol{G}_\ell)$ and $V_1^\ell = \mathrm{Im}\,(\boldsymbol{G}_\ell)$, we find that $\dim V_0^\ell = 2^\ell - \ell - 1$ and $\dim V_1^\ell = \ell + 1$. An explicit orthogonal basis for V_1^ℓ may be constructed as follows. Let $\boldsymbol{v}_0^0 = 1$ and let us assume that we already inductively found a set $\{\boldsymbol{v}_0^{\ell-1}, \ldots, \boldsymbol{v}_{\ell-1}^{\ell-1}\} \subseteq \mathbb{R}^{2^{\ell-1}}$. We then construct a new set $\{\boldsymbol{v}_0^\ell, \ldots, \boldsymbol{v}_{\ell-1}^\ell, \boldsymbol{v}_\ell^\ell\} \subseteq \mathbb{R}^{2^\ell}$ by putting

$$\boldsymbol{v}_k^\ell = \begin{pmatrix} \boldsymbol{v}_k^{\ell-1} \\ \boldsymbol{v}_k^{\ell-1} \end{pmatrix}, \quad \text{for } 0 \le k < \ell$$

and

$$\boldsymbol{v}_\ell^\ell = \begin{pmatrix} \boldsymbol{u}_{\ell-1} \\ -\boldsymbol{u}_{\ell-1} \end{pmatrix}, \quad \text{where } \boldsymbol{u}_{\ell-1} = \begin{pmatrix} 1 \\ \vdots \\ 1 \end{pmatrix} \in \mathbb{R}^{2^{\ell-1}}.$$

As an example, if $\ell = 1$, we find

$$\boldsymbol{v}_0^1 = \begin{pmatrix} 1 \\ 1 \end{pmatrix}, \quad \boldsymbol{v}_1^1 = \begin{pmatrix} 1 \\ -1 \end{pmatrix}$$

and, if $\ell = 2$, we find

$$\boldsymbol{v}_0^2 = \begin{pmatrix} 1 \\ 1 \\ 1 \\ 1 \end{pmatrix}, \quad \boldsymbol{v}_1^2 = \begin{pmatrix} 1 \\ -1 \\ 1 \\ -1 \end{pmatrix}, \quad \boldsymbol{v}_2^2 = \begin{pmatrix} 1 \\ 1 \\ -1 \\ -1 \end{pmatrix}.$$

As we will see later in chapter IV, these vectors $\boldsymbol{v}_k^\ell$ are exactly the i-th columns of the Walsh matrix $\boldsymbol{V}_\ell$ for $i = 0$ and $i = 2^j$ with $0 \le j < \ell$.

Let us now prove:

Proposition II.8. *With the previous notations, for any positive integer ℓ, the set $\{\boldsymbol{v}_0^\ell, \ldots, \boldsymbol{v}_\ell^\ell\}$ is an orthogonal basis for V_1^ℓ.*

Proof. For $\ell = 0$, the statement is obvious. So, let us assume it to hold true for length $0, \ldots, \ell - 1$ and let us prove it for length ℓ. In this case, if $0 \le k \ne k' < \ell$, we have

$$ {}^t\boldsymbol{v}_k^\ell \boldsymbol{v}_{k'}^\ell = \begin{pmatrix} {}^t\boldsymbol{v}_k^{\ell-1} & {}^t\boldsymbol{v}_k^{\ell-1} \end{pmatrix} \begin{pmatrix} \boldsymbol{v}_{k'}^{\ell-1} \\ \boldsymbol{v}_{k'}^{\ell-1} \end{pmatrix} = 2\, {}^t\boldsymbol{v}_k^{\ell-1} \boldsymbol{v}_{k'}^{\ell-1} = 0 $$

and

$$ {}^t\boldsymbol{v}_k^\ell \boldsymbol{v}_\ell^\ell = \begin{pmatrix} {}^t\boldsymbol{v}_k^{\ell-1} & {}^t\boldsymbol{v}_k^{\ell-1} \end{pmatrix} \begin{pmatrix} \boldsymbol{u}_{\ell-1} \\ -\boldsymbol{u}_{\ell-1} \end{pmatrix} = {}^t\boldsymbol{v}_k^{\ell-1} \boldsymbol{u}_{\ell-1} - {}^t\boldsymbol{v}_k^{\ell-1} \boldsymbol{u}_{\ell-1} = 0. $$

In particular, the set $\{\boldsymbol{v}_0^\ell, \ldots, , \boldsymbol{v}_\ell^\ell\}$ is clearly independent. In order to conclude, it thus remains to prove that all of the $\boldsymbol{v}_k^\ell$ belong to V_1^ℓ. We again argue by induction. If $k < \ell$, then

$$ \begin{aligned} \boldsymbol{G}_\ell \boldsymbol{v}_k^\ell &= \begin{pmatrix} \boldsymbol{G}_{\ell-1} + \boldsymbol{U}_{\ell-1} & \boldsymbol{G}_{\ell-1} - \boldsymbol{U}_{\ell-1} \\ \boldsymbol{G}_{\ell-1} - \boldsymbol{U}_{\ell-1} & \boldsymbol{G}_{\ell-1} + \boldsymbol{U}_{\ell-1} \end{pmatrix} \begin{pmatrix} \boldsymbol{v}_k^{\ell-1} \\ \boldsymbol{v}_k^{\ell-1} \end{pmatrix} \\ &= 2 \begin{pmatrix} \boldsymbol{G}_{\ell-1} \boldsymbol{v}_k^{\ell-1} \\ \boldsymbol{G}_{\ell-1} \boldsymbol{v}_k^{\ell-1} \end{pmatrix} = 2^\ell \begin{pmatrix} \boldsymbol{v}_k^{\ell-1} \\ \boldsymbol{v}_k^{\ell-1} \end{pmatrix} \\ &= 2^\ell \boldsymbol{v}_k^\ell. \end{aligned} $$

On the other hand, we also have

$$ \begin{aligned} \boldsymbol{G}_\ell \boldsymbol{v}_\ell^\ell &= \begin{pmatrix} \boldsymbol{G}_{\ell-1} + \boldsymbol{U}_{\ell-1} & \boldsymbol{G}_{\ell-1} - \boldsymbol{U}_{\ell-1} \\ \boldsymbol{G}_{\ell-1} - \boldsymbol{U}_{\ell-1} & \boldsymbol{G}_{\ell-1} + \boldsymbol{U}_{\ell-1} \end{pmatrix} \begin{pmatrix} \boldsymbol{u}_{\ell-1} \\ -\boldsymbol{u}_{\ell-1} \end{pmatrix} \\ &= 2 \begin{pmatrix} \boldsymbol{U}_{\ell-1} \boldsymbol{u}_{\ell-1} \\ -\boldsymbol{U}_{\ell-1} \boldsymbol{u}_{\ell-1} \end{pmatrix} = 2^\ell \begin{pmatrix} \boldsymbol{u}_{\ell-1} \\ -\boldsymbol{u}_{\ell-1} \end{pmatrix} \\ &= 2^\ell \boldsymbol{v}_\ell^\ell. \end{aligned} $$

This proves the assertion. □

It is clear that $\varepsilon^*(f) = 0$ and $\varepsilon^*(f) = 1$ if and only if $\boldsymbol{f} \in V_1^\ell$ and $\boldsymbol{f} \in V_0^\ell$, respectively.

For example, if $\ell = 2$, then $\boldsymbol{f} \in V_1^2$ if and only if

$$ \boldsymbol{f} \in \left\langle \begin{pmatrix} 1 \\ 1 \\ 1 \\ 1 \end{pmatrix}, \begin{pmatrix} 1 \\ -1 \\ 1 \\ -1 \end{pmatrix}, \begin{pmatrix} 1 \\ 1 \\ -1 \\ -1 \end{pmatrix} \right\rangle, $$

so

$$\begin{pmatrix} f(00) \\ f(01) \\ f(10) \\ f(11) \end{pmatrix} = \alpha \begin{pmatrix} 1 \\ 1 \\ 1 \\ 1 \end{pmatrix} + \beta \begin{pmatrix} 1 \\ -1 \\ 1 \\ -1 \end{pmatrix} + \gamma \begin{pmatrix} 1 \\ 1 \\ -1 \\ -1 \end{pmatrix}$$

or

$$\begin{aligned} f(00) &= \alpha + \beta + \gamma \\ f(01) &= \alpha - \beta + \gamma \\ f(10) &= \alpha + \beta - \gamma \\ f(11) &= \alpha - \beta - \gamma \end{aligned}$$

and this yields

$$f(00) + f(11) = f(01) + f(10).$$

The converse is true as well. Indeed, assume that $f(00) + f(11) = f(01) + f(10)$. We just saw that

$$\boldsymbol{f} \in \left\langle \begin{pmatrix} 1 \\ 1 \\ 1 \\ 1 \end{pmatrix}, \begin{pmatrix} 1 \\ -1 \\ 1 \\ -1 \end{pmatrix}, \begin{pmatrix} 1 \\ 1 \\ -1 \\ -1 \end{pmatrix} \right\rangle$$

if and only if we may find $\alpha, \beta, \gamma \in \mathbb{R}$ with

$$\begin{aligned} f(00) &= \alpha + \beta + \gamma \\ f(01) &= \alpha - \beta + \gamma \\ f(10) &= \alpha + \beta - \gamma \\ f(11) &= \alpha - \beta - \gamma. \end{aligned}$$

However, our assumption clearly implies the values

$$\alpha = \frac{f(00) + f(11)}{2}, \beta = \frac{f(00) - f(01)}{2}, \gamma = \frac{f(00) - f(10)}{2}$$

to do the trick.

4 Examples

Let us calculate the normalized epistasis of some typical (but rather extreme) functions using the techniques developed in the previous section.

For arbitrary ℓ, let *cons* be the constant function with norm $||cons|| = 1$, i.e., $cons = 2^{-\frac{\ell}{2}}\boldsymbol{u}_\ell$, where

$$\boldsymbol{u}_\ell = \begin{pmatrix} 1 \\ \vdots \\ 1 \end{pmatrix} \in \mathbb{R}^{2^\ell}.$$

Taking into account lemma II.2, it follows easily that

$${}^t\boldsymbol{u}_\ell \boldsymbol{G}_\ell \boldsymbol{u}_\ell = \sum_{i,j=0}^{2^\ell-1} g_{ij} = 2^{2\ell}$$

and

$$\varepsilon^*(cons) = \varepsilon^*(\boldsymbol{u}_\ell) = 1 - \frac{2^{-\ell}\,{}^t\boldsymbol{u}_\ell \boldsymbol{G}_\ell \boldsymbol{u}_\ell}{2^\ell} = 0.$$

Let us now consider the vectors

$$\boldsymbol{e}_\ell = \begin{pmatrix} 1 \\ 0 \\ \vdots \\ 0 \end{pmatrix}, \quad \boldsymbol{e}'_\ell = \begin{pmatrix} 0 \\ 0 \\ \vdots \\ 1 \end{pmatrix}$$

in $\mathbb{R}^{2^\ell}$ and let us denote by $\boldsymbol{0}_\ell \in \mathbb{R}^{2^\ell}$ the vector all of whose entries are 0.

The needle-in-a-haystack function *needle* (see section 5.2 of chapter I) centered at $t = 0$ is given by $needle(s) = \delta_{s,0}$, where δ denotes the "Kronecker delta", i.e., $needle(s) = 1$ if $s = 0$ and $needle(s) = 0$ elsewhere. So, the associated vector of *needle* is $\boldsymbol{e}_\ell$ and

$$\begin{aligned} {}^t\boldsymbol{e}_\ell \boldsymbol{G}_\ell \boldsymbol{e}_\ell &= ({}^t\boldsymbol{e}_{\ell-1}\ {}^t\boldsymbol{0}_{\ell-1}) \begin{pmatrix} \boldsymbol{G}_{\ell-1} + \boldsymbol{U}_{\ell-1} & \boldsymbol{G}_{\ell-1} - \boldsymbol{U}_{\ell-1} \\ \boldsymbol{G}_{\ell-1} - \boldsymbol{U}_{\ell-1} & \boldsymbol{G}_{\ell-1} + \boldsymbol{U}_{\ell-1} \end{pmatrix} \begin{pmatrix} \boldsymbol{e}_{\ell-1} \\ \boldsymbol{0}_{\ell-1} \end{pmatrix} \\ &= {}^t\boldsymbol{e}_{\ell-1}\boldsymbol{G}_{\ell-1}\boldsymbol{e}_{\ell-1} + {}^t\boldsymbol{e}_{\ell-1}\boldsymbol{U}_{\ell-1}\boldsymbol{e}_{\ell-1} = {}^t\boldsymbol{e}_{\ell-1}\boldsymbol{G}_{\ell-1}\boldsymbol{e}_{\ell-1} + 1 \\ &= {}^t\boldsymbol{e}_{\ell-2}\boldsymbol{G}_{\ell-2}\boldsymbol{e}_{\ell-2} + 2 = \cdots = {}^t\boldsymbol{e}_1\boldsymbol{G}_1\boldsymbol{e}_1 + \ell - 1 = \ell + 1. \end{aligned}$$

Then, the normalized epistasis of *needle* is

$$\varepsilon^*(needle) = 1 - \frac{\ell+1}{2^\ell}.$$

Finally, consider the so-called camel function *camel* (also defined in section 5.2, chapter I) with $camel(0) = camel(2^\ell - 1) = 1$ and $camel(s) = 0$ for $s \neq 0, 2^\ell - 1$. The associated vector is $\boldsymbol{c}_\ell = \boldsymbol{e}_\ell + \boldsymbol{e}'_\ell$ and

$$\begin{aligned}{}^t\boldsymbol{c}_\ell \boldsymbol{G}_\ell \boldsymbol{c}_\ell &= ({}^t\boldsymbol{e}_\ell + {}^t\boldsymbol{e}'_\ell)\boldsymbol{G}_\ell(\boldsymbol{e}_\ell + \boldsymbol{e}'_\ell) \\ &= {}^t\boldsymbol{e}_\ell \boldsymbol{G}_\ell \boldsymbol{e}_\ell + 2\,{}^t\boldsymbol{e}_\ell \boldsymbol{G}_\ell \boldsymbol{e}'_\ell + {}^t\boldsymbol{e}'_\ell \boldsymbol{G}_\ell \boldsymbol{e}'_\ell.\end{aligned}$$

The verification of ${}^t\boldsymbol{e}'_\ell \boldsymbol{G}_\ell \boldsymbol{e}'_\ell = \ell + 1$ is analogous to the calculation of ${}^t\boldsymbol{e}_\ell \boldsymbol{G}_\ell \boldsymbol{e}_\ell$. Using this, it follows that

$$\begin{aligned}{}^t\boldsymbol{e}_\ell \boldsymbol{G}_\ell \boldsymbol{e}'_\ell &= ({}^t\boldsymbol{e}_{\ell-1}, {}^t\boldsymbol{0}_{\ell-1}) \begin{pmatrix} \boldsymbol{G}_{\ell-1} + \boldsymbol{U}_{\ell-1} & \boldsymbol{G}_{\ell-1} - \boldsymbol{U}_{\ell-1} \\ \boldsymbol{G}_{\ell-1} - \boldsymbol{U}_{\ell-1} & \boldsymbol{G}_{\ell-1} + \boldsymbol{U}_{\ell-1} \end{pmatrix} \begin{pmatrix} \boldsymbol{0}_{\ell-1} \\ \boldsymbol{e}'_{\ell-1} \end{pmatrix} \\ &= {}^t\boldsymbol{e}_{\ell-1}\boldsymbol{G}_{\ell-1}\boldsymbol{e}'_{\ell-1} - {}^t\boldsymbol{e}_{\ell-1}\boldsymbol{U}_{\ell-1}\boldsymbol{e}'_{\ell-1} = {}^t\boldsymbol{e}_{\ell-1}\boldsymbol{G}_{\ell-1}\boldsymbol{e}'_{\ell-1} - 1 \\ &= {}^t\boldsymbol{e}_{\ell-2}\boldsymbol{G}_{\ell-2}\boldsymbol{e}'_{\ell-2} - 2 = \cdots = {}^t\boldsymbol{e}_1\boldsymbol{G}_1\boldsymbol{e}'_1 - (\ell - 1) = 1 - \ell.\end{aligned}$$

Finally, we have

$$ {}^t\boldsymbol{c}_\ell \boldsymbol{G}_\ell \boldsymbol{c}_\ell = 2(\ell + 1) + 2(1 - \ell) = 4, \tag{II.2}$$

hence the normalized epistasis of *camel* is given by

$$\varepsilon^*(camel) = 1 - \frac{4}{2^\ell \parallel \boldsymbol{c}_\ell \parallel^2} = 1 - \frac{4}{2^\ell 2} = 1 - \frac{1}{2^{\ell-1}}.$$

5 Extreme values

We have already seen that the normalized epistasis value ε^* takes values between 0 and 1. However, in most practical situations, one considers functions which only take positive values. For these functions, we will calculate the extreme values of the normalized epistasis. Let us also point out that maximal and minimal values of $\varepsilon^*(f)$ correspond to minimal and maximal values of $\gamma(f) = {}^t\boldsymbol{f}\boldsymbol{G}_\ell\boldsymbol{f}$, respectively, with $\|\boldsymbol{f}\| = 1$, where, of course, $0 \leq \gamma(f) \leq 2^\ell$.

5.1 The minimal value of normalized epistasis

First, observe that the theoretical *minimal* value $\varepsilon^*(f) = 0$ or, equivalently, the maximal value $\gamma(f) = 2^\ell$, may indeed be reached.

If $\ell = 1$, then $\dim V_1^1 = \ell + 1 = 2$, so $V_1^1 = \mathbb{R}^2$ and for any $\boldsymbol{f} \in \mathbb{R}^2$ with $\|\boldsymbol{f}\| = 1$, we find

$$\gamma(f) = {}^t\boldsymbol{f}\boldsymbol{G}_1\boldsymbol{f} = \begin{pmatrix} f_0 \, f_1 \end{pmatrix} \begin{pmatrix} 2 \; 0 \\ 0 \; 2 \end{pmatrix} \begin{pmatrix} f_0 \\ f_1 \end{pmatrix} = 2\left(f_0^2 + f_1^2\right) = 2.$$

If $\ell > 1$ and $\boldsymbol{f} = 2^{-\frac{\ell}{2}}\boldsymbol{u}_\ell$, then $\|\boldsymbol{f}\| = 1$, and using lemma II.2 it easily follows that

$$\gamma(f) = {}^t\boldsymbol{f}\boldsymbol{G}_\ell\boldsymbol{f} = \sum_{i,j=0}^{2^\ell-1} g_{ij} f_i f_j = 2^{-\ell} \sum_{i,j=0}^{2^\ell-1} g_{ij} = 2^\ell.$$

We will now show that $\varepsilon^*(f) = 0$ occurs exactly when f has minimal epistasis in the sense of Rawlins, i.e., when f is a first order function:

$$f(s) = \sum_{i=0}^{\ell-1} g_i(s_i) \qquad \text{for any } s = s_{\ell-1} \dots s_0.$$

Note also that $g_i(s_i)$ is often written as $g_i(s)$, with the implicit assumption that g_i only depends on the value of s at position i.)

This condition is easily seen to be equivalent to the existence of a vector $\boldsymbol{g} \in \mathbb{R}^{2\ell}$ such that $\boldsymbol{A}_\ell \boldsymbol{g} = \boldsymbol{f}$, where $\boldsymbol{A}_\ell = (a_{ij}) \in \mathbb{R}^{2^\ell} \times \mathbb{R}^{2\ell}$ is defined as follows: if we encode a 0 as 01 and a 1 as 10, then the i-th row of $\boldsymbol{A}_\ell$ will be the encoded version of the number $i - 1$ in binary notation.

For example, if $\ell = 2$, then

$$\boldsymbol{A}_2 = \begin{pmatrix} 0\;1\;0\;1 \\ 0\;1\;1\;0 \\ 1\;0\;0\;1 \\ 1\;0\;1\;0 \end{pmatrix}.$$

Alternatively, $\boldsymbol{A}_\ell$ may be defined by

$$a_{ij} = \begin{cases} 1 - (((i-1)\,\mathtt{div}^{\lceil \frac{2\ell-j+1}{2} \rceil - 1}\, 2) \bmod 2) & \text{if } j \text{ is even} \\ (((i-1)\,\mathtt{div}^{\lceil \frac{2\ell-j+1}{2} \rceil - 1}\, 2) \bmod 2) & \text{if } j \text{ is odd.} \end{cases}$$

Here, for any $x \in \mathbb{R}$, we let $\lceil x \rceil$ denote the smallest integer n with $n \geq x$ and $\mathtt{div}$ denotes integer division. Moreover $\mathtt{div}^k$ is inductively defined by

$$\begin{cases} n\,\mathtt{div}^0\, m = n \\ n\,\mathtt{div}^1\, m = n\,\mathtt{div}\, m \\ n\,\mathtt{div}^k\, m = (n\,\mathtt{div}^{k-1}\, m)\,\mathtt{div}\, m. \end{cases}$$

Taking into account proposition B.9, we know that a linear system $\boldsymbol{f} = \boldsymbol{A}_\ell \boldsymbol{g}$ has $\boldsymbol{A}_\ell^\dagger \boldsymbol{f}$ as a solution, whenever solutions exist. (The matrix $\boldsymbol{A}_\ell^\dagger$ is called the *generalized inverse* of $\boldsymbol{A}_\ell$.) So, it is clear that the existence of $\boldsymbol{g}$ such that $\boldsymbol{A}_\ell \boldsymbol{g} = \boldsymbol{f}$ is equivalent to

$$\boldsymbol{f} - \boldsymbol{A}_\ell \boldsymbol{A}_\ell^\dagger \boldsymbol{f} = \boldsymbol{0}.$$

We will see below that $\boldsymbol{E}_\ell = \boldsymbol{A}_\ell \boldsymbol{A}_\ell^\dagger$ and from this one easily deduces the following result:

Theorem II.9. *The following statements are equivalent:*

1. $\boldsymbol{f} = \boldsymbol{E}_\ell \boldsymbol{f}$,
2. f *is a first order function.*

To prove this theorem, we first show that the ranges of $\boldsymbol{A}_\ell$ and $\boldsymbol{E}_\ell$ are identical:

Lemma II.10. $\mathrm{Im}(\boldsymbol{A}_\ell) = \mathrm{Im}(\boldsymbol{E}_\ell)$.

Proof. If we denote by $\boldsymbol{a}_1^\ell, \boldsymbol{a}_2^\ell, \ldots, \boldsymbol{a}_{2\ell}^\ell$ the columns of $\boldsymbol{A}_\ell$ and by $\boldsymbol{g}_0^\ell, \boldsymbol{g}_1^\ell, \ldots, \boldsymbol{g}_{2^\ell-1}^\ell$ the columns of $\boldsymbol{G}_\ell$, we have to prove that

$$< \boldsymbol{a}_1^\ell, \boldsymbol{a}_2^\ell, \ldots, \boldsymbol{a}_{2\ell}^\ell > = < \boldsymbol{g}_0^\ell, \boldsymbol{g}_1^\ell, \ldots, \boldsymbol{g}_{2^\ell-1}^\ell > .$$

For $\ell = 1$, this is clear, since

$$< \boldsymbol{a}_1^1, \boldsymbol{a}_2^1 > = < (0,1), (1,0) > = < (2,0), (0,2) > = < \boldsymbol{g}_0^1, \boldsymbol{g}_1^1 > .$$

Let us now argue by induction, i.e., suppose that

$$< \boldsymbol{a}_1^{\ell-1}, \boldsymbol{a}_2^{\ell-1}, \ldots, \boldsymbol{a}_{2\ell-2}^{\ell-1} > = < \boldsymbol{g}_0^{\ell-1}, \boldsymbol{g}_1^{\ell-1}, \ldots, \boldsymbol{g}_{2^{\ell-1}-1}^{\ell-1} >$$

and let us prove the analogous result in the length ℓ case. First of all, note that applying elementary column operations to

$$\boldsymbol{G}_\ell = \begin{pmatrix} \boldsymbol{G}_{\ell-1} + \boldsymbol{U}_{\ell-1} & \boldsymbol{G}_{\ell-1} - \boldsymbol{U}_{\ell-1} \\ \boldsymbol{G}_{\ell-1} - \boldsymbol{U}_{\ell-1} & \boldsymbol{G}_{\ell-1} + \boldsymbol{U}_{\ell-1} \end{pmatrix}$$

easily yields that

$$\mathrm{Im}(\boldsymbol{G}_\ell) = \mathrm{Im}(\boldsymbol{G}'_\ell) = \mathrm{Im}(\boldsymbol{G}''_\ell),$$

where

$$\boldsymbol{G}'_\ell = \begin{pmatrix} \boldsymbol{G}_{\ell-1} & \boldsymbol{G}_{\ell-1} - \boldsymbol{U}_{\ell-1} \\ \boldsymbol{G}_{\ell-1} & \boldsymbol{G}_{\ell-1} + \boldsymbol{U}_{\ell-1} \end{pmatrix}$$

and

$$\boldsymbol{G}''_\ell = \begin{pmatrix} \boldsymbol{G}_{\ell-1} & -\boldsymbol{U}_{\ell-1} \\ \boldsymbol{G}_{\ell-1} & \boldsymbol{U}_{\ell-1} \end{pmatrix}.$$

On the other hand, note that

$$\boldsymbol{A}_\ell = \begin{pmatrix} 0\ 1 & \\ \vdots\ \vdots\ \boldsymbol{a}_1^{\ell-1} \cdots \boldsymbol{a}_{2\ell-2}^{\ell-1} \\ 0\ 1 & \\ 1\ 0 & \\ \vdots\ \vdots\ \boldsymbol{a}_1^{\ell-1} \cdots \boldsymbol{a}_{2\ell-2}^{\ell-1} \\ 1\ 0 & \end{pmatrix}$$

and

$$\boldsymbol{G}''_\ell = \begin{pmatrix} \boldsymbol{g}_0^{\ell-1} \cdots \boldsymbol{g}_{2^{\ell-1}-1}^{\ell-1} & -\boldsymbol{U}_{\ell-1} \\ \boldsymbol{g}_0^{\ell-1} \cdots \boldsymbol{g}_{2^{\ell-1}-1}^{\ell-1} & \boldsymbol{U}_{\ell-1} \end{pmatrix}.$$

If we take $i = 2^{\ell-1}, \dots, 2^\ell - 1$, the corresponding column of $\boldsymbol{G}''_\ell$ is of the form

$$\begin{pmatrix} -1 \\ \vdots \\ -1 \\ 1 \\ \vdots \\ 1 \end{pmatrix} = \begin{pmatrix} 0 \\ \vdots \\ 0 \\ 1 \\ \vdots \\ 1 \end{pmatrix} - \begin{pmatrix} 1 \\ \vdots \\ 1 \\ 0 \\ \vdots \\ 0 \end{pmatrix}.$$

Now, the i-th column of $\boldsymbol{G}''_\ell$, with $0 \le i < 2^{\ell-1}$, is of the form

$$\begin{pmatrix} \boldsymbol{g}_i^{\ell-1} \\ \boldsymbol{g}_i^{\ell-1} \end{pmatrix}.$$

But $\boldsymbol{g}_i^{\ell-1} \in < \boldsymbol{a}_1^{\ell-1}, \ldots, \boldsymbol{a}_{2\ell-2}^{\ell-1} >$, so

$$\begin{pmatrix} \boldsymbol{g}_i^{\ell-1} \\ \boldsymbol{g}_i^{\ell-1} \end{pmatrix} \in < \boldsymbol{a}_3^{\ell}, \ldots, \boldsymbol{a}_{2\ell}^{\ell} > .$$

Similarly each of the $2\ell - 2$ last columns of $\boldsymbol{A}_\ell$ is a linear combination of the first $2^{\ell-1} - 1$ columns of $\boldsymbol{G}_\ell''$.

Moreover, we know that $\boldsymbol{u}_{\ell-1} = \begin{pmatrix} 1 \\ \vdots \\ 1 \end{pmatrix}$ is a linear combination of the columns of $\boldsymbol{G}_\ell''$ as

$$\begin{pmatrix} 1 \\ \vdots \\ 1 \end{pmatrix} = \frac{1}{2} \left(\boldsymbol{g}_j^{\ell-1} + \boldsymbol{g}_{2^{\ell-1}-1-j}^{\ell-1} \right)$$

for $0 \leq j < 2^{\ell-1}$.

Finally, we have that

$$\begin{pmatrix} 0 \\ \vdots \\ 0 \\ 1 \\ \vdots \\ 1 \end{pmatrix} = \frac{1}{2} \begin{pmatrix} 1 \\ \vdots \\ 1 \\ 1 \\ \vdots \\ 1 \end{pmatrix} + \frac{1}{2} \begin{pmatrix} -1 \\ \vdots \\ -1 \\ 1 \\ \vdots \\ 1 \end{pmatrix},$$

and also

$$\begin{pmatrix} 1 \\ \vdots \\ 1 \\ 0 \\ \vdots \\ 0 \end{pmatrix} = \begin{pmatrix} 1 \\ \vdots \\ 1 \\ 1 \\ \vdots \\ 1 \end{pmatrix} - \begin{pmatrix} 0 \\ \vdots \\ 0 \\ 1 \\ \vdots \\ 1 \end{pmatrix}.$$

□

Proposition II.11. $\boldsymbol{E}_\ell = \boldsymbol{A}_\ell \boldsymbol{A}_\ell^\dagger$.

Proof. We know that $\boldsymbol{A}_\ell \boldsymbol{A}_\ell^\dagger$ is the orthogonal projection on the range $\mathrm{Im}(\boldsymbol{A}_\ell)$ and also that a linear map is an orthogonal projection if, and only if, its corresponding matrix is idempotent and symmetric (proposition B.28). Taking into account this and the previous result, we conclude that $\boldsymbol{E}_\ell = \boldsymbol{A}_\ell \boldsymbol{A}_\ell^\dagger$, because $\boldsymbol{E}_\ell$ is idempotent and symmetric and the orthogonal projection on a subspace is unique. □

An easier proof of theorem II.9 may be given as follows. For any $0 \le i < \ell$ define $h_i^\ell : \Omega \to \mathbb{R}$ by putting $h_i^\ell(s) = 1$ if $s_i = 1$ and $h_i^\ell(s) = 0$ elsewhere and denote by $\boldsymbol{h}_i^\ell$ the corresponding vector in $\mathbb{R}^{2^\ell}$. It is clear that for any $0 \le i < \ell - 1$, we have

$$\boldsymbol{h}_i^\ell = \begin{pmatrix} \boldsymbol{h}_i^{\ell-1} \\ \boldsymbol{h}_i^{\ell-1} \end{pmatrix},$$

while

$$\boldsymbol{h}_{\ell-1}^\ell = \begin{pmatrix} \mathbf{0}_{\ell-1} \\ \boldsymbol{u}_{\ell-1} \end{pmatrix} \quad \text{with } \mathbf{0}_i = \begin{pmatrix} 0 \\ \vdots \\ 0 \end{pmatrix} \in \mathbb{R}^{2^i},$$

for any $i \ge 0$.

So, a straightforward induction argument shows that the vectors $\boldsymbol{h}_0^\ell, \ldots, \boldsymbol{h}_{\ell-1}^\ell, \boldsymbol{u}_\ell$ belong to V_1^ℓ. It is also easy to see that they are linearly independent. Indeed, if $\sum \alpha_i \boldsymbol{h}_i^\ell + \beta \boldsymbol{u}_\ell = 0$, then, with $g = \sum \alpha_i h_i^\ell + \beta u_\ell$, we find $g(2^i) = \alpha_i + \beta = 0$ for $0 \le i < \ell$ and $g(0) = \beta = 0$, hence $\alpha_0 = \cdots = \alpha_{\ell-1} = \beta = 0$.

If $g_i : \Omega \to \mathbb{R}$ only depends on the i-th bit, i.e., $g_i(s) = a_i$ if $s_i = 1$ and $g_i(s) = b_i$ if $s_i = 0$, say, then we may write g_i as $a_i h_i^\ell + b_i \left(u_\ell - h_i^\ell\right)$.

So, if f is a first order function, then

$$\boldsymbol{f} \in \left\langle \boldsymbol{h}_0^\ell, \ldots, \boldsymbol{h}_{\ell-1}^\ell, \boldsymbol{u}_\ell \right\rangle = V_1^\ell,$$

i.e., $\varepsilon^*(f) = 0$.

Conversely, if $\boldsymbol{f} \in V_1^\ell$, then

$$f = \sum \alpha_i h_i^\ell + \beta u_\ell = \sum g_i,$$

where

$$g_0 = (\alpha_0 + \beta)\, h_0^\ell + \beta \left(u_\ell - h_0^\ell\right)$$

and

$$g_i = \alpha_i h_i^\ell \quad \text{for } 1 \leq i < \ell.$$

Example II.12. As an example, let us show that "linear" functions

$$f : \Omega \to \mathbb{R} : s \mapsto bs + a,$$

with $a, b \in \mathbb{R}$, have zero epistasis. For each $0 \leq i < \ell$, define the function g_i on $\{0, 1\}$ by

$$g_i(t) = 2^i bt + \frac{a}{\ell}.$$

Then, it is clear that

$$f(s) = b\sum_{i=0}^{\ell-1} 2^i s_i + a = \sum_{i=0}^{\ell-1}(bs_i 2^i + \frac{a}{\ell}) = \sum_{i=0}^{\ell-1} g_i(s_i).$$

5.2 The maximal value of normalized epistasis

Let us now take a look at the *maximal* value of $\varepsilon^*(f)$. We have already mentioned that $\varepsilon^*(f) \leq 1$. However, if we restrict to positive-valued functions, we claim that the maximal value of $\varepsilon^*(f)$ is $1 - \frac{1}{2^{\ell-1}}$. Note that we may, of course, assume that $||\boldsymbol{f}|| = 1$, and prove that the minimal value of $\gamma(f)$ is 2.

Let us first point out that $\gamma(f) = 2$ may effectively be reached, by choosing

$$\boldsymbol{f} = \begin{pmatrix} \alpha \\ 0 \\ \vdots \\ 0 \\ \alpha \end{pmatrix},$$

where $\alpha = \frac{\sqrt{2}}{2}$. This is clear since $f = \alpha camel$, where *camel* is the camel function defined in section 4. So, if we take into account calculation (II.2) in the same section, it follows that

$$\gamma(f) = \alpha({}^t\boldsymbol{c}_\ell \boldsymbol{G}_\ell \boldsymbol{c}_\ell)\alpha = 4\alpha^2 = 2.$$

Proposition II.13. *For any positive integer ℓ and any positive valued function $f:\Omega\to\mathbb{R}$, we have*

$$\varepsilon^*(f)\le 1-\frac{1}{2^{\ell-1}}.$$

Proof. Again we assume that $||\boldsymbol{f}||=1$, and show that $\gamma(f)\ge 2$. Since $\boldsymbol{G}_\ell$ is symmetric, there exists some orthogonal matrix $\boldsymbol{S}$ such that ${}^t\boldsymbol{S}\boldsymbol{G}_\ell\boldsymbol{S}=\boldsymbol{D}$ is a diagonal matrix (corollary B.40), whose diagonal entries are the eigenvalues of $\boldsymbol{G}_\ell$ (taking into account multiplicities), i.e.,

$$\boldsymbol{D}=\begin{pmatrix}2^\ell\boldsymbol{I}_{\ell+1} & 0\\ 0 & 0\end{pmatrix}\in M_{2^\ell}(\mathbb{Z}).$$

Let $\boldsymbol{g}={}^t\boldsymbol{S}\boldsymbol{f}$, then we thus find that

$$\gamma(f)=2^\ell\left(g_0^2+\cdots+g_\ell^2\right).$$

The matrix $\boldsymbol{S}$ may be constructed by choosing its columns to be an orthogonal basis consisting of normalized eigenvectors of $\boldsymbol{G}_\ell$. In particular, its first $\ell+1$ columns may be chosen to be the columns $2^{-\frac{\ell}{2}}\boldsymbol{v}_i^\ell$ $(i=0,\ldots,\ell)$, as $||\boldsymbol{v}_i^\ell||=2^{\frac{\ell}{2}}$ for all i.

Now, we have that

$$\gamma(f_0,\ldots,f_{2^\ell-1})=\gamma(f)=\sum_{i=0}^{\ell}\left({}^t\boldsymbol{v}_i^\ell\boldsymbol{f}\right)^2.$$

By construction (we temporarily add subscripts to γ to stress the dimension of its argument),

$$\begin{aligned}\gamma_\ell(f_0,\ldots,f_{2^\ell-1})=&\,\gamma_{\ell-1}(f_0+f_{2^{\ell-1}},\ldots,f_{2^{\ell-1}-1}+f_{2^\ell-1})\\ &+\left((f_0+f_1+\cdots+f_{2^{\ell-1}-1})-(f_{2^{\ell-1}}+\cdots+f_{2^\ell-1})\right)^2.\end{aligned}$$

Let us write

$$\hat{\boldsymbol{f}}=\begin{pmatrix}f_0+f_{2^{\ell-1}}\\ \vdots\\ f_{2^{\ell-1}-1}+f_{2^\ell-1}\end{pmatrix}\in\mathbb{R}^{2^{\ell-1}}.$$

Then

$$\begin{aligned}\|\hat{\boldsymbol{f}}\|^2 &= (f_0 + f_{2^{\ell-1}})^2 + \cdots + (f_{2^{\ell-1}-1} + f_{2^\ell-1})^2 \\ &= f_0^2 + \cdots + f_{2^\ell-1}^2 + 2\,(f_0 f_{2^{\ell-1}} + \cdots + f_{2^{\ell-1}-1} f_{2^\ell-1}) \\ &= 1 + 2\,(f_0 f_{2^{\ell-1}} + \cdots + f_{2^{\ell-1}-1} f_{2^\ell-1}) \\ &= a^2,\end{aligned}$$

for some positive real $a \geq 1$.

Put $f' = \frac{1}{a}\hat{f}$, then $\|\boldsymbol{f}'\| = 1$ and

$$\gamma_{\ell-1}(f') = \gamma_{\ell-1}(\frac{1}{a}\hat{f}) = \frac{1}{a^2}\gamma_{\ell-1}(\hat{f}).$$

Note that since f only takes positive values, so does f'. Now, let us assume for some positive integer ℓ that $\gamma_\ell(f) < 2$, for a positive $f : \Omega \to \mathbb{R}$ with $\|\boldsymbol{f}\| = 1$. Then

$$\begin{aligned}\gamma_{\ell-1}(\hat{f}) &\leq \gamma_{\ell-1}(\hat{f}) + ((f_0 + f_1 + \cdots + f_{2^{\ell-1}-1}) - (f_{2^{\ell-1}} + \cdots + f_{2^\ell-1}))^2 \\ &= \gamma_\ell(f) < 2.\end{aligned}$$

In particular,

$$\gamma_{\ell-1}(f') = \frac{1}{a^2}\gamma_{\ell-1}(\hat{f}) < \frac{2}{a^2} \leq 2$$

as well. But then, iterating this procedure, we find some positive $f : \{0,1\} \to \mathbb{R}$ with $\|\boldsymbol{f}\| = 1$, such that $\gamma_1(f) < 2$. This is impossible, however, as

$$\gamma_1(f) = 2.$$

This contradiction proves the assertion. □

We already pointed out above that the minimal value $\gamma(f) = 2$ may actually be reached for every ℓ. On the other hand, the example we gave is essentially the only one. Note, of course, that if $\ell = 1$ and $\|\boldsymbol{f}\| = 1$, then we always have $\gamma(f) = 2$.

So, for $\ell > 1$, define for any $0 \leq i < 2^{\ell-1}$ the vector $\boldsymbol{q}_i^\ell \in \mathbb{R}^{2^\ell}$ by $\left(\boldsymbol{q}_i^\ell\right)_i = \left(\boldsymbol{q}_i^\ell\right)_{\bar{\imath}} = \frac{\sqrt{2}}{2}$, and $\left(\boldsymbol{q}_i^\ell\right)_j = 0$ if $j \neq i, \bar{\imath}$ (where $\bar{\imath}$ denotes the binary complement of i).

Then:

Proposition II.14. *With notations as before, the following statements are equivalent for any $\ell \geq 2$ and any positive $\boldsymbol{f} \in \mathbb{R}^{2^\ell}$ with $\|\boldsymbol{f}\| = 1$:*

1. $\varepsilon^*(f) = 1 - \frac{1}{2^{\ell-1}}$;
2. *there exists some $0 \leq i < 2^{\ell-1}$ with $\boldsymbol{f} = \boldsymbol{q}_i^\ell$.*

Proof. Let us first verify the statement for $\ell = 2$. If $\gamma(f_0, f_1, f_2, f_3) = 2$ (with $\|\boldsymbol{f}\| = 1$), then $\gamma(f_0 + f_2, f_1 + f_3) + ((f_0 + f_1) - (f_2 + f_3))^2 = 2$. So, with notations as in the previous result,

$$2a^2 = a^2\gamma(f') = \gamma(\hat{f}) \leq 2,$$

hence, as $a \geq 1$, it follows that $a = 1$, so

$$f_0 f_2 + f_1 f_3 = 0 \tag{$*$}$$

and $\gamma(\hat{f}) = 2$, hence

$$f_0 + f_1 = f_2 + f_3. \tag{$**$}$$

Of course, $(*)$ is equivalent to $f_0 f_2 = f_1 f_3 = 0$, as f is positive. If $f_0 = 0$, then we cannot have $f_1 = 0$, as otherwise $f_2 = f_3 = 0$, by $(**)$, so $f_3 = 0$ and $\boldsymbol{f} = \boldsymbol{q}_1^2$. If $f_0 \neq 0$, then $f_2 = 0$ and $\boldsymbol{f} = \boldsymbol{q}_0^2$.

Let us now argue by induction, i.e., suppose the statement holds up to length $\ell - 1$, and let us see what happens for length ℓ.

If $\gamma_\ell(f) = 2$, then, as $\gamma_{\ell-1}(\hat{f}) \leq 2$, we again find $\gamma_{\ell-1}(\hat{f}) = 2$, and

$$f_0 + \cdots + f_{2^{\ell-1}-1} = f_{2^{\ell-1}} + \cdots + f_{2^\ell-1}.$$

From

$$2 \leq \gamma_{\ell-1}(f') = \frac{1}{a^2}\gamma_{\ell-1}(\hat{f}) = \frac{2}{a^2}$$

and $a \geq 1$, it follows that $\|\hat{\boldsymbol{f}}\| = 1$ and

$$f_0 f_{2^{\ell-1}} + \cdots + f_{2^{\ell-1}-1} f_{2^\ell-1} = 0,$$

i.e.,

$$f_0 f_{2^{\ell-1}} = \cdots = f_{2^{\ell-1}-1} f_{2^\ell-1} = 0.$$

By induction $\hat{\boldsymbol{f}} = \boldsymbol{f}' = \boldsymbol{g}_i^{\ell-1}$ for some $0 \leq i \leq 2^{\ell-2} - 1$, i.e.,

$$f_i + f_{2^{\ell-1}+i} = f_{2^{\ell-1}-i} + f_{2^\ell-i} = \frac{\sqrt{2}}{2}$$

and other entries vanish. Since $f_i f_{2^{\ell-1}+i} = f_{2^{\ell-1}-i} f_{2^\ell-i} = 0$, the same argument as above applies, showing that $\boldsymbol{f} = \boldsymbol{q}_i^\ell$ or $\boldsymbol{f} = \boldsymbol{g}_{2^{\ell-1}-i}^\ell$, indeed. □

Chapter III

Examples

This chapter is devoted to the explicit calculation of epistasis values for three classes of functions: generalized Royal Road functions, unitation functions, and template functions. Unlike the calculations in chapter IV, where we will work in the Walsh basis, the calculations in this chapter are *direct*, i.e., $\boldsymbol{f G_\ell f}$ is directly computed for the respective functions.

The Royal Road functions are of historical importance with regard to GA dynamics. They were one of the first "laboratory functions" especially designed to study the dynamics of GAs. Given the strong linkage between the bits in these functions, it seems justified to compute their epistasis and relate it to the problem difficulty they impose on a GA. We show in the first section of this chapter that within the parametrized class of generalized Royal Road functions, this problem difficulty is indeed directly related to the epistasis values.

The motivation for computing the epistasis of unitation functions is that these functions occur very frequently in GA theory because of their simple yet in some way flexible fitness assignment scheme. Moreover, they occur in GA problem difficulty research in the context of deception.

Template functions are a variation on the theme of the Royal Road functions. Their controlling parameter defines the length of the "template" whose presence in a chain yields its fitness value (see below for a precise definition). From their very definition, it seems obvious that the epistasis should increase when the template length does, and that this implies the function to be harder to optimize. The aim of the third section of this chapter is to show that this is indeed the case.

1 Royal Road functions

1.1 Generalized Royal Road functions of type I

Mimicking the constructions in [60] detailed in section 5.4 of chapter I, we introduce, for any pair of positive integers $m \leq n$, *generalized Royal Road functions* R^n_m *of Type I* through the schemata

$$H_i^{n,m} = \#^{(2^m i)} 1^{(2^m)} \#^{2^n - 2^m (i+1)},$$

where $0 \leq i < 2^{n-m}$. The value of R^n_m applied to a length $\ell = 2^n$ string s is given by

$$R^n_m(s) = \sum_{s \in H_i^{n,m}} c_i,$$

where, for any $0 \leq i < 2^{n-m}$, we put $c_i = 2^m$. Obviously, R^6_3 is Forrest and Mitchell's original Royal Road function R_1.

In order to calculate the normalized epistasis of the functions R^n_m, one first calculates

$$\gamma(R^n_m) = {}^t\boldsymbol{R}^n_m \boldsymbol{G}_{2^n} \boldsymbol{R}^n_m,$$

where, as before, $\boldsymbol{R}^n_m \in \mathbb{R}^{2^{2^n}}$ denotes the vector corresponding to R^n_m, i.e.,

$$\boldsymbol{R}^n_m = \begin{pmatrix} R^n_m(00\ldots0) \\ \vdots \\ R^n_m(11\ldots1) \end{pmatrix}.$$

To simplify the actual calculation, let us note that

$$\boldsymbol{R}^n_m = 2^m \sum_{0 \leq i < 2^{n-m}} \boldsymbol{h}_i^{n,m},$$

where, for any $0 \leq i < 2^{n-m}$, we denote by $\boldsymbol{h}_i^{n,m}$ the vector corresponding to the characteristic function $h_i^{n,m}$ of $H_i^{m,n}$, defined by

$$h_i^{n,m}(s) = \begin{cases} 1 & s \in H_i^{n,m} \\ 0 & s \notin H_i^{n,m}. \end{cases}$$

For example, if $n = m$, then

$$\boldsymbol{h}_0^{n,n} = \begin{pmatrix} 0 \\ 0 \\ \vdots \\ 0 \\ 1 \end{pmatrix}.$$

For each $\boldsymbol{r} \in \mathbb{R}^p$, let $\xi_q(\boldsymbol{r})$ denote the vector

$${}^t\Big(\boldsymbol{r}\,\boldsymbol{r} \cdots \boldsymbol{r}\Big) \in \mathbb{R}^{pq}.$$

If $n = m + 1$ we find, for $0 \leq i < 2^{n-m-1}$ (i.e., $i = 0$), that

$$\boldsymbol{h}_i^{n,n-1} = \begin{pmatrix} \xi_{2^{2^{n-1}}}(0) \\ \xi_{2^{2^{n-1}}}(0) \\ \vdots \\ \xi_{2^{2^{n-1}}}(0) \\ \xi_{2^{2^{n-1}}}(1) \end{pmatrix}$$

and for $2^{n-m-1} \leq i < 2^{n-m}$ (i.e. $i = 1$), that

$$\boldsymbol{h}_i^{n,n-1} = \xi_{2^{2^{n-1}}}(\begin{pmatrix} 0 \\ 0 \\ \vdots \\ 0 \\ 1 \end{pmatrix}).$$

In general, these vectors $\boldsymbol{h}_i^{n,m}$ may easily be constructed by induction, since we have, for $0 \leq i < 2^{n-m-1}$, that

$$\boldsymbol{h}_i^{n,m} = \begin{pmatrix} \xi_{2^{2^{n-1}}}(h_i^{n-1,m}(0)) \\ \xi_{2^{2^{n-1}}}(h_i^{n-1,m}(1)) \\ \vdots \\ \xi_{2^{2^{n-1}}}(h_i^{n-1,m}(2^{2^{n-1}} - 1)) \end{pmatrix}$$

and, for $2^{n-m-1} \leq i < 2^{n-m}$, that

$$\boldsymbol{h}_i^{n,m} = \xi_{2^{2^{n-1}}}(\boldsymbol{h}_{i-2^{n-m-1}}^{n-1,m}).$$

In order to better describe the structure of $\boldsymbol{G}_{2^n}$ and its interaction with the two different constructions of $\boldsymbol{h}_i^{n,m}$, define τ, a permutation on the rows of a column vector, by

$$\tau(\boldsymbol{h}_k^{n,m}) = \boldsymbol{h}_{k+2^{n-m-1}}^{n,m},$$

for all $0 \leq k < 2^{n-m-1}$. Clearly, $\tau = \tau^{-1}$, and we obtain

$$\boldsymbol{h}_k^{n,m} \boldsymbol{G}_{2^n} \boldsymbol{h}_k^{n,m} = \tau(\boldsymbol{h}_k^{n,m}) \boldsymbol{G}'_{2^n} \tau(\boldsymbol{h}_k^{n,m}),$$

where $\boldsymbol{G}'_{2^n} \in M_{2^{2^{n-1}}}(\mathbb{Z})$ is the result of applying τ both to the columns and rows of $\boldsymbol{G}_{2^n}$. If we denote by $\oplus$ the "exclusive or" operation, we may decompose $\boldsymbol{G}'_{2^n}$ as

$$\boldsymbol{G}'_{2^n} = \begin{pmatrix} \boldsymbol{L}_n^{0,0\oplus 0} & \boldsymbol{L}_n^{0,0\oplus 1} & \cdots & \boldsymbol{L}_n^{0,0\oplus 2^{2^{n-1}}-1} \\ \boldsymbol{L}_n^{1,1\oplus 0} & \boldsymbol{L}_n^{1,1\oplus 1} & \cdots & \boldsymbol{L}_n^{1,1\oplus 2^{2^{n-1}}-1} \\ \vdots & \vdots & \ddots & \vdots \\ \boldsymbol{L}_n^{2^{2^{n-1}}-1,2^{2^{n-1}}-1\oplus 0} & \boldsymbol{L}_n^{2^{2^{n-1}}-1,2^{2^{n-1}}-1\oplus 1} & \cdots & \boldsymbol{L}_n^{2^{2^{n-1}}-1,2^{2^{n-1}}-1\oplus 2^{2^{n-1}}-1} \end{pmatrix},$$

for some $\boldsymbol{L}_n^{\alpha,\beta} \in M_{2^{2^{n-1}}}(\mathbb{Z})$. Within $\boldsymbol{G}_{2^n}$, each matrix $\boldsymbol{L}_n^{\alpha,\beta}$ is dispersed by τ. Its elements are situated in a grid with top left element at row α and column $\alpha \oplus \beta$, and inter-element distance $2^{2^{n-1}} - 1$, as in

$$\begin{matrix} a & b & a & b \\ c & d & c & d \\ a & b & a & b \\ c & d & c & d \end{matrix}$$

for example. It is now clear that

$$\boldsymbol{L}_n^{\alpha,\beta} = \left(g_{\alpha+i2^{2^{n-1}},\,(\alpha\oplus\beta)+j2^{2^{n-1}}} \right)_{0\leq i,j<2^{2^{n-1}}}$$

provides an alternative definition of $\boldsymbol{L}_n^{\alpha,\beta}$. The following result shows the strong relationship between this matrix and $\boldsymbol{G}_{2^n}$:

Lemma III.1. *For all* $0 \leq \alpha, \beta < 2^{2^{n-1}}$, *we have*

$$\boldsymbol{L}_n^{\alpha,\beta} = \boldsymbol{G}_{2^{n-1}} + (2^{n-1} - 2\mathrm{d}_{0\beta}) \boldsymbol{U}_{2^{n-1}}.$$

Proof. The result follows from a short calculation (where we annotate g's to indicate the matrix they belong to):

$$\begin{aligned}
\left(\boldsymbol{L}_n^{\alpha,\beta}\right)_{ij} &= g^{(n)}_{\alpha+i2^{2^{n-1}},\,(\alpha\oplus\beta)+j2^{2^{n-1}}} \\
&= 2^n + 1 - 2\mathrm{d}_{\alpha+i2^{2^{n-1}},\,(\alpha\oplus\beta)+j2^{2^{n-1}}} \\
&= 2^n + 1 - 2\mathrm{d}_{2^{2^{n-1}}i,\,2^{2^{n-1}}j} - 2\mathrm{d}_{\alpha\oplus\beta,\,\alpha} \\
&= 2^{n-1} + 1 - 2\mathrm{d}_{ij} + 2^{n-1} - 2\mathrm{d}_{0\beta} \\
&= g^{(n-1)}_{ij} + 2^{n-1} - 2\mathrm{d}_{0\beta}.
\end{aligned}$$

So,

$$\boldsymbol{L}_n^{\alpha,\beta} = \boldsymbol{G}_{2^{n-1}} + (2^{n-1} - 2\mathrm{d}_{0\beta})\boldsymbol{U}_{2^{n-1}}.$$

□

The calculation of $\gamma(R^n_m)$ also depends on the following two lemmas:

Lemma III.2. *For all* $0 \le k < 2^{n-m}$, *we have*

$$^t\boldsymbol{h}_k^{n,m}\boldsymbol{G}_{2^n}\boldsymbol{h}_k^{n,m} = (2^m+1)\prod_{i=m+1}^{n} 2^{2^i}.$$

Proof. First note that the statement is obvious for $n = m$. Indeed, in this case,

$$^t\boldsymbol{h}_0^{n,n}\boldsymbol{G}_{2^n}\boldsymbol{h}_0^{n,n} = g_{2^n-1,\,2^n-1} = 2^n + 1.$$

Let us prove the general case by induction on n. So, pick $m < n$ and $0 \le k < 2^{n-m}$, and let us first assume that $2^{n-m-1} \le k < 2^{n-m}$. In this case, $\boldsymbol{h}_k^{n,m} = \xi_{2^{2^{n-1}}}(\boldsymbol{h}^{n-1,m}_{k-2^{n-m-1}})$. Note that

$$\begin{aligned}
&{}^t\boldsymbol{h}_k^{n,m}\left((\boldsymbol{G}_{2^n-2^{n-1}} - \boldsymbol{U}_{2^n-2^{n-1}}) \otimes \boldsymbol{U}_{2^{n-1}}\right)\boldsymbol{h}_k^{n,m} \\
&\quad= \sum_{0\le p<2^{2^{n-1}}}\ \sum_{0\le q<2^{2^{n-1}}} \left((g_{pq}-1)\,{}^t\boldsymbol{h}^{n-1,m}_{k-2^{n-m-1}}\boldsymbol{U}_{2^{n-1}}\boldsymbol{h}^{n-1,m}_{k-2^{n-m-1}}\right) \qquad \text{(III.1)} \\
&\quad= 0.
\end{aligned}$$

The last equality easily follows from the fact that for any $0 \le q < 2^{2^{n-1}}$, the binary complement $\bar{q}$ of q has the property that

$$(g_{pq}-1) + (g_{p\bar{q}}-1) = 2.2^{2^{n-1}} - 2(\mathrm{d}_{pq} + \mathrm{d}_{p\bar{q}}) = 0,$$

whence $\sum_q (g_{pq} - 1) = 0$. Moreover,

$$\begin{aligned}
{}^t\boldsymbol{h}_k^{n,m} \boldsymbol{G}_{2^n} \boldsymbol{h}_k^{n,m} &= {}^t\boldsymbol{h}_k^{n,m} \left(\boldsymbol{U}_{2^n-2^{n-1}} \otimes \boldsymbol{G}_{2^{n-1}} + (\boldsymbol{G}_{2^n-2^{n-1}} - \boldsymbol{U}_{2^n-2^{n-1}}) \otimes \boldsymbol{U}_{2^{n-1}}\right) \boldsymbol{h}_k^{n,m} \\
&= {}^t\boldsymbol{h}_k^{n,m} \left(\boldsymbol{U}_{2^n-2^{n-1}} \otimes \boldsymbol{G}_{2^{n-1}}\right) \boldsymbol{h}_k^{n,m} \\
&= 2^{2^{n-1}} 2^{2^{n-1}} {}^t\boldsymbol{h}_{k-2^{n-m-1}}^{n-1,m} \boldsymbol{G}_{2^{n-1}} \boldsymbol{h}_{k-2^{n-m-1}}^{n-1,m} \\
&= 2^{2^n} {}^t\boldsymbol{h}_{k-2^{n-m-1}}^{n-1,m} \boldsymbol{G}_{2^{n-1}} \boldsymbol{h}_{k-2^{n-m-1}}^{n-1,m}.
\end{aligned}$$

Next, if we assume that $0 \leq k < 2^{n-m-1}$, then it follows that

$$\begin{aligned}
{}^t\boldsymbol{h}_k^{n,m} \boldsymbol{G}_{2^n} \boldsymbol{h}_k^{n,m} &= {}^t\boldsymbol{h}_{k+2^{n-m-1}}^{n,m} \boldsymbol{G}'_{2^n} \boldsymbol{h}_{k+2^{n-m-1}}^{n,m} \\
&= \sum_{0 \leq \alpha < 2^{2^{n-1}}} \sum_{0 \leq \beta < 2^{2^{n-1}}} {}^t\boldsymbol{h}_k^{n-1,m} \boldsymbol{L}_n^{\alpha,\beta} \boldsymbol{h}_k^{n-1,m} \qquad \text{(III.2)} \\
&= \sum_{0 \leq \alpha < 2^{2^{n-1}}} \sum_{0 \leq \beta < 2^{2^{n-1}}} {}^t\boldsymbol{h}_k^{n-1,m} \left(\boldsymbol{G}_{2^{n-1}} + (2^{n-1} - 2\mathrm{d}_{0\beta}) \boldsymbol{U}_{2^{n-1}}\right) \boldsymbol{h}_k^{n-1,m} \\
&= 2^{2^{n-1}} 2^{2^{n-1}} {}^t\boldsymbol{h}_k^{n-1,m} \boldsymbol{G}_{2^{n-1}} \boldsymbol{h}_k^{n-1,m} \\
&\quad + 2^{2^{n-1}} \sum_{0 \leq \beta < 2^{2^{n-1}}} {}^t\boldsymbol{h}_k^{n-1,m} \boldsymbol{U}_{2^{n-1}} (2^{n-1} - 2\mathrm{d}_{0\beta}) \boldsymbol{h}_k^{n-1,m} \\
&= 2^{2^n} {}^t\boldsymbol{h}_k^{n-1,m} \boldsymbol{G}_{2^{n-1}} \boldsymbol{h}_k^{n-1,m},
\end{aligned}$$

since

$$\begin{aligned}
\sum_{0 \leq \beta < 2^{2^{n-1}}} (2^{n-1} - 2\mathrm{d}_{0\beta}) &= \sum_{0 \leq \beta < 2^{2^{n-1}-1}} (2^{n-1} - 2\mathrm{d}_{0\beta} + 2^{n-1} - 2\mathrm{d}_{0\bar{\beta}}) \\
&= \sum_{0 \leq \beta < 2^{2^{n-1}-1}} (2^n - 2(\mathrm{d}_{0\beta} + \mathrm{d}_{0\bar{\beta}})) \\
&= \sum_{0 \leq \beta < 2^{2^{n-1}-1}} (2^n - 2.2^{n-1}) = 0.
\end{aligned}$$

A straightforward induction argument proves the assertion. □

We will also need:

Lemma III.3. *For all $m < n$ and $0 \leq k \neq \ell < 2^{n-m}$, we have*

$$ {}^t\boldsymbol{h}_k^{n,m} \boldsymbol{G}_{2^n} \boldsymbol{h}_\ell^{n,m} = \prod_{i=m+1}^{n} 2^{2^i}. $$

Proof. Let us begin with the case $m = n-1$ and define the vectors $\boldsymbol{e}'$ and $\boldsymbol{u}$ in $\mathbb{R}^{2^{2^m}}$ as in section 4 of chapter II:

$$\boldsymbol{e}' = \begin{pmatrix} 0 \\ \vdots \\ 0 \\ 1 \end{pmatrix} = \begin{pmatrix} \xi_{2^{2^m}-1}(0) \\ 1 \end{pmatrix}, \qquad \boldsymbol{u} = \begin{pmatrix} 1 \\ 1 \\ \vdots \\ 1 \end{pmatrix} = \xi_{2^{2^m}}(1).$$

Clearly,

$$\boldsymbol{h}_0^{n,m} = \begin{pmatrix} \xi_{2^{2^n}-2^{2^m}}(0) \\ \boldsymbol{u} \end{pmatrix}, \qquad \boldsymbol{h}_1^{n,m} = \xi_{2^{2^m}}(\boldsymbol{e}').$$

Arguing as in the previous lemma, we have that

$$\begin{aligned} {}^t\boldsymbol{h}_0^{n,m} \boldsymbol{G}_{2^{m+1}} \boldsymbol{h}_1^{n,m} &= {}^t\boldsymbol{h}_0^{n,m} \left(\boldsymbol{U}_{2^m} \otimes \boldsymbol{G}_{2^m} + (\boldsymbol{G}_{2^m} - \boldsymbol{U}_{2^m}) \otimes \boldsymbol{U}_{2^m} \right) \boldsymbol{h}_1^{n,m} \\ &= {}^t\boldsymbol{h}_0^{n,m} \left(\boldsymbol{U}_{2^m} \otimes \boldsymbol{G}_{2^m} \right) \boldsymbol{h}_1^{n,m} \qquad \text{(by (III.1))} \\ &= 2^{2^m}\, {}^t\boldsymbol{h}_0^{n,m} \xi_{2^{2^n}-2^m}(\boldsymbol{G}_{2^m} \boldsymbol{e}') = 2^{2^m}\, {}^t\boldsymbol{u} \boldsymbol{G}_{2^m} \boldsymbol{e}' \\ &= 2^{2^m} \sum_{0 \le j < 2^{2^m}} g_{ij} = 2^{2^m} + 2^{2^m} = 2^{2^m+1}. \end{aligned}$$

Arguing inductively, exactly as in the proof of the previous lemma, we find

- if $2^{n-m-1} \le k \ne \ell < 2^{n-m}$,

$$ {}^t\boldsymbol{h}_k^{n,m} \boldsymbol{G}_{2^n} \boldsymbol{h}_\ell^{n,m} = 2^{2^n}\, {}^t\boldsymbol{h}_{k-2^{n-m-1}}^{n-1,m} \boldsymbol{G}_{2^{n-1}} \boldsymbol{h}_{\ell-2^{n-m-1}}^{n-1,m}, $$

- if $0 \le k \ne \ell < 2^{n-m-1}$,

$$ {}^t\boldsymbol{h}_k^{n,m} \boldsymbol{G}_{2^n} \boldsymbol{h}_\ell^{n,m} = 2^{2^n}\, {}^t\boldsymbol{h}_k^{n-1,m} \boldsymbol{G}_{2^{n-1}} \boldsymbol{h}_\ell^{n-1,m}. $$

Finally, if we assume that $0 \le k < 2^{n-m-1} \le \ell < 2^{n-m}$ then $\boldsymbol{h}_i^n$ can be written as

$$\boldsymbol{h}_i^n = \begin{pmatrix} \boldsymbol{u}_{i0}^n \\ \vdots \\ \boldsymbol{u}_{i(2^{2^{n-1}}-1)}^n \end{pmatrix},$$

where

$$\boldsymbol{u}_{pq}^n = \xi_{2^{2^{n-1}}}(\boldsymbol{h}_p^{n-1,m}(q)).$$

To conclude, note that

$$
\begin{aligned}
{}^t\boldsymbol{h}_k^{n,m}\boldsymbol{G}_{2^n}\boldsymbol{h}_\ell^{n,m} &= {}^t\boldsymbol{h}_k^{n,m}\left(\boldsymbol{U}_{2^{n-1}}\otimes\boldsymbol{G}_{2^{n-1}} + (\boldsymbol{G}_{2^{n-1}} - \boldsymbol{U}_{2^{n-1}})\otimes\boldsymbol{U}_{2^{n-1}}\right)\boldsymbol{h}_\ell^{n,m} \\
&= 2^{2^{n-1}}{}^t\boldsymbol{h}_k^{n,m}\xi_{2^{2^{n-1}}}(\boldsymbol{G}_{2^{n-1}}\boldsymbol{h}_{\ell-2^{n-m-1}}^{n-1,m}) \\
&= 2^{2^{n-1}}\sum_{0\le q<2^{2^{n-1}}}\boldsymbol{u}_{kq}^n\boldsymbol{G}_{2^{n-1}}\boldsymbol{h}_{\ell-2^{n-m-1}}^{n-1,m} \\
&= 2^{2^{n-1}}.\left|\left\{0\le q\le 2^{2^{n-1}}-1;\boldsymbol{u}_{pq}^n = \xi_{2^{2^{n-1}}}(1)\right\}\right|{}^t\xi_{2^{2^{n-1}}}(1)\boldsymbol{h}_{\ell-2^{n-m-1}}^{n-1,m} \\
&= 2^{2^n}.2^{2^{n-1}-2^m}.2^{2^{n-1}-2^m} \\
&= \prod_{i=m+1}^{n} 2^{2^i}.
\end{aligned}
$$

A straightforward induction argument finishes the proof. □

Finally, we obtain:

Proposition III.4. *For any pair of positive integers $m\le n$, we have*

$$\gamma(R_m^n) = 2^{n+m}(2^m+2^{n-m})\prod_{i=m+1}^{n}2^{2^i}.$$

Proof. It follows from the previous results that

$$
\begin{aligned}
\gamma(R_m^n) &= {}^t\boldsymbol{R}_m^n\boldsymbol{G}_{2^n}\boldsymbol{R}_m^n \\
&= 2^{2m}\left(\sum_{0\le i<2^{n-m}}{}^t\boldsymbol{h}_i^{n,m}\right)\boldsymbol{G}_{2^n}\left(\sum_{0\le i<2^{n-m}}\boldsymbol{h}_i^{n,m}\right) \\
&= 2^{2m}\sum_{0\le i<2^{n-m}}{}^t\boldsymbol{h}_i^{n,m}\boldsymbol{G}_{2^n}\boldsymbol{h}_i^{n,m} + 2^{2m}\sum_{0\le i\ne j<2^{n-m}}{}^t\boldsymbol{h}_i^{n,m}\boldsymbol{G}_{2^n}\boldsymbol{h}_j^{n,m} \\
&= 2^{n+m}(2^m+1)\prod_{i=m+1}^{n}2^{2^i} + 2^{n+m}(2^{n-m}-1)\prod_{i=m+1}^{n}2^{2^i} \\
&= 2^{n+m}(2^m+2^{n-m})\prod_{i=m+1}^{n}2^{2^i}.
\end{aligned}
$$

□

It remains to calculate the norm of the generalized Royal Road Functions. This will be realized through the following straightforward lemmas.

Lemma III.5. *For all* $0 \le k < 2^{n-m}$*, we have*

$$ {}^t\boldsymbol{h}_k^{n,m}\boldsymbol{h}_k^{n,m} = \prod_{i=m+1}^{n} 2^{2^{i-1}} = \prod_{i=m}^{n-1} 2^{2^i}. $$

Proof. Let us begin with the case $n = m$. Then $k = 0$ and

$$ {}^t\boldsymbol{h}_0^{n,n}\boldsymbol{h}_0^{n,n} = 1. $$

If $n = m + 1$, we have $k = 0$ or $k = 1$, so

$$ {}^t\boldsymbol{h}_0^{n,n-1}\boldsymbol{h}_0^{n,n-1} = {}^t\xi_{2^{2^{n-1}}}(1)\xi_{2^{2^{n-1}}}(1) = 2^{2^{n-1}} $$

and

$$ {}^t\boldsymbol{h}_1^{n,n-1}\boldsymbol{h}_1^{n,n-1} = {}^t\xi_{2^{2^{n-1}}}(\boldsymbol{e}')\xi_{2^{2^{n-1}}}(\boldsymbol{e}') = 2^{2^{n-1}}. $$

In the general case, we proceed by induction. Indeed, if $0 \le k < 2^{n-m-1}$, then

$$ \boldsymbol{h}_k^{n,m} = \begin{pmatrix} \xi_{2^{2^{n-1}}}(\boldsymbol{h}_k^{n-1,m}(0)) \\ \xi_{2^{2^{n-1}}}(\boldsymbol{h}_k^{n-1,m}(1)) \\ \vdots \\ \xi_{2^{2^{n-1}}}(\boldsymbol{h}_k^{n-1,m}(2^{2^{n-1}}-1)) \end{pmatrix} $$

so,

$$ {}^t\boldsymbol{h}_k^{n,m}\boldsymbol{h}_k^{n,m} = 2^{2^{n-1}}\,{}^t\boldsymbol{h}_k^{n-1,m}\boldsymbol{h}_k^{n-1,m}. $$

Similarly, if $2^{n-m-1} \le k < 2^{n-m}$, then $\boldsymbol{h}_k^{n,m} = \xi_{2^{2^{n-1}}}(\boldsymbol{h}_{k-2^{n-m-1}}^{n-1,m})$, hence

$$ {}^t\boldsymbol{h}_k^{n,m}\boldsymbol{h}_k^{n,m} = 2^{2^{n-1}}\,{}^t\boldsymbol{h}_{k-2^{n-m-1}}^{n-1,m}\boldsymbol{h}_{k-2^{n-m-1}}^{n-1,m}. $$

In both cases, we find by induction that ${}^t\boldsymbol{h}_k^{n,m}\boldsymbol{h}_k^{n,m} = \prod_{i=m+1}^{n} 2^{2^{i-1}}$. □

Lemma III.6. *For all* $n \ge 0$ *and* $0 \le k \ne \ell < 2^{n-m}$*, we have*

$$ {}^t\boldsymbol{h}_k^{n,m}\boldsymbol{h}_\ell^{n,m} = \begin{cases} 1 & \text{if } m = n-1 \\ \prod_{i=m+2}^{n} 2^{2^{i-1}} & \text{if } m < n-1. \end{cases} $$

Proof. For $m = n - 1$, the statement is obvious. For arbitrary $m \le n$, we argue by induction. We consider three cases:

- If $2^{n-m-1} \leq k \neq \ell < 2^{n-m}$, then
$$ {}^t\boldsymbol{h}_k^{n,m}\boldsymbol{h}_\ell^{n,m} = 2^{2^{n-1}}\,{}^t\boldsymbol{h}_{k-2^{n-m-1}}^{n-1,m}\boldsymbol{h}_{\ell-2^{n-m-1}}^{n-1,m}. $$
- If $0 \leq k \neq \ell < 2^{n-m-1}$, then
$$ {}^t\boldsymbol{h}_k^{n,m}\boldsymbol{h}_\ell^{n,m} = 2^{2^{n-1}}\,{}^t\boldsymbol{h}_k^{n-1,m}\boldsymbol{h}_\ell^{n-1,m}. $$
- Finally, if $0 \leq k < 2^{n-m-1} \leq \ell < 2^{n-m}$, then the number of 1's in the vector $\boldsymbol{h}_i^{n,m}$ is $\prod_{j=m+1}^{n} 2^{2^{j-1}}$. It follows that
$$ \begin{aligned} {}^t\boldsymbol{h}_k^{n,m}\boldsymbol{h}_\ell^{n,m} &= \begin{pmatrix} \xi_{2^{2^{n-1}}}(h_k^{n-1,m}(0)) \\ \xi_{2^{2^{n-1}}}(h_k^{n-1,m}(1)) \\ \vdots \\ \xi_{2^{2^{n-1}}}(h_k^{n-1,m}(2^{2^{n-1}}-1)) \end{pmatrix} \xi_{2^{2^{n-1}}}(\boldsymbol{h}_{\ell-2^{n-m-1}}^{n-1,m}) \\ &= \left|\left\{0 \leq i < 2^{2^{n-1}};\, h_k^{n-1,m}(i) = 1\right\}\right| (1,\ldots,1)\,\boldsymbol{h}_{\ell-2^{n-m-1}}^{n-1,m} \\ &= \prod_{i=m+1}^{n-1} 2^{2^{i-1}} \left|\left\{0 \leq i < 2^{2^{n-1}};\, h_k^{n-1,m}(i) = 1\right\}\right| \\ &= \left(\prod_{i=m+1}^{n-1} 2^{2^{i-1}}\right) {}^t\boldsymbol{h}_k^{n-1,m}\boldsymbol{h}_k^{n-1,m} = \prod_{i=m+2}^{n-1} 2^{2^{i-1}}. \end{aligned} $$

An easy induction argument completes the proof. □

Proposition III.7. *The norm of the Royal Road function R_m^n is*
$$ \|\boldsymbol{R}_m^n\|^2 = 2^{n+m}(2^{2^m} + 2^{n-m} - 1) \prod_{i=m+2}^{n} 2^{2^{i-1}}. $$

Proof. This easily follows from
$$ \begin{aligned} {}^t\boldsymbol{R}_m^n\boldsymbol{R}_m^n &= 2^{2m}\Big(\sum_{0\leq i<2^{n-m}} {}^t\boldsymbol{h}_i^{n,m}\Big)\Big(\sum_{0\leq i<2^{n-m}} \boldsymbol{h}_i^{n,m}\Big) \\ &= 2^{2m} \sum_{0\leq i<2^{n-m}} {}^t\boldsymbol{h}_i^{n,m}\boldsymbol{h}_i^{n,m} + 2^{2m} \sum_{0\leq i\neq j<2^{n-m}} {}^t\boldsymbol{h}_i^{n,m}\boldsymbol{h}_j^{n,m} \\ &= 2^{n+m} \prod_{i=m+1}^{n} 2^{2^{i-1}} + (2^{n-m}-1)2^{n+m} \prod_{i=m+2}^{n} 2^{2^{i-1}} \\ &= 2^{n+m}(2^{2^m} + 2^{n-m} - 1) \prod_{i=m+2}^{n} 2^{2^{i-1}}. \end{aligned} $$
□

Theorem III.8. *The epistasis of the Royal Road function R^n_m is given by*

$$\varepsilon^*(R^n_m) = \frac{2^{2^m} - 2^m - 1}{2^{2^m} + 2^{n-m} - 1}.$$

Proof. The epistasis of the R^n_m Royal Road function is

$$\begin{aligned}\varepsilon^*(R^n_m) &= 1 - \frac{\gamma(R^m_n)}{2^{2^n}\|R^m_n\|^2} \\ &= 1 - \frac{2^{n+m}(2^m + 2^{n-m})\prod_{i=m+1}^{n} 2^{2^i}}{2^{2^n}2^{n+m}(2^{2^m} + 2^{n-m} - 1)\prod_{i=m+2}^{n} 2^{2^{i-1}}} \\ &= \frac{2^{2^m} - 2^m - 1}{2^{2^m} + 2^{n-m} - 1}.\end{aligned}$$

□

Applying this formula to the "classical" Royal Road function $R_1 = R^6_3$ gives a high epistatic value of

$$\varepsilon^*(R_1) = \frac{2^{2^3} - 2^3 - 1}{2^{2^3} + 2^3 - 1} = \frac{256 - 8 - 1}{263} = 0.93916.$$

For general values of n, the minimal and maximal epistatic values are given by

$$\varepsilon^*(R^n_0) = \frac{2^{2^0} - 2^0 - 1}{2^{2^0} + 2^0 - 1} = 0$$

and

$$\varepsilon^*(R^n_n) = \frac{2^{2^n} - 2^n - 1}{2^{2^n} + 2^0 - 1} = 1 - \frac{2^n + 1}{2^{2^n}} \approx 1 - \frac{1}{2^{2^n - n}},$$

which is close to the maximal possible value $1 - \frac{1}{2^{2^n-1}}$ of normalized epistasis over length 2^n strings.

1.2 Generalized Royal Road functions of type II

Inspired in the constructions of [60], we also introduce *generalized Royal Road functions of type II*, as follows. For any subset $T \subseteq \{1, \ldots, n\}$, define R^n_T by

$$R^n_T(s) = \sum_{\substack{s \in H^{n,m}_i \\ m \in T}} c^{m,n}_i,$$

where $c_i^{m,n} = 2^m$, for any $0 \le i < 2^{n-m}$.

The associated vector $\boldsymbol{R}_T^n$ can be written as a sum of vectors $\boldsymbol{h}_i^{n,m}$:

$$\boldsymbol{R}_T^n = \sum_{m\in T} \left(2^m \sum_{0\le i<2^{n-m}} \boldsymbol{h}_i^{n,m} \right).$$

Lemma III.9. *For all positive integers $n = p > q \ge 0$ and any $0 \le b < 2^{n-q}$, we have*

$$^t\boldsymbol{h}_0^{n,n}\boldsymbol{G}_{2^n}\boldsymbol{h}_b^{n,q} = (2^q+1)\prod_{i=q+1}^{n} 2^{2^{i-1}} = (2^q+1)2^{2^n-2^q}.$$

Proof. We prove this result by induction on n. Let us begin with $n = q+1$, then

$$\begin{aligned}
^t\boldsymbol{h}_0^{n,n}\boldsymbol{G}_{2^n}\boldsymbol{h}_0^{n,q} &= {}^t\boldsymbol{h}_0^{n,n}[\boldsymbol{U}_{2^{n-1}}\otimes \boldsymbol{G}_{2^{n-1}} + (\boldsymbol{G}_{2^{n-1}} - \boldsymbol{U}_{2^{n-1}})\otimes \boldsymbol{U}_{2^{n-1}}]\boldsymbol{h}_0^{n,q} \\
&= (0,\dots,0,1)\boldsymbol{G}_{2^{n-1}}\begin{pmatrix}1\\ \vdots\\ 1\end{pmatrix} + 2^{2^{n-1}}(0,\dots,0,1)\boldsymbol{U}_{2^{n-1}}\begin{pmatrix}1\\ \vdots\\ 1\end{pmatrix} \\
&= 2^{2^q} + 2^{2^q}2^q \\
&= 2^{2^q}(2^q+1)
\end{aligned}$$

and

$$\begin{aligned}
^t\boldsymbol{h}_0^{n,n}\boldsymbol{G}_{2^n}\boldsymbol{h}_1^{n,q} &= {}^t(\tau(\boldsymbol{h}_0^{n,n}))\boldsymbol{G}'_{2^n} + \tau(\boldsymbol{h}_1^{n,q}) \\
&= {}^t\boldsymbol{h}_0^{n,n}\boldsymbol{G}'_{2^n}\boldsymbol{h}_0^{n,q} \\
&= (0,\dots,0,1)\boldsymbol{L}_n^{2^{2^{n-1}}-1,0}\begin{pmatrix}1\\ \vdots\\ 1\end{pmatrix} \\
&= (0,\dots,0,1)[\boldsymbol{G}_{2^{n-1}} + 2^{n-1}\boldsymbol{U}_{2^{n-1}}]\begin{pmatrix}1\\ \vdots\\ 1\end{pmatrix} \\
&= 2^{2^q}(2^q+1).
\end{aligned}$$

Let us suppose the lemma to be true for all $n < q$. If $0 \le b < 2^{n-q-1}$, we can argue as in lemma III.2 and obtain that

$$
\begin{aligned}
{}^t\boldsymbol{h}_0^{n,n}\boldsymbol{G}_{2^n}\boldsymbol{h}_b^{n,q} &= \sum_{\beta=0}^{2^{2^{n-1}}-1} {}^t\boldsymbol{h}_0^{n-1,n-1}\boldsymbol{L}_n^{2^{2^{n-1}}-1,\beta}\boldsymbol{h}_b^{n,q} \\
&= 2^{2^{n-1}}\,{}^t\boldsymbol{h}_0^{n-1,n-1}\boldsymbol{G}_{2^{n-1}}\boldsymbol{h}_b^{n-1,q} \\
&= (2^q+1)\prod_{i=q+1}^{n} 2^{2^{i-1}}
\end{aligned}
$$

as well as

$$
\begin{aligned}
{}^t\boldsymbol{h}_0^{n,n}\boldsymbol{G}_{2^n}\boldsymbol{h}_b^{n,q} &= {}^t\boldsymbol{h}_0^{n,n}[\boldsymbol{U}_{2^{n-1}}\otimes\boldsymbol{G}_{2^{n-1}} + (\boldsymbol{G}_{2^{n-1}} - \boldsymbol{U}_{2^{n-1}})\otimes\boldsymbol{U}_{2^{n-1}}]\boldsymbol{h}_b^{n,q} \\
&= 2^{2^n}\,{}^t\boldsymbol{h}_0^{n-1,n-1}\boldsymbol{G}_{2^{n-1}}\boldsymbol{h}_{b-2^{n-q-1}}^{n-1,q} \\
&= (2^q+1)\prod_{i=q+1}^{n} 2^{2^{i-1}}.
\end{aligned}
$$

This finishes the proof. □

In order to simplify the calculations, we will use the following notation

$$
\begin{aligned}
\hat{a} &= a \mod 2^{n-p-1}, \\
\hat{b} &= b \mod 2^{n-q-1}.
\end{aligned}
$$

Lemma III.10. *For any positive integers $n > p > q \geq 0$, we have*

1. *for all $0 \leq a < 2^{n-p-1}$ and $0 \leq b < 2^{n-q-1}$,*

$$
{}^t\boldsymbol{h}_a^{n,p}\boldsymbol{G}_{2^n}\boldsymbol{h}_b^{n,q} = 2^{2^n}\,{}^t\boldsymbol{h}_a^{n-1,p}\boldsymbol{G}_{2^{n-1}}\boldsymbol{h}_b^{n-1,q},
$$

2. *for all $2^{n-p-1} \leq a < 2^{n-p}$ and $2^{n-q-1} \leq b < 2^{n-q}$,*

$$
{}^t\boldsymbol{h}_a^{n,p}\boldsymbol{G}_{2^n}\boldsymbol{h}_b^{n,q} = 2^{2^n}\,{}^t\boldsymbol{h}_{\hat{a}}^{n-1,p}\boldsymbol{G}_{2^{n-1}}\boldsymbol{h}_{\hat{b}}^{n-1,q},
$$

3. *for all $0 \leq a < 2^{n-p-1}$ and $2^{n-q-1} \leq b < 2^{n-q}$,*

$$
{}^t\boldsymbol{h}_a^{n,p}\boldsymbol{G}_{2^n}\boldsymbol{h}_b^{n,q} = 2^{2^{n+1}-2^p-2^q}.
$$

Proof. We will treat the three cases in the statement separately.

1. To prove the first statement, use the matrix $\boldsymbol{L}_n^{\alpha,\beta}$ and apply lemma III.2 to deduce

$$\begin{aligned}{}^t\boldsymbol{h}_a^{n,p}\boldsymbol{G}_{2^n}\boldsymbol{h}_b^{n,q} &= \sum_\alpha\sum_\beta {}^t\boldsymbol{h}_a^{n-1,p}\boldsymbol{L}_n^{\alpha,\beta}\boldsymbol{h}_b^{n-1,q}\\ &= 2^{2^n}\,{}^t\boldsymbol{h}_a^{n-1,p}\boldsymbol{G}_{2^{n-1}}\boldsymbol{h}_b^{n-1,q}.\end{aligned}$$

2. The second case follows from

$$\begin{aligned}{}^t\boldsymbol{h}_a^{n,p}\boldsymbol{G}_{2^n}\boldsymbol{h}_b^{n,q} &= {}^t\boldsymbol{h}_a^{n,p}[\boldsymbol{U}_{2^{n-1}}\otimes\boldsymbol{G}_{2^{n-1}} + (\boldsymbol{G}_{2^{n-1}} - \boldsymbol{U}_{2^{n-1}})\otimes\boldsymbol{U}_{2^{n-1}}]\boldsymbol{h}_b^{n,q}\\ &= {}^t\boldsymbol{h}_a^{n,p}[\boldsymbol{U}_{2^{n-1}}\otimes\boldsymbol{G}_{2^{n-1}}]\boldsymbol{h}_b^{n,q}\\ &= 2^{2^{n-1}}2^{2^{n-1}}\,{}^t\boldsymbol{h}_{a-2^{n-p-1}}^{n-1,p}\boldsymbol{G}_{2^{n-1}}\boldsymbol{h}_{b-2^{n-q-1}}^{n-1,q}\\ &= 2^{2^n}\,{}^t\boldsymbol{h}_{\hat{a}}^{n-1,p}\boldsymbol{G}_{2^{n-1}}\boldsymbol{h}_{\hat{b}}^{n-1,q}.\end{aligned}$$

3. The last identity may be verified as follows:

$$\begin{aligned}{}^t\boldsymbol{h}_a^{n,p}\boldsymbol{G}_{2^n}\boldsymbol{h}_b^{n,q} &= {}^t\boldsymbol{h}_a^{n,p}[\boldsymbol{U}_{2^{n-1}}\otimes\boldsymbol{G}_{2^{n-1}} + (\boldsymbol{G}_{2^{n-1}} - \boldsymbol{U}_{2^{n-1}})\otimes\boldsymbol{U}_{2^{n-1}}]\boldsymbol{h}_b^{n,q}\\ &= {}^t\boldsymbol{h}_a^{n,p}[\boldsymbol{U}_{2^{n-1}}\otimes\boldsymbol{G}_{2^{n-1}}]\boldsymbol{h}_b^{n,q}\\ &= {}^t\boldsymbol{h}_a^{n,p}\begin{pmatrix}\boldsymbol{G}_{2^{n-1}}\boldsymbol{h}_{\hat{b}}^{n-1,q}\\ \vdots\\ \boldsymbol{G}_{2^{n-1}}\boldsymbol{h}_{\hat{b}}^{n-1,q}\end{pmatrix}\\ &= 2^{2^n-2^p}\,{}^t\begin{pmatrix}1\\ \vdots\\ 1\end{pmatrix}\boldsymbol{G}_{2^{n-1}}\boldsymbol{h}_{\hat{b}}^{n-1,q}\\ &= 2^{2^{n+1}-2^p-2^q}.\end{aligned}$$

□

To calculate $\gamma(\boldsymbol{R}_T^n)$ we will apply the previous lemmas and recursion on n. So,

$$\begin{aligned}\sum_{\substack{0\le a<2^{n-p}\\ 0\le b<2^{n-q}}} {}^t\boldsymbol{h}_a^{n,p}\boldsymbol{G}_{2^n}\boldsymbol{h}_b^{n,q} &= \sum_{0\le b<2^{n-q}} {}^t\boldsymbol{h}_0^{n,p}\boldsymbol{G}_{2^n}\boldsymbol{h}_b^{n,q} + \sum_{\substack{0<a<2^{n-p}\\ 0\le b<2^{n-q}}} {}^t\boldsymbol{h}_a^{n,p}\boldsymbol{G}_{2^n}\boldsymbol{h}_b^{n,q}\\ &= \sum_{0\le b<2^{n-q}} 2^{2^n-2^p}\,{}^t\boldsymbol{h}_0^{n,n}\boldsymbol{G}_{2^n}\boldsymbol{h}_b^{n,q} + \sum_{\substack{0\le a<2^{n-p}\\ 0\le b<2^{n-q}}} {}^t\boldsymbol{h}_a^{n,p}\boldsymbol{G}_{2^n}\boldsymbol{h}_b^{n,q}\\ &= 2^{n-q}(2^q+1)2^{2^{n+1}-2^p-2^q} + 2^{n-q}(2^{n-p}-1)2^{2^{n+1}-2^p-2^q}\\ &= 2^{2^{n+1}-2^p-2^q+n-q}(2^q+2^{n-p}).\end{aligned}$$

Using this, we find, denoting R^n_T by $R^n_{p,q}$ when $T = \{p, p+1, \ldots, q\}$,

$$\begin{aligned}\gamma(\boldsymbol{R}^n_{p,q}) &= \gamma(\boldsymbol{R}^n_p) + \gamma(\boldsymbol{R}^n_q) + 2^{1+p+q}2^{2^{n+1}-2^p-2^q+n-q}(2^q + 2^{n-p}) \\ &= \gamma(\boldsymbol{R}^n_p) + \gamma(\boldsymbol{R}^n_q) + 2^{1+2^{n+1}-2^p-2^q+n+p}(2^q + 2^{n-p}).\end{aligned}$$

To finish the calculation of the normalized epistasis of the Royal Road functions of type II, it remains to determine their norm. We need the following result.

Lemma III.11. *For any positive integers $n \geq p > q \geq 0$, we have*

1. *for $n = p$ and for all $0 \leq b < 2^{n-q}$,*

$$^t\boldsymbol{h}^{n,p}_0 \boldsymbol{h}^{n,q}_b = {}^t\boldsymbol{h}^{n,n}_0 \boldsymbol{h}^{n,q}_b = 1,$$

2. *for $n > p$, for all $0 \leq a < 2^{n-p-1}$, $0 \leq b < 2^{n-q-1}$ and also for all $2^{n-p-1} \leq a < 2^{n-p}$, $2^{n-q-1} \leq b < 2^{n-q}$,*

$$^t\boldsymbol{h}^{n,p}_a \boldsymbol{h}^{n,q}_b = 2^{2^{n-1}}\, {}^t\boldsymbol{h}^{n-1,p}_{\hat{a}} \boldsymbol{h}^{n-1,q}_{\hat{b}},$$

3. *for $n > p$ and for all $0 \leq a < 2^{n-p-1}$, $2^{n-q-1} \leq b < 2^{n-q}$,*

$$^t\boldsymbol{h}^{n,p}_a \boldsymbol{h}^{n,q}_b = 2^{2^n-2^p-2^q}.$$

Assuming $n > p$, it easily follows that

$$\begin{aligned}\sum_{\substack{0 \leq a < 2^{n-p} \\ 0 \leq b < 2^{n-q}}} {}^t\boldsymbol{h}^{n,p}_a \boldsymbol{h}^{n,q}_b &= \sum_{0 \leq b < 2^{n-q}} {}^t\boldsymbol{h}^{n,p}_0 \boldsymbol{h}^{n,q}_b + \sum_{\substack{0 < a < 2^{n-p} \\ 0 \leq b < 2^{n-q}}} {}^t\boldsymbol{h}^{n,p}_a \boldsymbol{h}^{n,q}_b \\ &= 2^{2^n-2^p}2^{n-q} + 2^{n-q}(2^{n-p}-1)2^{2^n-2^p-2^q} \\ &= 2^{2^n-2^p-2^q+n-q}(2^{2^q} + 2^{n-p} - 1).\end{aligned}$$

The norm $\|\boldsymbol{R}^n_{p,q}\|$ may now be given by

$$\begin{aligned}\|\boldsymbol{R}^n_{p,q}\|^2 &= \|\boldsymbol{R}^n_p\|^2 + \|\boldsymbol{R}^n_q\|^2 + 2^{1+p+q}2^{2^n-2^p-2^q+n-q}(2^{2^q} + 2^{n-p} - 1) \\ &= \|\boldsymbol{R}^n_p\|^2 + \|\boldsymbol{R}^n_q\|^2 + 2^{1+2^n-2^p-2^q+n+p}(2^{2^q} + 2^{n-p} - 1).\end{aligned}$$

Finally, if $T \subseteq \{1, \ldots, n\}$, then the normalized epistasis of the corresponding generalized Royal Road function of type II is

$$\varepsilon^*(R^n_T) = \frac{\sum_{p \in T} A_p + 2\sum_{p<q \in T} 2^{p-2^p-2^q}(2^{2^q} - 2^q - 1)}{\sum_{p \in T} B_p + 2\sum_{p<q \in T} 2^{p-2^p-2^q}(2^{2^q} - 2^{n-p} - 1)},$$

where

$$A_p = 2^{-(2^{n+1}-p)}(2^{2^p} - 2^p - 1)$$

and

$$B_p = 2^{-(2^{p+1}-p)}(2^{2^p} + 2^{n-p} - 1).$$

For $R^n_{p,q}$, we obtain

$$\varepsilon^*(R^n_{p,q}) = \frac{\sum_{i=p}^{q}(2^{-(2^{n+1}-i)}(2^{2^i-2^i-1})) + 2\sum_{i=p+1}^{q}\sum_{j=p}^{i-1}(2^{i-2^i-2^j}(2^{2^j-2^j-1}))}{\sum_{i=p}^{q}(2^{-(2^{i+1}-i)}(2^{2^i+2^{n-i}-1})) + 2\sum_{i=p+1}^{q}\sum_{j=p}^{i-1}(2^{i-2^i-2^j}(2^{2^j+2^{n-i}-1}))}.$$

For the "classical" Royal Road function R_2, we find

$$\varepsilon^*(R_2) = \varepsilon^*(R^6_{3,6}) \approx 0.93983.$$

In particular, note that $\varepsilon^*(R_1) < \varepsilon^*(R_2)$.

1.3 Some experimental results

Although for general functions, (high) epistasis and problem difficulty for GAs are hardly related, it appears that within fixed length classes of generalized Royal Road functions, there is a nice correlation between these two values. This is shown in table III.1. We compare the average number of generations needed to optimize generalized Royal Road functions, used as a measure of problem difficulty, and the normalized epistasis. We consider length 64 strings and use a generational GA with binary tournament selection, a population of size 100, one-point crossover with probability 0.8 and ordinary mutation at rate 1/64. We stop the algorithm when the optimum is discovered by the GA.

It is interesting to note that both normalized epistasis and Jones and Forrest's fitness distance correlation (section 7.1 of chapter I) yield the same ordering for generalized Royal Road functions. This can be observed in table III.2 where functions are ordered by their fitness distance correlation.

For a comparison between and more details about the practical aspects of both metrics, we refer to [48, 67, 76]. Neither should be seen as the definitive problem difficulty predictor. They only open the research for further classifications of fitness functions. Briefly, we can say that normalized epistasis recognizes the simplest

f	$\varepsilon^*(f)$	mean	std dev
R_0^6	0.000	58.7	6.92
R_1^6	0.028	75.6	18.6
R_2^6	0.355	130	51.2
R_3^6	0.939	551	279
R_4^6	0.999	$\gg 5000$	

Table III.1: Epistasis and problem difficulty of generalized Royal Road functions of type I. We show the mean and standard deviation over 100 independent runs of the number of generations of the GA until the optimum is reached. The GA is detailed above.

f	$\varepsilon^*(f)$
R_1^4	0.091
$R_{1,4}^4$	0.188
R_2^5	0.478
$R_{2,5}^5$	0.527
R_3^6	0.939
$R_{3,6}^6$	0.940

Table III.2: Epistasis of generalized Royal Road functions. The functions are ordered by their fitness distance correlation.

additive functions, where no interactions between bits are present, whereas fitness distance correlation computes the deviation from linearity. In particular, both metrics are unable to detect more than one class of fitness functions.

2 Unitation functions

2.1 Generalities

For any string $s = s_{\ell-1} \dots s_0 \in \Omega = \{0,1\}^\ell$, let us denote by $u(s)$ the Hamming distance d_{s0}, where $0 = 0 \dots 0$ denotes the zero-string, i.e., $u(s)$ is the number of bits in s with value 1. For example,

$$u(101101) = 4.$$

We call $u(s)$ the *unitation* of s. A function

$$f : \Omega \to \mathbb{R}$$

is said to be a *unitation function*, cf. section 5.3 of chapter I, if we may find some real-valued function $h : \{0, \dots, \ell\} \to \mathbb{R}$ such that $f(s) = h(u(s))$ for all $s \in \Omega$.

2.2 Matrix formulation

Suppose that $f = h \circ u$ and define

$$\boldsymbol{h} = \begin{pmatrix} h(0) \\ \vdots \\ h(\ell) \end{pmatrix} = \begin{pmatrix} f(0\dots00) \\ f(0\dots01) \\ f(0\dots11) \\ \vdots \\ f(1\dots11) \end{pmatrix} \in \mathbb{R}^{\ell+1}.$$

We will use, as before, the notation $f_0, \dots, f_{2^\ell-1}$ for the components of $\boldsymbol{f}$ and $h_0, \dots, h_\ell$ for the components of $\boldsymbol{h}$. In this way, each of the vectors

$$\boldsymbol{f} = \begin{pmatrix} f_0 \\ \vdots \\ f_{2^\ell-1} \end{pmatrix}, \qquad \boldsymbol{h} = \begin{pmatrix} h_0 \\ \vdots \\ h_\ell \end{pmatrix}$$

completely determines the function f.

Let us inductively define, for any positive integer $\ell > 1$, the $2^\ell \times (\ell + 1)$ matrix $\boldsymbol{A}_\ell$ by

$$\boldsymbol{A}_\ell = \begin{pmatrix} \boldsymbol{A}_{\ell-1} & \boldsymbol{0}_{2^{\ell-1}} \\ \boldsymbol{0}_{2^{\ell-1}} & \boldsymbol{A}_{\ell-1} \end{pmatrix},$$

where $\boldsymbol{0}_\ell$ is the zero-vector of length ℓ and $\boldsymbol{A}_1$ is the two-dimensional identity matrix.

So,

$$\boldsymbol{A}_1 = \begin{pmatrix} 1 & 0 \\ 0 & 1 \end{pmatrix}, \qquad \boldsymbol{A}_2 = \begin{pmatrix} 1 & 0 & 0 \\ 0 & 1 & 0 \\ 0 & 1 & 0 \\ 0 & 0 & 1 \end{pmatrix},$$

and so on.

If we denote by $\boldsymbol{O}_\ell$ the $2^\ell \times (\ell+1)$-dimensional zero-matrix and by $\boldsymbol{I}_\ell$ the ℓ-dimensional identity matrix, this clearly implies

$$\boldsymbol{A}_\ell = \begin{pmatrix} \boldsymbol{A}_{\ell-1} & \boldsymbol{O}_{\ell-1} \\ \boldsymbol{O}_{\ell-1} & \boldsymbol{A}_{\ell-1} \end{pmatrix} \begin{pmatrix} \boldsymbol{I}_\ell & \mathbf{0}_\ell \\ \mathbf{0}_\ell & \boldsymbol{I}_\ell \end{pmatrix}, \tag{III.3}$$

as one easily verifies.

With these notations, it is easy to see that

$$\boldsymbol{f} = \boldsymbol{A}_\ell \boldsymbol{h}.$$

2.3 The epistasis of a unitation function

From chapter II we know that

$$\varepsilon^*(f) = \varepsilon^2\left(\frac{\boldsymbol{f}}{||\boldsymbol{f}||}\right) = 1 - \frac{1}{2^\ell}\frac{{}^t\boldsymbol{f}\,\boldsymbol{G}_\ell\,\boldsymbol{f}}{||\boldsymbol{f}||^2}.$$

Since, in the present context, $\boldsymbol{f}$ is completely determined by the vector $\boldsymbol{h}$, we want to determine a square matrix $\boldsymbol{B}_\ell \in M_{\ell+1}(\mathbb{R})$, with the property that

$${}^t\boldsymbol{f}\,\boldsymbol{G}_\ell\,\boldsymbol{f} = {}^t\boldsymbol{h}\,\boldsymbol{B}_\ell\,\boldsymbol{h}.$$

The above definition of normalized epistasis will then simplify to

$$\varepsilon^*(f) = 1 - \frac{1}{2^\ell}\frac{{}^t\boldsymbol{h}\,\boldsymbol{B}_\ell\,\boldsymbol{h}}{\sum_{p=0}^{\ell}\binom{\ell}{p}h_p^2}.$$

2.4 The matrix $\boldsymbol{B}_\ell$

Of course, from the relation $\boldsymbol{f} = \boldsymbol{A}_\ell \boldsymbol{h}$, it obviously follows that

$$\boldsymbol{B}_\ell = {}^t\boldsymbol{A}_\ell \boldsymbol{G}_\ell \boldsymbol{A}_\ell.$$

Let us use this relation to calculate $\boldsymbol{B}_\ell = (b^\ell_{pq})$ explicitly.

First, note that (III.3) and the induction formula

$$\boldsymbol{G}_\ell = \begin{pmatrix} \boldsymbol{G}_{\ell-1} + \boldsymbol{U}_{\ell-1} & \boldsymbol{G}_{\ell-1} - \boldsymbol{U}_{\ell-1} \\ \boldsymbol{G}_{\ell-1} - \boldsymbol{U}_{\ell-1} & \boldsymbol{G}_{\ell-1} + \boldsymbol{U}_{\ell-1} \end{pmatrix}$$

for $\boldsymbol{G}_\ell$ imply that

$$\begin{aligned} \boldsymbol{B}_\ell &= {}^t\boldsymbol{A}_\ell \boldsymbol{G}_\ell \boldsymbol{A}_\ell \\ &= \begin{pmatrix} \boldsymbol{I}_\ell & {}^t\boldsymbol{0}_\ell \\ {}^t\boldsymbol{0}_\ell & \boldsymbol{I}_\ell \end{pmatrix} \begin{pmatrix} {}^t\boldsymbol{A}_{\ell-1}(\boldsymbol{G}_{\ell-1} + \boldsymbol{U}_{\ell-1})\boldsymbol{A}_{\ell-1} & {}^t\boldsymbol{A}_{\ell-1}(\boldsymbol{G}_{\ell-1} - \boldsymbol{U}_{\ell-1})\boldsymbol{A}_{\ell-1} \\ {}^t\boldsymbol{A}_{\ell-1}(\boldsymbol{G}_{\ell-1} - \boldsymbol{U}_{\ell-1})\boldsymbol{A}_{\ell-1} & {}^t\boldsymbol{A}_{\ell-1}(\boldsymbol{G}_{\ell-1} + \boldsymbol{U}_{\ell-1})\boldsymbol{A}_{\ell-1} \end{pmatrix} \begin{pmatrix} \boldsymbol{I}_\ell & \boldsymbol{0}_\ell \\ \boldsymbol{0}_\ell & \boldsymbol{I}_\ell \end{pmatrix} \\ &= \boldsymbol{B'}_\ell + \boldsymbol{B''}_\ell, \end{aligned}$$

with

$$\boldsymbol{B'}_\ell = \begin{pmatrix} \boldsymbol{I}_\ell & {}^t\boldsymbol{0}_\ell \\ {}^t\boldsymbol{0}_\ell & \boldsymbol{I}_\ell \end{pmatrix} \begin{pmatrix} {}^t\boldsymbol{A}_{\ell-1}\boldsymbol{U}_{\ell-1}\boldsymbol{A}_{\ell-1} & -{}^t\boldsymbol{A}_{\ell-1}\boldsymbol{U}_{\ell-1}\boldsymbol{A}_{\ell-1} \\ -{}^t\boldsymbol{A}_{\ell-1}\boldsymbol{U}_{\ell-1}\boldsymbol{A}_{\ell-1} & {}^t\boldsymbol{A}_{\ell-1}\boldsymbol{U}_{\ell-1}\boldsymbol{A}_{\ell-1} \end{pmatrix} \begin{pmatrix} \boldsymbol{I}_\ell & \boldsymbol{0}_\ell \\ \boldsymbol{0}_\ell & \boldsymbol{I}_\ell \end{pmatrix}$$

and

$$\boldsymbol{B''}_\ell = \begin{pmatrix} \boldsymbol{I}_\ell & {}^t\boldsymbol{0}_\ell \\ {}^t\boldsymbol{0}_\ell & \boldsymbol{I}_\ell \end{pmatrix} \begin{pmatrix} \boldsymbol{B}_{\ell-1} & \boldsymbol{B}_{\ell-1} \\ \boldsymbol{B}_{\ell-1} & \boldsymbol{B}_{\ell-1} \end{pmatrix} \begin{pmatrix} \boldsymbol{I}_\ell & \boldsymbol{0}_\ell \\ \boldsymbol{0}_\ell & \boldsymbol{I}_\ell \end{pmatrix}.$$

In order to calculate $\boldsymbol{B'}_\ell$, we need the following result:

Lemma III.12. *For any positive integer ℓ, consider the matrix*

$$\boldsymbol{C}_\ell = (c_{pq}) = {}^t\boldsymbol{A}_\ell \boldsymbol{U}_\ell \boldsymbol{A}_\ell \in M_{\ell+1}(\mathbb{R}).$$

Then, for any $0 \leq p, q \leq \ell$, we have

$$c_{pq} = \binom{\ell}{p}\binom{\ell}{q}.$$

Proof. Let us argue by induction on ℓ. For $\ell = 1$, the matrix $\boldsymbol{A}_\ell$ is the two-dimensional identity matrix, hence the assertion is obviously true. Assume the statement to be correct for $1, \ldots, \ell - 1$, and let us verify it in dimension ℓ. Put

$$\boldsymbol{v}_\ell = \begin{pmatrix} \binom{\ell}{0} \\ \binom{\ell}{1} \\ \vdots \\ \binom{\ell}{\ell} \end{pmatrix} \in \mathbb{R}^{\ell+1}.$$

Then

$$\begin{aligned}
\boldsymbol{C}_\ell &= {}^t\boldsymbol{A}_\ell \boldsymbol{U}_\ell \boldsymbol{A}_\ell \\
&= \begin{pmatrix} \boldsymbol{I}_\ell & {}^t\boldsymbol{0}_\ell \\ {}^t\boldsymbol{0}_\ell & \boldsymbol{I}_\ell \end{pmatrix} \begin{pmatrix} {}^t\boldsymbol{A}_{\ell-1} & {}^t\boldsymbol{O}_{\ell-1} \\ {}^t\boldsymbol{O}_{\ell-1} & {}^t\boldsymbol{A}_{\ell-1} \end{pmatrix} \begin{pmatrix} \boldsymbol{U}_{\ell-1} & \boldsymbol{U}_{\ell-1} \\ \boldsymbol{U}_{\ell-1} & \boldsymbol{U}_{\ell-1} \end{pmatrix} \begin{pmatrix} \boldsymbol{A}_{\ell-1} & \boldsymbol{O}_{\ell-1} \\ \boldsymbol{O}_{\ell-1} & \boldsymbol{A}_{\ell-1} \end{pmatrix} \begin{pmatrix} \boldsymbol{I}_\ell & \boldsymbol{0}_\ell \\ \boldsymbol{0}_\ell & \boldsymbol{I}_\ell \end{pmatrix} \\
&= \begin{pmatrix} \boldsymbol{I}_\ell & {}^t\boldsymbol{0}_\ell \\ {}^t\boldsymbol{0}_\ell & \boldsymbol{I}_\ell \end{pmatrix} \begin{pmatrix} \boldsymbol{C}_{\ell-1} & \boldsymbol{C}_{\ell-1} \\ \boldsymbol{C}_{\ell-1} & \boldsymbol{C}_{\ell-1} \end{pmatrix} \begin{pmatrix} \boldsymbol{I}_\ell & {}^t\boldsymbol{0}_\ell \\ {}^t\boldsymbol{0}_\ell & \boldsymbol{I}_\ell \end{pmatrix} \\
&= \begin{pmatrix} \boldsymbol{I}_\ell & {}^t\boldsymbol{0}_\ell \\ {}^t\boldsymbol{0}_\ell & \boldsymbol{I}_\ell \end{pmatrix} \begin{pmatrix} \boldsymbol{v}_{\ell-1} \\ \boldsymbol{v}_{\ell-1} \end{pmatrix} \left({}^t\boldsymbol{v}_{\ell-1}, {}^t\boldsymbol{v}_{\ell-1} \right) \begin{pmatrix} \boldsymbol{I}_\ell & \boldsymbol{0}_\ell \\ \boldsymbol{0}_\ell & \boldsymbol{I}_\ell \end{pmatrix}.
\end{aligned}$$

Since

$$\begin{pmatrix} \boldsymbol{I}_\ell & {}^t\boldsymbol{0}_\ell \\ {}^t\boldsymbol{0}_\ell & \boldsymbol{I}_\ell \end{pmatrix} \begin{pmatrix} \boldsymbol{v}_{\ell-1} \\ \boldsymbol{v}_{\ell-1} \end{pmatrix} = \begin{pmatrix} \binom{\ell-1}{0} \\ \binom{\ell-1}{1} + \binom{\ell-1}{0} \\ \vdots \\ \binom{\ell-1}{\ell-1} + \binom{\ell-1}{\ell-2} \\ \binom{\ell-1}{\ell-1} \end{pmatrix} = \boldsymbol{v}_\ell,$$

we find $\boldsymbol{C}_\ell = \boldsymbol{v}_\ell {}^t\boldsymbol{v}_\ell$, whence the assertion. □

It now follows that

$$\begin{aligned}
\boldsymbol{B}'_\ell &= \begin{pmatrix} \boldsymbol{I}_\ell & {}^t\boldsymbol{0}_\ell \\ {}^t\boldsymbol{0}_\ell & \boldsymbol{I}_\ell \end{pmatrix} \begin{pmatrix} \boldsymbol{C}_{\ell-1} & -\boldsymbol{C}_{\ell-1} \\ -\boldsymbol{C}_{\ell-1} & \boldsymbol{C}_{\ell-1} \end{pmatrix} \begin{pmatrix} \boldsymbol{I}_\ell & \boldsymbol{0}_\ell \\ \boldsymbol{0}_\ell & \boldsymbol{I}_\ell \end{pmatrix} \\
&= \begin{pmatrix} \boldsymbol{I}_\ell & {}^t\boldsymbol{0}_\ell \\ {}^t\boldsymbol{0}_\ell & \boldsymbol{I}_\ell \end{pmatrix} \begin{pmatrix} \boldsymbol{v}_{\ell-1} \\ -\boldsymbol{v}_{\ell-1} \end{pmatrix} \left({}^t\boldsymbol{v}_{\ell-1}, -{}^t\boldsymbol{v}_{\ell-1} \right) \begin{pmatrix} \boldsymbol{I}_\ell & \boldsymbol{0}_\ell \\ \boldsymbol{0}_\ell & \boldsymbol{I}_\ell \end{pmatrix}
\end{aligned}$$

and, as one easily verifies that

$$\begin{pmatrix} \boldsymbol{I}_\ell & {}^t\boldsymbol{0}_\ell \\ {}^t\boldsymbol{0}_\ell & \boldsymbol{I}_\ell \end{pmatrix} \begin{pmatrix} \boldsymbol{v}_{\ell-1} \\ -\boldsymbol{v}_{\ell-1} \end{pmatrix} = \begin{pmatrix} \binom{\ell}{0} \\ \binom{\ell}{1}\frac{\ell-2}{\ell} \\ \vdots \\ \binom{\ell}{p}\frac{\ell-2p}{\ell} \\ \vdots \\ -\binom{\ell}{\ell} \end{pmatrix},$$

we find that

$$\boldsymbol{B}_\ell^{'} = (b_{\ell,pq}^{'}) = \left(\binom{\ell}{p}\binom{\ell}{q}\frac{(\ell-2p)}{\ell}\frac{(\ell-2q)}{\ell}\right).$$

The next result calculates the matrix $\boldsymbol{B}_\ell^{''} = (b_{\ell,pq}^{''})$:

Lemma III.13. *The components $b_{\ell,pq}^{''}$ of the matrix $\boldsymbol{B}_\ell^{''}$ are determined by*

1. $b_{\ell,00}^{''} = b_{00}^{\ell-1}$, $b_{\ell,\ell\ell}^{''} = b_{\ell-1,\ell-1}^{\ell-1}$, $b_{\ell,0\ell}^{''} = b_{0,\ell-1}^{\ell-1}$,
2. *if* $1 \le q < \ell$, *then* $b_{\ell,0q}^{''} = b_{0q}^{\ell-1} + b_{0,q-1}^{\ell-1}$ *and* $b_{\ell,\ell q}^{''} = b_{\ell-1,q}^{\ell-1} + b_{\ell-1,q-1}^{\ell-1}$,
3. *if* $1 \le p, q < \ell$, *then* $b_{\ell,pq}^{''} = b_{pq}^{\ell-1} + b_{p,q-1}^{\ell-1} + b_{p-1,q}^{\ell-1} + b_{p-1,q-1}^{\ell-1}$.

Proof. This is a straightforward consequence of

$$\boldsymbol{B}_\ell^{''} = \begin{pmatrix} \boldsymbol{I}_\ell & {}^t\boldsymbol{0}_\ell \\ {}^t\boldsymbol{0}_\ell & \boldsymbol{I}_\ell \end{pmatrix} \begin{pmatrix} \boldsymbol{B}_{\ell-1} & \boldsymbol{B}_{\ell-1} \\ \boldsymbol{B}_{\ell-1} & \boldsymbol{B}_{\ell-1} \end{pmatrix} \begin{pmatrix} \boldsymbol{I}_\ell & \boldsymbol{0}_\ell \\ \boldsymbol{0}_\ell & \boldsymbol{I}_\ell \end{pmatrix}.$$

□

Using the previous result and the fact that $\boldsymbol{B}_\ell = \boldsymbol{B}_\ell^{'} + \boldsymbol{B}_\ell^{''}$, we obtain:

Corollary III.14. *The components b_{pq}^{ℓ} of the matrix $\boldsymbol{B}_\ell$ are determined by*

1. $b_{00}^{\ell} = b_{\ell\ell}^{\ell} = 1 + \ell$, $b_{0\ell}^{\ell} = b_{\ell 0}^{\ell} = 1 - \ell$,
2. *if* $1 \le q < \ell$, *then we have*

$$b_{0q}^{\ell} = b_{0q}^{\ell-1} + b_{0q-1}^{\ell-1} + \binom{\ell}{q}\frac{\ell-2q}{\ell}$$

and

$$b_{\ell q}^{\ell} = b_{\ell-1,q}^{\ell-1} + b_{\ell-1,q-1}^{\ell-1} - \binom{\ell}{q}\frac{\ell-2q}{\ell},$$

3. *if* $1 \leq p, q < \ell$, *then we have*

$$b^\ell_{pq} = b^{\ell-1}_{pq} + b^{\ell-1}_{p-1,q} + b^{\ell-1}_{p,q-1} + b^{\ell-1}_{p-1,q-1} + \binom{\ell}{p}\binom{\ell}{q}\frac{(\ell-2p)}{\ell}\frac{(\ell-2q)}{\ell}.$$

We may now finally prove:

Proposition III.15. *For any* $0 \leq p, q \leq \ell$, *the component* b^ℓ_{pq} *of the matrix* $\boldsymbol{B}_\ell$ *is given by*

$$b^\ell_{pq} = \binom{\ell}{p}\binom{\ell}{q}\left(1 + \frac{(\ell-2p)(\ell-2q)}{\ell}\right).$$

Proof. The previous result may be written as

$$b^\ell_{pq} = \sum_{i,j=0}^{1} b^{\ell-1}_{p-i,q-j} + \left(\binom{\ell-1}{p} - \binom{\ell-1}{p-1}\right)\left(\binom{\ell-1}{q} - \binom{\ell-1}{q-1}\right).$$

Iterating, we obtain

$$b^\ell_{pq} = \binom{\ell}{p}\binom{\ell}{q} b^0_{00} + \sum_{r=0}^{\ell-1} B^\ell_{rp} B^\ell_{rq},$$

where

$$B^\ell_{rp} = A^\ell_{rp} - A^\ell_{r,p-1},$$

with

$$A^\ell_{rp} = \sum_{i=0}^{r}\binom{r}{i}\binom{\ell-r-1}{p-i}.$$

Since it is easy to see that, for any $r \leq \ell$, we have

$$A^\ell_{rp} = \binom{\ell-1}{p},$$

it follows that

$$B^\ell_{rp} = \binom{\ell-1}{p} - \binom{\ell-1}{p-1},$$

which finishes the proof. □

We thus have proved:

Theorem III.16. *Let* $f = h \circ u : \Omega \to \mathbb{R}$ *be a unitation function with associated function* $h : \{0, \ldots, \ell\} \to \mathbb{R}$. *Then the normalized epistasis of* f *is given by*

$$\varepsilon^*(f) = 1 - \frac{1}{2^\ell}\frac{\sum_{p,q=0}^{\ell}\binom{\ell}{p}\binom{\ell}{q}\left(1 + \frac{(\ell-2p)(\ell-2q)}{\ell}\right)h(p)h(q)}{\sum_{p=0}^{\ell}\binom{\ell}{p}h(p)^2}. \qquad \text{(III.4)}$$

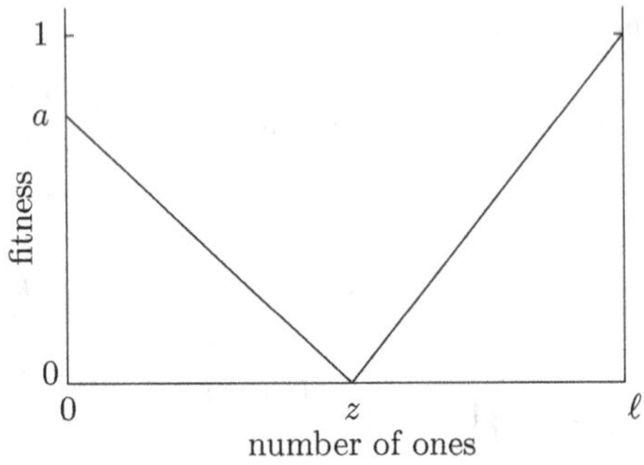

Figure III.1: Pictorial representation of the unitation function $g_{z,a}$.

2.5 Experimental results

Let us now link the normalized epistasis of a unitation function f, as calculated in the previous section, to the performance of a standard GA on f. As a measure of problem difficulty, we again use the number of generations needed to reach the optimum. The GA has the same characteristics as the one used for the Royal Road functions: it is generational, with a population of size 100, and it uses binary tournament selection, one-point crossover at rate 0.8 and ordinary mutation at rate $1/\ell$. We present the mean and standard deviation computed over 100 independent runs.

Example 1

Just as in [14] and similar to a construction in section 5.3 of chapter I, we define for any integer $0 \leq z \leq \ell$ and any $0 \leq a \leq 1$ the unitation function $g_{z,a}$ by

$$g_{z,a}(s) = \begin{cases} \frac{a}{z}(z - u(s)) & \text{if } u(s) \leq z, \\ \frac{1}{\ell - z}(u(s) - z) & \text{otherwise.} \end{cases}$$

The function is depicted in figure III.1.

Due to the presence of the local optimum a, it is clear that deception will contribute to slowing down the GA. For high values of a in combination with values of z much

a	$\varepsilon^*(g_{z,a})$	mean	std dev
0.1	0.219	97.53	10.26
0.2	0.253	99.52	11.20
0.3	0.284	97.61	11.83
0.4	0.309	96.25	10.25
0.5	0.329	96.50	9.906
0.6	0.344	97.49	10.18
0.7	0.355	147.8	487.7
0.8	0.362	444.0	1250
0.9	0.365	885.1	1796

Table III.3: Problem difficulty compared to epistasis for $g_{z,a}$, with $\ell = 100$ and $z = 50$. Note that for $a \geq 0.7$, the distribution of number of generations adheres more to a log-normal than to a normal distribution.

greater than 0.5, it may also cause the GA to not always finding the global optimum. We refer to [14] for a detailed study of this phenomenon.

Fixing $\ell = 100$, $z = 50$ and comparing epistasis values and problem difficulty for varying values of $0 \leq a \leq 1$, we are led to table III.3, from which it becomes clear that the epistasis increases with a increasing. The GA, however, does not distinguish between low values of a, say $0 \leq a \leq 0.6$. Only when a becomes significantly large, deception starts to play its role of slowing down the GA.

Example 2

For any $0 \leq z \leq \ell$ and $0 \leq a \leq 1$, define the unitation function $f_{z,a}$ by

$$f_{z,a}(s) = \begin{cases} a\frac{u(s)}{z} & \text{if } u(s) \leq z \\ a + (1-a)\frac{u(s)-z}{\ell - z} & \text{if } z < u(s) \leq \ell. \end{cases}$$

This function is depicted in figure III.2.

The inflection point z has fitness value $f_{z,a}(z) = a$. Obviously, when $z = 0$ or $z = \ell$, or when $a = \frac{z}{\ell}$, we are in the linear case and $\varepsilon^*(f_{z,a}) = 0$. In the table below, still with $\ell = 100$, we compare epistasis and problem difficulty for $z = 75$ and varying a.

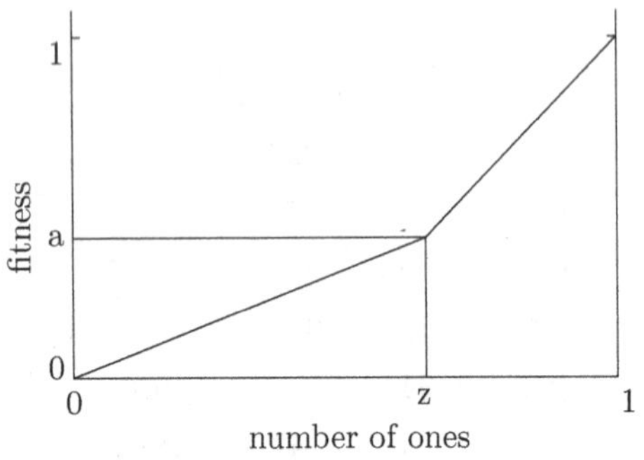

Figure III.2: Pictorial representation of the unitation function $f_{z,a}$.

We first note that our GA sees no difference between any of the functions with $0 < a < 1$, because it uses tournament selection: the functions are monotonically and strictly increasing, and the selection operator only observes that $f_{z,a}(s) > f_{z,a}(t)$ when $u(s) > u(t)$. The cases $a = 0$ and $a = 1$ are clearly different. In the former, the GA needs to discover, by accident, a string with 75 ones in it before it can pick up a signal. In the latter, it is fully guided and it has reached the optimum when it hits a string with 75 ones.

The epistasis values reflect the difficulties at $a = 0$, but do not distinguish between values near one and exactly one. Moreover, it cannot know that we are using tournament selection, and that the divergence from linearity in the range $0 < a < 1$ is completely irrelevant. To demonstrate the difference with fitness proportional selection, we present in table III.5 epistasis and problem difficulty for $f_{z,a}$ with $\ell = 30$ and $z = 25$ (we chose a smaller string length because the GA with proportional selection is not able to solve any of the functions within a reasonable amount of time because of too little selective pressure).

The change in selection operator has not modified the situation for the extreme points $a = 0$ and $a = 1$ (very hard and very easy). For the intermediate values, however, the problem difficulty now increases nicely with a increasing, while the epistasis values first decrease and then increase again.

a	$\varepsilon^*(f_{z,a})$	mean	std dev
0.0	0.999998	$\gg 50000$	
0.1	6.2×10^{-8}	97.95	10.28
0.2	1.1×10^{-8}	96.97	11.18
0.3	3.4×10^{-9}	97.47	8.948
0.4	1.1×10^{-9}	97.29	9.539
0.5	3.7×10^{-10}	97.43	10.17
0.6	9.3×10^{-11}	97.05	10.30
0.7	7.6×10^{-12}	97.68	10.02
0.8	5.8×10^{-12}	97.96	10.18
0.9	4.1×10^{-11}	97.20	9.635
1.0	9.3×10^{-11}	14.17	3.225

Table III.4: Problem difficulty compared to epistasis for $f_{z,a}$, with $\ell = 100$ and $z = 75$.

3 Template functions

3.1 Basic properties

The "template functions" we are about to consider calculate the fitness of a string of length ℓ, by sliding a fixed string t of length $n \leq \ell$ (the "template") over it. Each time an occurrence of t in s is found, a fixed amount a is added to the fitness of s. For convenience's sake, we will assume throughout that $a = 1$ and that t is the length n string $11\ldots11$. So, the template functions depend only on the parameters ℓ and n and will be denoted by T_ℓ^n. For example,

$$T_\ell^2(1^\ell) = T_\ell^2(11\ldots11) = \ell - 1,$$

because 11 may be found $\ell - 1$ times in the length ℓ string $11\ldots11$, whereas

$$T_\ell^3(01110\ldots011) = 1.$$

It seems reasonable to expect that increasing the length n of the template will also increase the epistasis of T_ℓ^n, in view of the strong linkage between the different loci. The calculations below will make these statements more precise.

a	$\varepsilon^*(f_{z,a})$	mean	std dev
0.0	0.9995	7009	3578
0.1	3.7×10^{-4}	38.29	12.52
0.2	6.9×10^{-5}	45.38	13.99
0.3	2.2×10^{-5}	51.29	17.73
0.4	8.1×10^{-6}	61.12	24.43
0.5	3.1×10^{-6}	85.32	30.63
0.6	1.0×10^{-6}	144.9	78.43
0.7	2.5×10^{-7}	384.0	320.7
0.8	1.2×10^{-8}	868.1	877.7
0.9	3.8×10^{-8}	1818	1851
1.0	1.9×10^{-7}	13.71	7.822

Table III.5: Problem difficulty compared to epistasis for $f_{z,a}$, with $\ell = 30$ and $z = 25$. The GA now uses fitness proportional selection. Note: from $a \approx 0.6$ onwards, and for $a = 0$, the distribution of the number of generations to hit the optimum adheres more closely to a log-normal distribution than to a normal one. This explains the excessive standard deviations. For $a = 0$, a maximum limit of 10,000 generations must also be taken into account.

The main purpose of this section is to explicitly calculate the normalized epistasis of the functions T_ℓ^n. The first step of this consists in evaluating $\|\boldsymbol{T}_\ell^n\|$ where $\boldsymbol{T}_\ell^n \in \mathbb{R}^{2^\ell}$ denotes the vector corresponding to T_ℓ^n, i.e.,

$$\boldsymbol{T}_\ell^n = \begin{pmatrix} T_\ell^n(00\ldots0) \\ \vdots \\ T_\ell^n(11\ldots1) \end{pmatrix}.$$

To simplify this calculation, let us first note that for any $n \leq \ell$, we have

$$\boldsymbol{T}_\ell^n = \begin{pmatrix} \boldsymbol{T}_{\ell-1}^n \\ \boldsymbol{T}_{\ell-1}^n + \boldsymbol{D}_{\ell-1}^n \end{pmatrix}$$

with

$$\boldsymbol{D}^n_{\ell-1} = \begin{pmatrix} \mathbf{0}_{\ell-2} \\ \vdots \\ \mathbf{0}_{\ell-n} \\ \boldsymbol{u}_{\ell-n} \end{pmatrix} \in \mathbb{R}^{2^{\ell-1}}.$$

Here, as before, we denote for any positive integer i by $\mathbf{0}_i$ the zero vector in $\mathbb{R}^{2^i}$ and by $\boldsymbol{u}_i$ the vector in $\mathbb{R}^{2^i}$ all of whose entries have value 1. For example, if $n = 2$,

$$\boldsymbol{T}^2_1 = \begin{pmatrix} 0 \\ 0 \end{pmatrix}, \ \boldsymbol{T}^2_2 = \begin{pmatrix} 0 \\ 0 \\ 0 \\ 1 \end{pmatrix} \text{ and } \boldsymbol{T}^2_3 = \begin{pmatrix} 0 \\ 0 \\ 0 \\ 1 \\ 0 \\ 0 \\ 1 \\ 2 \end{pmatrix}.$$

We need to know more about the structure of $\boldsymbol{T}^n_\ell$. An easy induction argument yields:

Lemma III.17. *For any $1 \leq i \leq n$, we have*

$$\boldsymbol{T}^n_{n+i} = \begin{pmatrix} \boldsymbol{T}^n_{n+i-1} \\ \boldsymbol{T}^n_{n+i-2} \\ \vdots \\ \boldsymbol{T}^n_n \\ \widetilde{\mathbf{0}}^i_n \\ \boldsymbol{u}_{i-1} \\ 2\boldsymbol{u}_{i-2} \\ \vdots \\ i\boldsymbol{u}_0 \\ i+1 \end{pmatrix},$$

where $\widetilde{\mathbf{0}}^i_n = {}^t(0, \ldots, 0) \in \mathbb{R}^{2^n - 2^i}$.

Using this, we may prove:

Lemma III.18. *For any $1 \leq i \leq n$, we have*

$$ {}^t\boldsymbol{T}^n_{n+i}\boldsymbol{D}^n_{n+i} = 2^{i+1} - 1. $$

Proof. It suffices to note that ${}^t\boldsymbol{T}^n_{n+i}\boldsymbol{D}^n_{n+i}$ is the sum of the 2^{i+1} last components of $\boldsymbol{T}^n_{n+i}$. By using the previous result, we then have

$$ \begin{aligned} {}^t\boldsymbol{T}^n_{n+i}\boldsymbol{D}^n_{n+i} &= (i+1) + 2^0\, i + 2^1(i-1) + \cdots + 2^{i-1}\,1 + 2^i\,0 \\ &= (i+1) + \sum_{k=0}^{i} 2^k(i-k). \end{aligned} $$

Since $\sum_{k=0}^{i} k2^k = 2 + (i-1)2^{i+1}$, it easily follows that ${}^t\boldsymbol{T}^n_{n+i}\boldsymbol{D}^n_{n+i} = 2^{i+1} - 1$. $\square$

From this one deduces:

Lemma III.19. *For any $0 \leq i \leq n$, we have*

$$ \left\|\boldsymbol{T}^n_{n+i}\right\|^2 = 2^i(3i-1) + 2. $$

Proof. Let us again argue by induction on i. The statement obviously holds true for $i = 0$. In the general case, we have

$$ \begin{aligned} \left\|\boldsymbol{T}^n_{n+i}\right\|^2 = {}^t\boldsymbol{T}^n_{n+i}\boldsymbol{T}^n_{n+i} &= \left({}^t\boldsymbol{T}^n_{n+i-1}, {}^t\boldsymbol{T}^n_{n+i-1} + {}^t\boldsymbol{D}^n_{n+i-1}\right)\begin{pmatrix} \boldsymbol{T}^n_{n+i-1} \\ \boldsymbol{T}^n_{n+i-1} + \boldsymbol{D}^n_{n+i-1} \end{pmatrix} \\ &= 2\left\|\boldsymbol{T}^n_{n+i-1}\right\|^2 + 2\,{}^t\boldsymbol{T}^n_{n+i-1}\boldsymbol{D}^n_{n+i-1} + \left\|\boldsymbol{D}^n_{n+i-1}\right\|^2, \end{aligned} $$

where

$$ \begin{aligned} \left\|\boldsymbol{D}^n_{n+i-1}\right\|^2 = {}^t\boldsymbol{D}^n_{n+i-1}\boldsymbol{D}^n_{n+i-1} &= \left({}^t\boldsymbol{0}_{n+i-2}, \ldots, {}^t\boldsymbol{0}_i, {}^t\boldsymbol{u}_i\right)\begin{pmatrix} \boldsymbol{0}_{n+i-2} \\ \vdots \\ \boldsymbol{0}_i \\ \boldsymbol{u}_i \end{pmatrix} \\ &= \left\|\boldsymbol{u}_i\right\|^2 = 2^i. \end{aligned} $$

Using the induction hypothesis and the previous lemma, it thus follows by induction that

$$\left\|\boldsymbol{T}^n_{n+i}\right\|^2 = 2\left(2^{i-1}(3(i-1)-1)+2\right) + 2\left(2^i - 1\right) + 2^i = 2^i(3i-1)+2.$$

□

Using this, we may prove:

Lemma III.20. *For any $n \leq \ell$, the sum of the components of $\boldsymbol{T}^n_\ell$ is given by*

$${}^t\boldsymbol{T}^n_\ell \boldsymbol{u}_\ell = 2^{\ell-n}(\ell-n+1).$$

Proof. Let us use $a_{\ell,n}$ to denote the sum of the components of $\boldsymbol{T}^n_\ell$. Then

$$\begin{aligned} a_{\ell,n} &= {}^t\boldsymbol{T}^n_\ell \boldsymbol{u}_\ell = \left({}^t\boldsymbol{T}^n_{\ell-1}, {}^t\boldsymbol{T}^n_{\ell-1} + {}^t\boldsymbol{D}^n_{\ell-1}\right)\begin{pmatrix}\boldsymbol{u}_{\ell-1}\\ \boldsymbol{u}_{\ell-1}\end{pmatrix} \\ &= 2{}^t\boldsymbol{T}^n_{\ell-1}\boldsymbol{u}_{\ell-1} + {}^t\boldsymbol{D}^n_{\ell-1}\boldsymbol{u}_{\ell-1} = 2a_{\ell-1,n} + \mathrm{Tr}(\boldsymbol{D}^n_{\ell-1}) \\ &= 2a_{\ell-1,n} + 2^{\ell-n}. \end{aligned}$$

It thus follows that

$$a_{\ell,n} = 2a_{\ell-1,n} + 2^{\ell-n} = 2^2 a_{\ell-2,n} + 22^{\ell-n} = \cdots = 2^p a_{\ell-p,n} + p2^{\ell-n}.$$

Putting $p = \ell - n$, we obtain

$$a_{\ell,n} = 2^{\ell-n} a_{n,n} + (\ell-n)2^{\ell-n} = 2^{\ell-n}\,{}^t\boldsymbol{T}^n_n \boldsymbol{u}_n + 2^{\ell-n}(\ell-n) = 2^{\ell-n}(\ell-n+1),$$

which proves the assertion. □

Combining these lemmas yields:

Proposition III.21. *For any $n \leq \ell$, we have*

$$\|\boldsymbol{T}^n_\ell\|^2 = \begin{cases} 2^{\ell-n}(3(\ell-n)-1)+2 & \text{if } n \leq \ell \leq 2n, \\ 2^{\ell-n}(3(\ell-n)-1) + 2^{\ell-2n}(2+(\ell-2n)(\ell-2n-1)) & \text{if } 2n \leq \ell. \end{cases}$$

Proof. The first case ($n \leq \ell \leq 2n$) is just lemma III.19. In the second case ($2n \leq \ell$), we again apply induction on ℓ. First, note that

$$\begin{aligned}\|\boldsymbol{T}_\ell^n\|^2 &= {}^t\boldsymbol{T}_\ell^n \boldsymbol{T}_\ell^n = \left({}^t\boldsymbol{T}_{\ell-1}^n, {}^t\boldsymbol{T}_{\ell-1}^n + {}^t\boldsymbol{D}_{\ell-1}^n\right) \begin{pmatrix} \boldsymbol{T}_{\ell-1}^n \\ \boldsymbol{T}_{\ell-1}^n + \boldsymbol{D}_{\ell-1}^n \end{pmatrix} \\ &= 2\left\|\boldsymbol{T}_{\ell-1}^n\right\|^2 + 2{}^t\boldsymbol{T}_{\ell-1}^n \boldsymbol{D}_{\ell-1}^n + \left\|\boldsymbol{D}_{\ell-1}^n\right\|^2, \end{aligned} \tag{III.5}$$

with $\left\|\boldsymbol{D}_{\ell-1}^n\right\|^2 = \|\boldsymbol{u}_{\ell-n}\|^2 = 2^{\ell-n}$. On the other hand,

$$\begin{aligned}{}^t\boldsymbol{T}_{\ell-1}^n \boldsymbol{D}_{\ell-1}^n &= \left({}^t\boldsymbol{T}_{\ell-2}^n, {}^t\boldsymbol{T}_{\ell-2}^n + {}^t\boldsymbol{D}_{\ell-2}^n\right) \begin{pmatrix} \boldsymbol{0}_{\ell-2} \\ \vdots \\ \boldsymbol{0}_{\ell-n} \\ \boldsymbol{u}_{\ell-n} \end{pmatrix} \\ &= {}^t\boldsymbol{T}_{\ell-2}^n \begin{pmatrix} \boldsymbol{0}_{\ell-3} \\ \vdots \\ \boldsymbol{0}_{\ell-n} \\ \boldsymbol{u}_{\ell-n} \end{pmatrix} + {}^t\boldsymbol{D}_{\ell-2}^n \begin{pmatrix} \boldsymbol{0}_{\ell-3} \\ \vdots \\ \boldsymbol{0}_{\ell-n} \\ \boldsymbol{u}_{\ell-n} \end{pmatrix} \\ &\vdots \\ &= {}^t\boldsymbol{T}_{\ell-n+1}^n \begin{pmatrix} \boldsymbol{0}_{\ell-n} \\ \boldsymbol{u}_{\ell-n} \end{pmatrix} + \sum_{k=2}^{n-1} {}^t\boldsymbol{D}_{\ell-k}^n \begin{pmatrix} \boldsymbol{0}_{\ell-k-1} \\ \vdots \\ \boldsymbol{0}_{\ell-n} \\ \boldsymbol{u}_{\ell-n} \end{pmatrix}. \end{aligned}$$

Again using the recursive form of the template functions, we obtain that

$$\begin{aligned}
{}^t\boldsymbol{T}^n_{\ell-1}\boldsymbol{D}^n_{\ell-1} &= \left({}^t\boldsymbol{T}^n_{\ell-n}, {}^t\boldsymbol{T}^n_{\ell-n} + {}^t\boldsymbol{D}^n_{\ell-n}\right)\begin{pmatrix}\mathbf{0}_{\ell-n}\\ \boldsymbol{u}_{\ell-n}\end{pmatrix}\\
&\quad+\sum_{k=2}^{n-1}\left({}^t\mathbf{0}_{\ell-k-1},\ldots,{}^t\mathbf{0}_{\ell-k-n+1},{}^t\boldsymbol{u}_{\ell-k-n+1}\right)\begin{pmatrix}\mathbf{0}_{\ell-k-1}\\ \vdots\\ \mathbf{0}_{\ell-n}\\ \boldsymbol{u}_{\ell-n}\end{pmatrix}\\
&= {}^t\boldsymbol{T}^n_{\ell-n}\boldsymbol{u}_{\ell-n} + \|\boldsymbol{u}_{\ell-2n+1}\|^2 + \sum_{k=2}^{n-1}\|\boldsymbol{u}_{\ell-k-n+1}\|^2\\
&= {}^t\boldsymbol{T}^n_{\ell-n}\boldsymbol{u}_{\ell-n} + 2^{\ell-2n+1} + 2^{\ell-n+1}\sum_{k=2}^{n-1}2^{-k}\\
&= {}^t\boldsymbol{T}^n_{\ell-n}\boldsymbol{u}_{\ell-n} + 2^{\ell-2n}(2^n-2)\\
&= 2^{\ell-2n}(\ell-2n-1+2^n).
\end{aligned}$$

Substituting this in (III.5) and denoting $\|\boldsymbol{T}^n_\ell\|^2$ by $b_{\ell,n}$, we obtain

$$\begin{aligned}
b_{\ell,n} &= 2b_{\ell-1,n} + 2^{\ell-2n+1}(\ell-2n+3.2^{n-1}-1)\\
&= 2^2b_{\ell-2,n} + 2^{\ell-2n+1}\left(2(\ell-2n+3.2^{n-1})-(1+2)\right)\\
&\;\;\vdots\\
&= 2^pb_{\ell-p,n} + 2^{\ell-2n+1}\left(p(\ell-2n+3.2^{n-1})-\sum_{k=1}^{p}k\right).
\end{aligned}$$

Putting $p=\ell-2n$ finally yields

$$b_{\ell,n} = 2^{\ell-2n}b_{2n,n} + 2^{\ell-2n+1}(\ell-2n)\left(\ell-2n+3.2^{n-1}-\frac{\ell-2n+1}{2}\right).$$

Since lemma III.19 implies $b_{2n,n}=2^n(3n-1)+2$, the result now easily follows. □

3.2 Epistasis of template functions

In order to determine the normalized epistasis of template functions, it remains to calculate $\gamma(T_\ell^n) = {}^t\boldsymbol{T}_\ell^n \boldsymbol{G}_\ell \boldsymbol{T}_\ell^n$. To realize this, let us first note that, for all $n \leq \ell$,

$$\begin{aligned}
\gamma(T_\ell^n) &= {}^t\boldsymbol{T}_\ell^n \boldsymbol{G}_\ell \boldsymbol{T}_\ell^n \\
&= \left({}^t\boldsymbol{T}_{\ell-1}^n, {}^t\boldsymbol{T}_{\ell-1}^n + {}^t\boldsymbol{D}_{\ell-1}^n\right) \begin{pmatrix} \boldsymbol{G}_{\ell-1} + \boldsymbol{U}_{\ell-1} & \boldsymbol{G}_{\ell-1} - \boldsymbol{U}_{\ell-1} \\ \boldsymbol{G}_{\ell-1} - \boldsymbol{U}_{\ell-1} & \boldsymbol{G}_{\ell-1} + \boldsymbol{U}_{\ell-1} \end{pmatrix} \begin{pmatrix} \boldsymbol{T}_{\ell-1}^n \\ \boldsymbol{T}_{\ell-1}^n + \boldsymbol{D}_{\ell-1}^n \end{pmatrix} \\
&= 4\,{}^t\boldsymbol{T}_{\ell-1}^n \boldsymbol{G}_{\ell-1} \boldsymbol{T}_{\ell-1}^n + 4\,{}^t\boldsymbol{T}_{\ell-1}^n \boldsymbol{G}_{\ell-1} \boldsymbol{D}_{\ell-1}^n + {}^t\boldsymbol{D}_{\ell-1}^n \boldsymbol{G}_{\ell-1} \boldsymbol{D}_{\ell-1}^n + {}^t\boldsymbol{D}_{\ell-1}^n \boldsymbol{U}_{\ell-1} \boldsymbol{D}_{\ell-1}^n,
\end{aligned}$$

where the last terms are

$$\begin{aligned}
{}^t\boldsymbol{D}_{\ell-1}^n \boldsymbol{U}_{\ell-1} \boldsymbol{D}_{\ell-1}^n &= \left({}^t\mathbf{0}_{\ell-2}, \ldots, {}^t\mathbf{0}_{\ell-n}, {}^t\boldsymbol{u}_{\ell-n}\right) \begin{pmatrix} 1 \cdots 1 \\ \vdots \ddots \vdots \\ 1 \cdots 1 \end{pmatrix} \begin{pmatrix} \mathbf{0}_{\ell-2} \\ \vdots \\ \mathbf{0}_{\ell-n} \\ \boldsymbol{u}_{\ell-n} \end{pmatrix} \\
&= 2^{\ell-n} \left\|\boldsymbol{u}_{\ell-n}\right\|^2 = 4^{\ell-n},
\end{aligned}$$

and

$$\begin{aligned}
{}^t\boldsymbol{D}_{\ell-1}^n \boldsymbol{G}_{\ell-1} \boldsymbol{D}_{\ell-1}^n &= \left({}^t\mathbf{0}_{\ell-2}, \ldots, {}^t\mathbf{0}_{\ell-n}, {}^t\boldsymbol{u}_{\ell-n}\right) \begin{pmatrix} \boldsymbol{G}_{\ell-2} + \boldsymbol{U}_{\ell-2} & \boldsymbol{G}_{\ell-2} - \boldsymbol{U}_{\ell-2} \\ \boldsymbol{G}_{\ell-2} - \boldsymbol{U}_{\ell-2} & \boldsymbol{G}_{\ell-2} + \boldsymbol{U}_{\ell-2} \end{pmatrix} \begin{pmatrix} \mathbf{0}_{\ell-2} \\ \vdots \\ \mathbf{0}_{\ell-n} \\ \boldsymbol{u}_{\ell-n} \end{pmatrix} \\
&= \left({}^t\mathbf{0}_{\ell-3}, \ldots, {}^t\mathbf{0}_{\ell-n}, {}^t\boldsymbol{u}_{\ell-n}\right) \left(\boldsymbol{G}_{\ell-2} + \boldsymbol{U}_{\ell-2}\right) \begin{pmatrix} \mathbf{0}_{\ell-3} \\ \vdots \\ \mathbf{0}_{\ell-n} \\ \boldsymbol{u}_{\ell-n} \end{pmatrix} \\
&\;\;\vdots \\
&= {}^t\boldsymbol{u}_{\ell-n} \boldsymbol{G}_{\ell-n} \boldsymbol{u}_{\ell-n} + (n-1)4^{\ell-n} \\
&= \sum_{i,j=0}^{2^{\ell-n}-1} g_{ij} + (n-1)4^{\ell-n} = n4^{\ell-n}.
\end{aligned}$$

Let us put

$$\alpha_k = \left({}^t\mathbf{0}_{\ell-k-1}, \ldots, {}^t\mathbf{0}_{\ell-n}, {}^t\boldsymbol{u}_{\ell-n}\right) \boldsymbol{G}_{\ell-k}\boldsymbol{D}^n_{\ell-k}$$
$$\beta_k = \left({}^t\mathbf{0}_{\ell-k-1}, \ldots, {}^t\mathbf{0}_{\ell-n}, {}^t\boldsymbol{u}_{\ell-n}\right) \boldsymbol{G}_{\ell-k}\boldsymbol{T}^n_{\ell-k},$$

then the calculation of the second term in the above expression for $\gamma(T^n_\ell)$ depends on the following result:

Lemma III.22. *For any* $1 \le k \le \ell - n + 1$, *we have*

$$\alpha_k = (n+1-k)2^{k-1}4^{\ell-n-k+1}$$
$$\beta_k = 2\beta_{k+1} + (n+1-k)2^k 4^{\ell-n-k}.$$

Proof. We calculate α_k by recursion. As α_k is equal to

$$\left({}^t\mathbf{0}_{\ell-k-1}, \ldots, {}^t\mathbf{0}_{\ell-n}, {}^t\boldsymbol{u}_{\ell-n}\right) \begin{pmatrix} \boldsymbol{G}_{\ell-k-1} + \boldsymbol{U}_{\ell-k-1} & \boldsymbol{G}_{\ell-k-1} - \boldsymbol{U}_{\ell-k-1} \\ \boldsymbol{G}_{\ell-k-1} - \boldsymbol{U}_{\ell-k-1} & \boldsymbol{G}_{\ell-k-1} + \boldsymbol{U}_{\ell-k-1} \end{pmatrix} \begin{pmatrix} \mathbf{0}_{\ell-k-1} \\ \vdots \\ \mathbf{0}_{\ell-n-k+1} \\ \boldsymbol{u}_{\ell-n-k+1} \end{pmatrix},$$

clearly,

$$\begin{aligned}
\alpha_k &= \left({}^t\mathbf{0}_{\ell-k-2}, \ldots, {}^t\mathbf{0}_{\ell-n}, {}^t\boldsymbol{u}_{\ell-n}\right) \left(\boldsymbol{G}_{\ell-k-1} + \boldsymbol{U}_{\ell-k-1}\right) \begin{pmatrix} \mathbf{0}_{\ell-k-2} \\ \vdots \\ \mathbf{0}_{\ell-n-k+1} \\ \boldsymbol{u}_{\ell-n-k+1} \end{pmatrix} \\
&= \left({}^t\mathbf{0}_{\ell-k-2}, \ldots, {}^t\mathbf{0}_{\ell-n}, {}^t\boldsymbol{u}_{\ell-n}\right) \boldsymbol{G}_{\ell-k-1} \begin{pmatrix} \mathbf{0}_{\ell-k-2} \\ \vdots \\ \mathbf{0}_{\ell-n-k+1} \\ \boldsymbol{u}_{\ell-n-k+1} \end{pmatrix} + 2^{\ell-n}2^{\ell-n-k+1} \\
&\ \ \vdots \\
&= {}^t\boldsymbol{u}_{\ell-n}\boldsymbol{G}_{\ell-n} \begin{pmatrix} \mathbf{0}_{\ell-n-1} \\ \vdots \\ \mathbf{0}_{\ell-n-k+1} \\ \boldsymbol{u}_{\ell-n-k+1} \end{pmatrix} + (n-k)2^{k-1}4^{\ell-n-k+1},
\end{aligned}$$

if $\ell - n - k + 1 \geq 0$. Now, again using the recursive formula for $\boldsymbol{G}_{\ell-n}$, we have

$$\alpha_k = 2{}^t\boldsymbol{u}_{\ell-n-1}\boldsymbol{G}_{\ell-n-1}\begin{pmatrix} \boldsymbol{0}_{\ell-n-2} \\ \vdots \\ \boldsymbol{0}_{\ell-n-k+1} \\ \boldsymbol{u}_{\ell-n-k+1} \end{pmatrix} + (n-k)2^{k-1}4^{\ell-n-k+1}$$

$$\vdots$$

$$= 2^{k-1}{}^t\boldsymbol{u}_{\ell-n-k+1}\boldsymbol{G}_{\ell-n-k+1}\boldsymbol{u}_{\ell-n-k+1} + (n-k)2^{k-1}4^{\ell-n-k+1}$$

$$= 2^{k-1}\sum_{i,j=0}^{2^{\ell-n+1-k}-1} g_{ij} + (n-k)2^{k-1}4^{\ell-n-k+1} = (n+1-k)2^{k-1}4^{\ell-n-k+1}.$$

A similar argument yields the expression for β. □

From this, it follows:

Lemma III.23. *For all* $2n < \ell$,

$${}^t\boldsymbol{T}^n_{\ell-1}\boldsymbol{G}_{\ell-1}\boldsymbol{D}^n_{\ell-1} = 4^{\ell-n-1}(n^2 - 3n + 2\ell).$$

Proof. Using the symmetry of $\boldsymbol{G}_\ell$ and the previous result, we obtain

$${}^t\boldsymbol{T}^n_{\ell-1}\boldsymbol{G}_{\ell-1}\boldsymbol{D}^n_{\ell-1} = \beta_1 = 2\beta_2 + 4^{\ell-n-1}2n = \cdots = 2^i\beta_{i+1} + 4^{\ell-n-1}i(2n+1-i).$$

In particular, if $i = n-2$, then

$$\begin{aligned} {}^t\boldsymbol{T}^n_{\ell-1}\boldsymbol{G}_{\ell-1}\boldsymbol{D}^n_{\ell-1} &= 2^{n-2}\beta_{n-1} + 4^{\ell-n-1}(n-2)(n+3) \\ &= 2^{n-2}\left({}^t\boldsymbol{0}_{\ell-n}, {}^t\boldsymbol{u}_{\ell-n}\right)\boldsymbol{G}_{\ell-n+1}\boldsymbol{T}^n_{\ell-n+1} + 4^{\ell-n-1}(n-2)(n+3) \\ &= 2^{n-2}\left(2{}^t\boldsymbol{u}_{\ell-n}\boldsymbol{G}_{\ell-n}\boldsymbol{T}^n_{\ell-n} + {}^t\boldsymbol{u}_{\ell-n}\boldsymbol{G}_{\ell-n}\boldsymbol{D}^n_{\ell-n}\right. \\ &\quad \left. + {}^t\boldsymbol{u}_{\ell-n}\boldsymbol{U}_{\ell-n}\boldsymbol{D}^n_{\ell-n}\right) + 4^{\ell-n-1}(n-2)(n+3). \end{aligned}$$

As ${}^t\boldsymbol{u}_{\ell-n}\boldsymbol{G}_{\ell-n} = \left(\sum_{j=0}^{2^{\ell-n}-1} g_{0j}\right){}^t\boldsymbol{u}_{\ell-n} = 2^{\ell-n}\,{}^t\boldsymbol{u}_{\ell-n}$ and ${}^t\boldsymbol{u}_{\ell-n}\boldsymbol{U}_{\ell-n} = 2^{\ell-n}\,{}^t\boldsymbol{u}_{\ell-n}$, using this, and lemma III.20, we have

$$\begin{aligned} {}^t\boldsymbol{T}^n_{\ell-1}\boldsymbol{G}_{\ell-1}\boldsymbol{D}^n_{\ell-1} &= 2^{\ell-1}\left({}^t\boldsymbol{T}^n_{\ell-n}\boldsymbol{u}_{\ell-n} + \mathrm{Tr}(\boldsymbol{D}^n_{\ell-n})\right) + 4^{\ell-n-1}(n-2)(n+3) \\ &= 4^{\ell-n-1}(n^2 - 3n + 2\ell). \end{aligned}$$

□

Combining the previous results, we obtain

$$\begin{aligned}\gamma(T_\ell^n) &= {}^t\boldsymbol{T}_\ell^n \boldsymbol{G}_\ell \boldsymbol{T}_\ell^n = 4\,{}^t\boldsymbol{T}_{\ell-1}^n \boldsymbol{G}_{\ell-1} \boldsymbol{T}_{\ell-1}^n + 4\,{}^t\boldsymbol{T}_{\ell-1}^n \boldsymbol{G}_{\ell-1} \boldsymbol{D}_{\ell-1}^n \\ &\quad + {}^t\boldsymbol{D}_{\ell-1}^n \boldsymbol{G}_{\ell-1} \boldsymbol{D}_{\ell-1}^n + {}^t\boldsymbol{D}_{\ell-1}^n \boldsymbol{U}_{\ell-1} \boldsymbol{D}_{\ell-1}^n \\ &= 4\gamma(T_{\ell-1}^n) + 4^{\ell-n}(n^2 - 3n + 2\ell) + n4^{\ell-n} + 4^{\ell-n} \\ &= 4\gamma(T_{\ell-1}^n) + 4^{\ell-n}(n^2 + 2\,(\ell - n) + 1),\end{aligned}$$

if $\ell - 2n + 1 \geq 0$. Let us denote $\gamma(T_\ell^n)$ by $c_{\ell,n}$, then

$$\begin{aligned}c_{\ell,n} &= 4c_{\ell-1,n} + 4^{\ell-n}(n^2 - 2n + 2\ell + 1) \\ &= 4^2 c_{\ell-2,n} + 4^{\ell-n}\left(2(n^2 - 2n + 2\ell + 1) - 2\right) \\ &\;\;\vdots \\ &= 4^p c_{\ell-p,n} + 4^{\ell-n}\left(p(n^2 - 2n + 2\ell + 1) - p(p-1)\right),\end{aligned} \tag{III.6}$$

with $2n < \ell$. In particular, for $p = \ell - 2n + 1$,

$$c_{\ell,n} = 4^{\ell-2n+1} c_{2n-1,n} + 4^{\ell-n}(\ell - 2n + 1)(n^2 + \ell + 1).$$

It remains to calculate $c_{2n-1,n} = \gamma(T_{2n-1}^n)$. We will do this through the following result:

Lemma III.24. *For all* $0 \leq i \leq n$, *we have*

$$\gamma(T_{n+i}^n) = 4^i(n(i+1)^2 + 1 + \frac{i}{3}(4 - i^2)).$$

Proof. For $i = 0$, we have

$$\gamma(T_n^n) = {}^t\boldsymbol{T}_n^n \boldsymbol{G}_n \boldsymbol{T}_n^n = (0, \ldots, 0, 1)\, \boldsymbol{G}_n {}^t(0, \ldots, 0, 1) = g_{2^n-1, 2^n-1} = n + 1,$$

and the statement is true. In the general case, we proceed by induction on i. Indeed, with notations as before,

$$\begin{aligned}\gamma(T_{n+i}^n) &= c_{n+i,n} = {}^t\boldsymbol{T}_{n+i}^n \boldsymbol{G}_{n+i} \boldsymbol{T}_{n+i}^n \\ &= 4\,{}^t\boldsymbol{T}_{n+i-1}^n \boldsymbol{G}_{n+i-1} \boldsymbol{T}_{n+i-1}^n + 4\,{}^t\boldsymbol{T}_{n+i-1}^n \boldsymbol{G}_{n+i-1} \boldsymbol{D}_{n+i-1}^n \\ &\quad + {}^t\boldsymbol{D}_{n+i-1}^n \boldsymbol{G}_{n+i-1} \boldsymbol{D}_{n+i-1}^n + {}^t\boldsymbol{D}_{n+i-1}^n \boldsymbol{U}_{n+i-1} \boldsymbol{D}_{n+i-1}^n \\ &= 4c_{n+i-1,n} + 4\beta_1 + 4^i n + 4^i,\end{aligned}$$

with $\beta_1 = {}^t\boldsymbol{T}^n_{n+i-1}\boldsymbol{G}_{n+i-1}\boldsymbol{D}^n_{n+i-1}$ (note that $\ell = n+i$).
Arguing recursively, as in the proof of lemma III.23, we obtain

$$c_{n+i,n} = 4c_{n+i-1,n} + 4.2^{i-1}\beta_i + 4^{i-1}(i-1)(2n+1-(i-1)) + (n+1)4^i.$$

But

$$\begin{aligned}
\beta_i &= \left({}^t\boldsymbol{0}_{n-1},\ldots,{}^t\boldsymbol{0}_i,{}^t\boldsymbol{u}_i\right)\boldsymbol{G}_n\boldsymbol{T}^n_n \\
&= \left({}^t\boldsymbol{0}_{n-2},\ldots,{}^t\boldsymbol{0}_i,{}^t\boldsymbol{u}_i\right)(\boldsymbol{G}_{n-1}+\boldsymbol{U}_{n-1})\,{}^t(0,\ldots,0,1) \\
&= \left({}^t\boldsymbol{0}_{n-2},\ldots,{}^t\boldsymbol{0}_i,{}^t\boldsymbol{u}_i\right)\boldsymbol{G}_{n-1}{}^t(0,\ldots,0,1) + \|\boldsymbol{u}_i\|^2 \\
&\;\;\vdots \\
&= \left({}^t\boldsymbol{0}_i,{}^t\boldsymbol{u}_i\right)\boldsymbol{G}_{i+1}{}^t(0,\ldots,0,1) + (n-i-1)\|\boldsymbol{u}_i\|^2 \\
&= {}^t\boldsymbol{u}_i\boldsymbol{G}_i{}^t(0,\ldots,0,1) + (n-i)\|\boldsymbol{u}_i\|^2 = (n-i+1)2^i.
\end{aligned}$$

So, the expression of $c_{n+i,n}$ reduces to

$$\begin{aligned}
c_{n+i,n} &= 4c_{n+i-1,n} + 4^i\left((2n+1)i - i^2 + (n+1)\right) \\
&= 4^2c_{n+i-2,n} + 4^i\left((2n+1)(i+(i-1)) - \left(i^2+(i-1)^2\right) + 2(n+1)\right) \\
&\;\;\vdots \\
&= 4^pc_{n+i-p,n} + 4^i\left((2n+1)\sum_{k=0}^{p-1}(i-k) - \sum_{k=0}^{p-1}(i-k)^2 + p(n+1)\right).
\end{aligned}$$

Using the fact that $\sum_{k=0}^{p-1}k = p\frac{p-1}{2}$ and $\sum_{k=0}^{p-1}k^2 = \frac{1}{6}\left(2p^3-3p^2+p\right)$, it follows that

$$c_{n+i,n} = 4^pc_{n+i-p,n} + 4^ip\left(n(2i-p+2) + i(p-i) + \frac{1}{3}(4-p^2)\right).$$

Finally, for $p = i$, we have

$$\begin{aligned}
c_{n+i,n} &= 4^ic_{n,n} + 4^ii\left(n(i+2) + \frac{1}{3}(4-i^2)\right) \\
&= 4^i(n+1) + 4^ii\left(n(i+2) + \frac{1}{3}(4-i^2)\right) \\
&= 4^i\left(n(i+1)^2 + 1 + \frac{i}{3}(4-i^2)\right).
\end{aligned}$$

This proves our assertion. □

Combining these lemmas yields:

Proposition III.25. *For any pair of integers $n \leq \ell$, we have*

$$\gamma(T_\ell^n) = \begin{cases} 4^{\ell-n}\left(1 + n(\ell-n+1)^2 + \frac{\ell-n}{3}\left(4-(\ell-n)^2\right)\right) & \text{if } n \leq \ell \leq 2n, \\ 4^{\ell-n}\left((\ell-2n)(n^2+\ell+2) + \frac{n}{3}(2n^2+7) + 2n^2+1\right) & \text{if } 2n \leq \ell. \end{cases}$$

Proof. The expression for $n \leq \ell \leq 2n$ has been proved in lemma III.24. On the other hand, the case $2n \leq \ell$ easily follows from the expression of $c_{\ell,n}$ given in (III.6) with $p = \ell - 2n$. □

Combining the previous results finally yields the epistasis of the template function T_ℓ^n:

Theorem III.26. *The epistasis of the template function T_ℓ^n is given by*

$$\varepsilon^*(T_\ell^n) = \begin{cases} 1 - \frac{1+n(\ell-n+1)^2+\frac{(\ell-n)}{3}\left(4-(\ell-n)^2\right)}{2^n\left(3(\ell-n)-1+2^{n-\ell+1}\right)} & \text{if } n \leq \ell \leq 2n \\ \\ 1 - \frac{(\ell-2n)(n^2+\ell+2)+\frac{n}{3}(2n^2+7)+2n^2+1}{2^n(3(\ell-n)-1)+(\ell-2n)^2+2(n+1)-\ell} & \text{if } 2n \leq \ell. \end{cases}$$

Applying the previous formula to the case $n = \ell$, gives a *high* epistatic value

$$\varepsilon^*(T_\ell^\ell) = 1 - \frac{1+\ell}{2^\ell}.$$

On the other hand, for general values of ℓ, the minimal value for epistasis is easily seen to be given by the case $n = 1$:

$$\varepsilon^*(T_\ell^1) = 1 - \frac{(\ell-2)(\ell+3)+6}{2(3(\ell-1)-1)+(\ell-2)^2+4-\ell} = 1 - \frac{\ell^2+\ell}{\ell^2+\ell} = 0.$$

Indeed, in this case, T_ℓ^1 just counts the number of $1's$ in a string and $T_\ell^1 = \sum_{i=0}^{\ell-1} g_i$, where $g_i(s) = \delta_{1,s_i}$ is the Kronecker function with value 1 when $s_i = 1$ and zero elsewhere.

3.3 Experimental results

This section shows, with some explicit runs, that for template functions the epistasis measure is a nice indicator of problem difficulty. We again measure the latter by counting the number of generations required to first hit the optimum. We also stick to our generational GA with binary tournament selection, one-point crossover at rate 0.8, mutation at a rate of one over the string length, and a population size of 100 in the first experiment, and of the size of the string length in the second experiment, where the latter is varied.

We first fix the string length ℓ and calculate both the epistasis and problem difficulty for the template functions T_ℓ^n, with $\ell = 100$ and $1 \leq n \leq 17$. The results are shown in figure III.3. As expected, the epistasis strongly correlates with problem difficulty, i.e., as we increase the size n of the template, we notice an increase of both epistasis and average number of generations needed to reach the optimum. The reason for not going beyond $n = 17$ is that for large n, template functions become needle-in-a-haystack problems; the probability of randomly obtaining a sequence of n ones becomes exponentially small with n increasing.

A second experiment, detailed in figure III.4, is to fix the size of the template $n = 10$ and calculate the epistasis for different values of ℓ (of course, $n \leq \ell$). Here, a *negative* correlation between epistasis and problem difficulty is observed, which can be motivated as follows. As the string length increases, the template length becomes proportionally smaller. This results in (slightly) smaller epistasis values. The effect on the GA is a higher number of generations required to find the optimum: the template is smaller, and as a result, fewer mutations occur near a template, and fewer crossovers combine parts of templates.

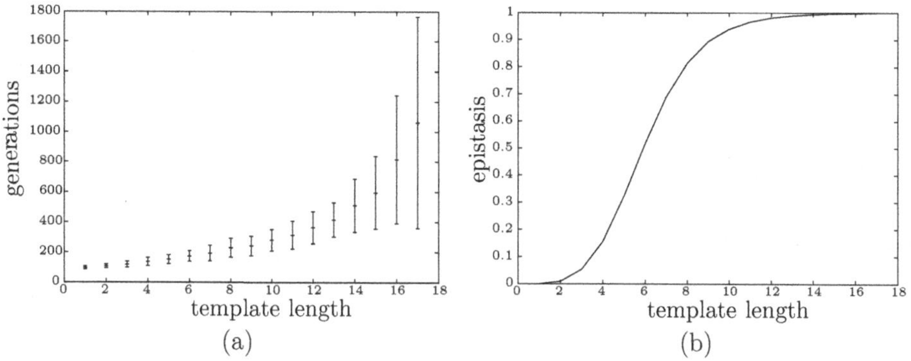

Figure III.3: Problem difficulty and epistasis for increasing template lengths $1 \leq n \leq 17$ and a fixed string length $\ell = 100$. In plot (a) we show, computed over 100 independent runs, the mean and standard devation of the number of generations to hit the optimum. The GA is described in the text; it is the same as the one used for the Royal Road and unitation functions. In plot (b), we computed the epistasis using the explicit formula of theorem III.26.

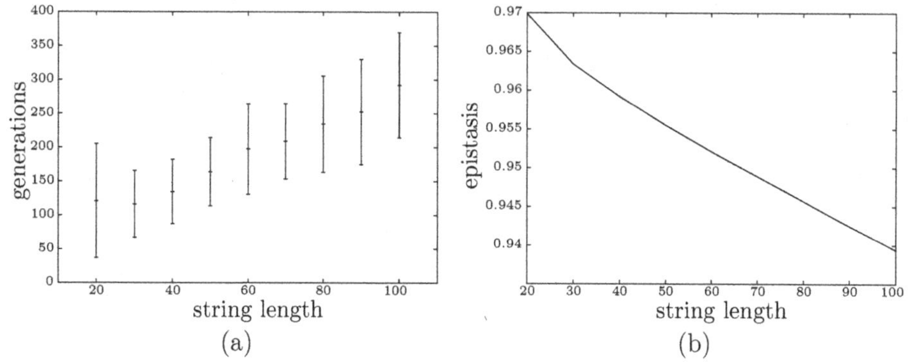

Figure III.4: Problem difficulty and epistasis for increasing string lengths and a fixed template length $n = 10$ (same set-up as in figure III.3)

Chapter IV

Walsh transforms

Just as the ordinary Fourier transform is extremely well-suited to study periodicity and density properties of real or complex valued functions, its binary counterpart, the *Walsh* or *Walsh–Hadamard* transform [6], appears to play a fundamental role in the analysis of GAs. In particular, as the Walsh transform and its associated Walsh coefficients essentially describe averages of the function with respect to certain well-determined elementary schemata, they allow for an elegant and rather practical description of its epistasis.

In this chapter we first develop the basic machinery of Walsh transforms, both in their classical and matrix setting, and show how they relate to schema averages. The classical setting in sections 1 and 2 largely follows Goldberg [24, 25] and Heckendorn and Whitley [33]; section 3 is based on Suys [95]. We then apply this to the study of epistasis, showing how the "Walsh point of view" provides an alternative and much more straightforward treatment of some of the examples considered in the previous chapter.

1 The Walsh transform

1.1 Walsh functions

As before, we let Ω denote the set $\{0,1\}^\ell$, which we identify with the set of length ℓ binary strings $s = s_{\ell-1}\dots s_0$. Let us consider an arbitrary function

$$f : \Omega \to \mathbb{R} : s \mapsto f(s).$$

Our aim is to define functions ψ_t, $t \in \Omega$, the so-called *Walsh functions*, for which f may canonically be written as

$$f(s) = \sum_{t\in\Omega} v_t \psi_t(s).$$

As before, we will sometimes identify elements $t \in \Omega$ with their numerical value, viewing t as the binary expansion of some non-negative integer $0 \le t < 2^\ell$. So,

$$f(s) = \sum_{t\in\Omega} v_t\psi_t(s) = \sum_{t=0}^{2^\ell-1} v_t\psi_t(s).$$

The t-th Walsh function ψ_t is defined as follows:

$$\psi_t(s) = \begin{cases} +1 & \text{if the number of loci where both } s \text{ and } t \text{ have value 1 is even,} \\ -1 & \text{if the number of loci where both } s \text{ and } t \text{ have value 1 is odd.} \end{cases}$$

So, for example, $\psi_{11010}(01110) = 1$. Also, with $\ell = 5$, we have $\psi_{19}(9) = -1$, as 9 is identified with 01001 and 19 with 10011.

Alternative ways of defining ψ_t may be given as follows.

First, define the conjunction $\wedge$ in Ω by putting for any $s = s_{\ell-1}\dots s_0$ and $t = t_{\ell-1}\dots t_0$ in Ω,

$$s \wedge t = (s_{\ell-1} \wedge t_{\ell-1})\dots(s_0 \wedge t_0),$$

where $x \wedge y$ is just the product xy, for any $x, y \in \{0,1\}$. If we let $u : \Omega \to \mathbb{N}$ denote the unitation function, i.e., for any $s \in \Omega$ we let $u(s)$ denote the number of loci with value 1 in s, then it is easy to see that

$$\psi_t(s) = (-1)^{u(s\wedge t)}.$$

Next, note that we may define the scalar product $s \cdot t$ of $s, t \in \Omega$ in the obvious way:

$$s \cdot t = \sum_{i=0}^{\ell-1} s_i t_i.$$

It then follows that

$$\psi_t(s) = (-1)^{s \cdot t} = \prod_{i=0}^{\ell-1} (-1)^{s_i t_i}.$$

Let us point out that what we just defined are actually the so-called *Hadamard functions*. The "correct" definition of the Walsh functions should be

$$\psi_t(s) = \prod_{i=0}^{\ell-1} (-1)^{s_i t_{(\ell-1)-i}},$$

which is, of course, just a permutation of the Hadamard functions. In most practical applications this definition is usually easier, as it allows for a "fast Fourier"-like implementation, for example. The Hadamard definition, however, is of more use in theoretical applications in view of its recursive properties, as we will see below.

1.2 Properties of Walsh functions

Let us mention some first properties of the Walsh functions.

a. *Symmetry*: From the very definition of the Walsh functions, it follows that

$$\forall s, t \in \Omega : \psi_t(s) = \psi_s(t).$$

b. *Exclusive-or property*: Define the exclusive-or of $s, t \in \Omega$ to be

$$s \oplus t = (s_{\ell-1} \oplus t_{\ell-1}) \ldots (s_0 \oplus t_0),$$

where, for each $x, y \in \{0, 1\}$, we put

$$x \oplus y = \begin{cases} 1 & \text{if } x \neq y \\ 0 & \text{if } x = y. \end{cases}$$

(Viewing x and y as elements of $\mathbb{Z}/2\mathbb{Z}$, i.e., working modulo 2, this is of course just the sum of x and y). For any $s, t, t' \in \Omega$ we now have

$$\prod(-1)^{s_i t_i}\prod(-1)^{s_i t'_i} = \prod(-1)^{s_i(t_i+t'_i)} = \prod(-1)^{s_i(t_i \oplus t'_i)}.$$

It then follows that

$$\forall s, t, t' \in \Omega : \psi_t(s)\psi_{t'}(s) = \psi_{t\oplus t'}(s).$$

c. *Conjunction compression*: It is easy to verify that

$$\forall s, t \in \Omega : \psi_t(s) = \psi_{s\wedge t}(s).$$

Indeed, it suffices to note that

$$\psi_{s\wedge t}(s) = (-1)^{u(s\wedge(s\wedge t))} = (-1)^{u(s\wedge t)} = \psi_t(s).$$

As a first corollary of these properties, let us mention:

Lemma IV.1. *For all $t, t' \in \Omega$, we have*

$$\sum_{s\in\Omega}\psi_t(s)\psi_{t'}(s) = 2^\ell\,\delta_{t,t'}\,,$$

where $\delta_{t,t'}$ is the Kronecker delta.

Proof. We have

$$\sum_{s\in\Omega}\psi_t(s)\psi_{t'}(s) = \sum_{s\in\Omega}\psi_{t\oplus t'}(s).$$

If $t = t'$, then $t \oplus t' = 0\ldots0$, so $\psi_{t\oplus t'}(s) = 1$ for any $s \in \Omega$. In this case, the above sum is equal to 2^ℓ.

If $t \neq t'$, then we claim that the above sum is equal to 0. Of course, since $t \neq t'$ is equivalent to $t \oplus t' \neq 0\ldots0$, it suffices to check that if $t \neq 0$ then $\sum_s \psi_t(s) = 0$. Since $t = t_{\ell-1}\ldots t_0 \neq 0\ldots0$, at least one component t_i has value 1. Let us suppose $t_0 = 1$ for notational convenience. Any $s \in \Omega$ may be written as $s = \hat{s}s_0$ with $\hat{s} \in \Omega_{\ell-1} = \{0,1\}^{\ell-1}$. In particular, $t = \hat{t}1$ for some $\hat{t} \in \Omega_{\ell-1}$. Then

$$\begin{aligned}\sum_{s\in\Omega}\psi_t(s) &= \sum_{s\in\Omega}(-1)^{\hat{s}\cdot\hat{t}}(-1)^{s_0}\\ &= \sum_{\substack{\hat{s}\in\Omega_{\ell-1}\\(s_0=0)}}(-1)^{\hat{s}\cdot\hat{t}} - \sum_{\substack{\hat{s}\in\Omega_{\ell-1}\\(s_0=1)}}(-1)^{\hat{s}\cdot\hat{t}} = 0.\end{aligned}$$

□

If we put $t' = 0$ in the expression of the previous lemma, we directly obtain:

Corollary IV.2. *Except for $\psi_0 \equiv 1$, all Walsh functions have a zero average:*

$$\sum_{s\in\Omega} \psi_t(s) = \begin{cases} 2^\ell & \text{if } t = 0 \\ 0 & \text{if } t \neq 0. \end{cases}$$

For any $f : \Omega \to \mathbb{R}$, let us now define $v_t \in \mathbb{R}$ by

$$v_t = \frac{1}{2^\ell} \sum_{s\in\Omega} f(s)\psi_t(s).$$

We call v_t the *t-th Walsh coefficient* of f. As an example, since $\psi_0 \equiv 1$, clearly v_0 is the average value of f.

Proposition IV.3. *With the above definitions, for any $f : \Omega \to \mathbb{R}$, we have*

$$f(s) = \sum_{t\in\Omega} v_t\psi_t(s) \quad \forall s \in \Omega.$$

Proof. We have

$$\begin{aligned}
\sum_{t\in\Omega} v_t\psi_t(s) &= \sum_{t\in\Omega} \left(\frac{1}{2^\ell} \sum_{t'\in\Omega} f(t')\psi_t(t') \right) \psi_t(s) \\
&= \frac{1}{2^\ell} \sum_{t'\in\Omega} f(t') \sum_{t\in\Omega} \psi_t(t')\psi_t(s) \\
&= \frac{1}{2^\ell} \sum_{t'\in\Omega} f(t') \sum_{t\in\Omega} \psi_{t'}(t)\psi_s(t) \\
&= \frac{1}{2^\ell} \sum_{t'\in\Omega} f(t')\, 2^\ell\, \delta_{t',s} \\
&= f(s).
\end{aligned}$$

This proves our claim. □

Corollary IV.4. *The Walsh functions $\{\psi_t;\ t \in \Omega\}$ form a basis for the vector space of real-valued functions on Ω.*

Proof. The previous result shows that the ψ_t form a set of generators, so, it remains to prove that they are linearly independent.

Suppose that $\sum_{t\in\Omega} a_t\psi_t = 0$ for some $a_t \in \mathbb{R}$, i.e., for all $s \in \Omega$ we have

$$\sum_{t\in\Omega} a_t\psi_t(s) = 0.$$

Then, for any $t' \in \Omega$ we have

$$\begin{aligned} 0 = \sum_{s\in\Omega}\left(\sum_{t\in\Omega} a_t\psi_t(s)\right)\psi_{t'}(s) &= \sum_{t\in\Omega} a_t \sum_{s\in\Omega} \psi_t(s)\psi_{t'}(s) \\ &= \sum_{t\in\Omega} a_t\, 2^\ell\, \delta_{t,t'} = 2^\ell\, a_{t'}. \end{aligned}$$

So, for any $t' \in \Omega$, we have $a_{t'} = 0$, proving that the ψ_t are linearly independent indeed. □

1.3 The Walsh matrix

With notations as before, let us introduce the matrix

$$\boldsymbol{V} = \boldsymbol{V}_\ell = (\psi_t(s))_{s,t\in\Omega} \in M_{2^\ell}(\mathbb{Z}).$$

Note that this implies $\boldsymbol{V}$ to be symmetric.

The elements v_t associated to an arbitrary function $f : \Omega \to \mathbb{R}$ may be collected into a vector

$$\boldsymbol{v} = \begin{pmatrix} v_{0\ldots0} \\ \vdots \\ v_{1\ldots1} \end{pmatrix} = \begin{pmatrix} v_0 \\ \vdots \\ v_{2^\ell-1} \end{pmatrix} \in \mathbb{R}^{2^\ell}.$$

In a similar way, f yields a vector

$$\boldsymbol{f} = \begin{pmatrix} f(0\ldots0) \\ \vdots \\ f(1\ldots1) \end{pmatrix} = \begin{pmatrix} f(0) \\ \vdots \\ f(2^\ell-1) \end{pmatrix} \in \mathbb{R}^{2^\ell}.$$

These data are related through:

Proposition IV.5. *With the above definitions, we have*

$$\boldsymbol{f} = \boldsymbol{V}\boldsymbol{v}.$$

Proof. This is just a rewriting in matrix form of the fact that

$$f(s) = \sum_{t\in\Omega} v_t \psi_t(s) \quad \forall s \in \Omega.$$

□

The matrix $\boldsymbol{V}_\ell$ is easy to calculate for small values of ℓ. Indeed, we have

$$\begin{aligned}
\boldsymbol{V}_0 &= (1), \\
\boldsymbol{V}_1 &= \begin{pmatrix} 1 & 1 \\ 1 & -1 \end{pmatrix}, \\
\boldsymbol{V}_2 &= \begin{pmatrix} 1 & 1 & 1 & 1 \\ 1 & -1 & 1 & -1 \\ 1 & 1 & -1 & -1 \\ 1 & -1 & -1 & 1 \end{pmatrix},
\end{aligned}$$

as one easily verifies.

For arbitrary values of ℓ, one may calculate $\boldsymbol{V}_\ell$ by recursion:

Proposition IV.6. *For any $\ell \geq 1$, we have*

$$\boldsymbol{V}_\ell = \begin{pmatrix} \boldsymbol{V}_{\ell-1} & \boldsymbol{V}_{\ell-1} \\ \boldsymbol{V}_{\ell-1} & -\boldsymbol{V}_{\ell-1} \end{pmatrix}.$$

Proof. This is a straightforward consequence of the fact that for any $s = s_{\ell-1}\hat{s}$, $t = t_{\ell-1}\hat{t} \in \Omega$, with $\hat{s}, \hat{t} \in \Omega_{\ell-1}$, we have

$$\begin{aligned}
\psi_t(s) = (-1)^{s\cdot t} &= (-1)^{s_{\ell-1}t_{\ell-1}}(-1)^{\hat{s}\cdot\hat{t}} \\
&= (-1)^{s_{\ell-1}t_{\ell-1}}\psi_{\hat{t}}(\hat{s}).
\end{aligned}$$

Note that $(-1)^{s_{\ell-1}t_{\ell-1}}$ always has value $+1$, except when $s_{\ell-1} = t_{\ell-1} = 1$. □

Corollary IV.7. *For any positive integer ℓ, we have*

$$\boldsymbol{V}_\ell^2 = 2^\ell \boldsymbol{I}_\ell.$$

Proof. This may easily be proved by induction. Note that the result is obvious for $\ell = 1$, so let us assume the statement to be correct up to $\ell - 1$. We then have:

$$\boldsymbol{V}_\ell^2 = \begin{pmatrix} \boldsymbol{V}_{\ell-1} & \boldsymbol{V}_{\ell-1} \\ \boldsymbol{V}_{\ell-1} & -\boldsymbol{V}_{\ell-1} \end{pmatrix} \begin{pmatrix} \boldsymbol{V}_{\ell-1} & \boldsymbol{V}_{\ell-1} \\ \boldsymbol{V}_{\ell-1} & -\boldsymbol{V}_{\ell-1} \end{pmatrix} = \begin{pmatrix} 2\boldsymbol{V}_{\ell-1}^2 & \boldsymbol{O}_{\ell-1} \\ \boldsymbol{O}_{\ell-1} & 2\boldsymbol{V}_{\ell-1}^2 \end{pmatrix}$$
$$= \begin{pmatrix} 2 \cdot 2^{\ell-1}\boldsymbol{I}_{\ell-1} & \boldsymbol{O}_{\ell-1} \\ \boldsymbol{O}_{\ell-1} & 2 \cdot 2^{\ell-1}\boldsymbol{I}_{\ell-1} \end{pmatrix} = 2^\ell \begin{pmatrix} \boldsymbol{I}_{\ell-1} & \boldsymbol{O}_{\ell-1} \\ \boldsymbol{O}_{\ell-1} & \boldsymbol{I}_{\ell-1} \end{pmatrix} = 2^\ell \boldsymbol{I}_\ell,$$

where $\boldsymbol{O}_{\ell-1}$ denotes the $2^{\ell-1}$-dimensional square matrix all of whose entries are zero. Alternatively, the (s,t) entry of $\boldsymbol{V}_\ell^2$ is equal to

$$\sum_{t' \in \Omega_\ell} \psi_{t'}(s)\psi_t(t') = \sum_{t' \in \Omega_\ell} \psi_s(t')\psi_t(t') = 2^\ell \, \delta_{s,t},$$

by lemma IV.1. □

In practice, it is somewhat easier to work with a modified version of the matrix $\boldsymbol{V}_\ell$. Indeed, if we introduce the *Walsh matrix* $\boldsymbol{W}_\ell = 2^{-\ell/2}\boldsymbol{V}_\ell$, then we have

$$\boldsymbol{W}_\ell^2 = \boldsymbol{I}_\ell,$$

in view of the previous result. If we define for any $f : \Omega \to \mathbb{R}$ with associated vector $\boldsymbol{f} \in \mathbb{R}^{2^\ell}$,

$$\boldsymbol{w} = \boldsymbol{W}_\ell \boldsymbol{f},$$

then it is obvious that $\boldsymbol{f}$ (and hence f!) may be recovered from $\boldsymbol{w}$ by

$$\boldsymbol{f} = \boldsymbol{W}_\ell \boldsymbol{w}.$$

The components $w_s = w_s(f)$ of the vector $\boldsymbol{w}$ are also called *Walsh coefficients* of f. It follows that $v_s = 2^{-\ell/2} w_s$ for each $s \in \Omega$.

In particular, the relation

$$f(s) = \sum_{t \in \Omega_\ell} v_t \psi_t(s)$$

may be rewritten as

$$f(s) = 2^{-\ell/2} \sum_{t \in \Omega_\ell} w_t \psi_t(s).$$

Note also that $\boldsymbol{W}_\ell$ satisfies the recursion relation

$$\boldsymbol{W}_\ell = 2^{-\frac{1}{2}} \begin{pmatrix} \boldsymbol{W}_{\ell-1} & \boldsymbol{W}_{\ell-1} \\ \boldsymbol{W}_{\ell-1} & -\boldsymbol{W}_{\ell-1} \end{pmatrix},$$

in view of the analogous relation for $\boldsymbol{V}_\ell$.

2 Link with schema averages

Corollary IV.2, which states that

$$\sum_{s\in\Omega} \psi_t(s) = \begin{cases} 2^\ell & \text{if } t = 0 \\ 0 & \text{if } t \neq 0, \end{cases}$$

is sometimes referred to as the *balanced sum theorem.* Of course, this just says that the sum of the elements in an arbitrary row (or column) of the matrix $\boldsymbol{V}_\ell$ is 0, except for the first one, where this sum has value 2^ℓ. This also holds for the matrix $\boldsymbol{W}_\ell$ but here, of course, the non-zero sum has value $2^\ell \cdot 2^{-\ell/2} = 2^{\ell/2}$.

Observe that in the above sum s takes values in the whole of Ω. One may wonder what happens if one restricts s to a schema (section 4, chapter 1) H in Ω.

We will need some preliminaries and notations first.

Define functions $\alpha, \beta : \{0,1,\#\}^\ell \to \{0,1\}^\ell = \Omega$ as follows. Let $H = h_{\ell-1}\ldots h_0$ be a schema, i.e., an element of $\{0,1,\#\}^\ell$. We then put

$$\alpha(H)_i = \begin{cases} 0 & \text{if } h_i = \# \\ 1 & \text{if } h_i = 0 \text{ or } h_i = 1, \end{cases}$$

$$\beta(H)_i = \begin{cases} 0 & \text{if } h_i = \# \text{ or } h_i = 0 \\ 1 & \text{if } h_i = 1. \end{cases}$$

For example, if $H = \#10\#1$, then $\alpha(H) = 01101$ and $\beta(H) = 01001$.

Let

$$J(H) = \{t \in \Omega;\ \forall 0 \le i < \ell,\ h_i = \# \Rightarrow t_i = 0\},$$

i.e., $J(H)$ is the set of all strings which have a 0 where H has a #. It is fairly easy to see that

$$t \in J(H) \Leftrightarrow t \wedge \overline{\alpha(H)} = 0,$$

where for any $s \in \Omega$, we denote by $\bar{s}$ its binary complement. For example, if $s = 010011$ then $\bar{s} = 101100$.

Denoting as before by $|H|$ the cardinality of H (the number of strings "in" or "satisfying" H), i.e., $|H| = 2^{\ell - o(H)}$, where $o(H)$ denotes the order of H, we may prove:

Theorem IV.8 (Balanced sum theorem for hyperplanes). *For any schema H and any $t \in \Omega$, we have*

$$\sum_{s \in H} \psi_t(s) = \begin{cases} 0 & \textit{if } t \notin J(H) \\ \psi_t(\beta(H))|H| & \textit{if } t \in J(H). \end{cases}$$

Note that if $H = \# \ldots \#$, i.e., if $H = \Omega$, then $|H| = 2^\ell$ and $J(H) = \{0 \ldots 0\} = \{0\}$. Moreover, $\beta(H) = 0 \ldots 0$, so $\psi_t(\beta(H)) = 1$ for all $t \in \Omega$. We thus recover the fact that

$$\sum_{s \in \Omega} \psi_t(s) = \begin{cases} 0 & \text{if } t \neq 0 \\ 2^\ell & \text{if } t = 0. \end{cases}$$

To prove the theorem, we will need:

Lemma IV.9. *For any $s \in \Omega$ and any hyperplane H, we have*

$$s \in H \Leftrightarrow s \wedge \alpha(H) = \beta(H).$$

Proof. Let us first assume $s \in H$. For each bit position i, we have to consider three cases:

(a) $h_i = 0$. Then $\alpha(H)_i = 1$ and as $s_i = 0$ as well, we have

$$(s \wedge \alpha(H))_i = 0 \wedge 1 = 0 = \beta(H)_i.$$

(b) $h_i = 1$. Then $\alpha(H)_i = 1$ and as $s_i = 1$ as well, we have

$$(s \wedge \alpha(H))_i = 1 \wedge 1 = 1 = \beta(H)_i.$$

(c) $h_i = \#$. Then $\alpha(H)_i = 0$, so

$$(s \wedge \alpha(H))_i = s_i \wedge 0 = 0 = \beta(H)_i.$$

Conversely, if $s \wedge \alpha(H) = \beta(H)$, then we again consider three cases:

(a) $h_i = 0$. Then $\alpha(H)_i = 1$ and $\beta(H)_i = 0$. From $s \wedge \alpha(H) = \beta(H)$, it follows that $s_i \wedge 1 = 0$, so $s_i = 0$.

(b) $h_i = 1$. Then $\alpha(H)_i = 1$ and $\beta(H)_i = 1$. In this case we have $s_i \wedge 1 = 1$, so $s_i = 1$.

(c) $h_i = \#$. Then $\alpha(H)_i = 0$ and $\beta(H)_i = 0$, so for all s_i we have $s_i \wedge 0 = 0$.

□

Proof. (of the theorem)

Case 1: $t \in J(H)$, i.e., $t \wedge \overline{\alpha(H)} = 0$. Since for any $s \in \Omega$, obviously,

$$s = (s \wedge \overline{\alpha(H)}) \oplus (s \wedge \alpha(H)),$$

we obtain that

$$\sum_{s \in H} \psi_t(s) = \sum_{s \in H} \psi_t(s \wedge \overline{\alpha(H)})\, \psi_t(s \wedge \alpha(H)).$$

Now, by conjunction compression,

$$\psi_t(s \wedge \overline{\alpha(H)}) = \psi_t(s \wedge \overline{\alpha(H)} \wedge t) = \psi_t(0) = 1.$$

On the other hand, for each $s \in H$, we have $\psi_t(s \wedge \alpha(H)) = \psi_t(\beta(H))$ thanks to the previous lemma, so the above sum reduces to

$$\sum_{s \in H} \psi_t(s \wedge \alpha(H)) = \sum_{s \in H} \psi_t(\beta(H)) = |H|\, \psi_t(\beta(H)).$$

Case 2: $t \notin J(H)$, i.e., $t \wedge \overline{\alpha(H)} \neq 0$. In this case, there exists some bit position i such that $t_i = 1$ and i is not one of the fixed positions of H. For notational convenience, assume $i = 0$. In particular, we then have $h_0 = \#$. Any $s \in \Omega$ may be written as $s = \hat{s}s_0$, with $\hat{s} \in \Omega_{\ell-1}$. Let $t = \hat{t}1$, then

$$\sum_{s\in H} \psi_t(s) = \sum_{s\in H} (-1)^{\hat{s}\cdot\hat{t}}(-1)^{s_0}.$$

Put $H_0 = \{s \in H;\ s_0 = 0\}$ and $H_1 = \{s \in H;\ s_0 = 1\}$. It is clear that H_0 and H_1 correspond, bijectively, as $\hat{s}0 \in H_0 \Leftrightarrow \hat{s}1 \in H_1$ since $h_0 = \#$. The above sum thus reduces to

$$\begin{aligned}\sum_{s\in H} (-1)^{\hat{s}\cdot\hat{t}}(-1)^{s_0} &= \sum_{s\in H_0} (-1)^{\hat{s}\cdot\hat{t}}(-1)^{s_0} + \sum_{s\in H_1} (-1)^{\hat{s}\cdot\hat{t}}(-1)^{s_0} \\ &= \sum_{\hat{s}\in \hat{H}} (-1)^{\hat{s}\cdot\hat{t}} - \sum_{\hat{s}\in \hat{H}} (-1)^{\hat{s}\cdot\hat{t}} = 0,\end{aligned}$$

where $\hat{H} \in \{0, 1, \#\}^{\ell-1}$ is defined by $H = \#\hat{H}$. This proves the assertion. □

In section 3.2 of chapter VI, we prove the same theorem for strings over multary alphabets (theorem VI.11). The proof of this more general theorem is much shorter than the one shown here, which is based on Heckendorn and Whitley [33].

Corollary IV.10 (Hyperplane averaging theorem). *The average of f over any schema H is given by*

$$f(H) = \frac{1}{|H|}\sum_{s\in H} f(s) = \sum_{t\in J(H)} v_t\psi_t(\beta(H)).$$

Proof. We have

$$\frac{1}{|H|}\sum_{s\in H} f(s) = \frac{1}{|H|}\sum_{s\in H}\sum_{t\in\Omega_\ell} v_t\psi_t(s) = \frac{1}{|H|}\sum_{t\in\Omega_\ell} v_t \sum_{s\in H}\psi_t(s).$$

By the previous result, the sum $\sum_{s\in H}\psi_t(s)$ is only non-zero when $t \in J(H)$, so,

$$\begin{aligned}f(H) &= \frac{1}{|H|}\sum_{t\in J(H)} v_t \sum_{s\in H}\psi_t(s) = \frac{1}{|H|}\sum_{t\in J(H)} v_t\psi_t(\beta(H))|H| \\ &= \sum_{t\in J(H)} v_t\psi_t(\beta(H)).\end{aligned}$$

□

Note that the corollary may also be written as

$$f(H) = 2^{-\ell/2} \sum_{t \in J(H)} w_t \psi_t(\beta(H)).$$

Let us apply the foregoing to some concrete examples.

First, since $\beta(\#\ldots\#) = 0$ and since $\#\ldots\#$ corresponds to the whole space Ω, we recover

$$f(\Omega) = f(\#\ldots\#) = 2^{-\ell/2}\, w_0,$$

i.e., w_0 is essentially just the average of f.

Next, let us consider schemata of the form $H = \#\ldots\#a\#\ldots\#$, with $a \in \{0,1\}$ at position i. In this case, for all $t \in \Omega$, we find

$$\psi_t(\beta(H)) = (-1)^{a \cdot t_i},$$

as $\beta(H) = 0\ldots0\,a\,0\ldots0$. If $t \in J(H)$, then t is necessarily of the form $t = 0\ldots0\,t_i\,0\ldots0$.

We distinguish two cases:

(i) If $t_i = 0$ then $t = 0$, and we find

$$\psi_t(\beta(H)) = \psi_0(\beta(H)) = 1.$$

(ii) If $t_i = 1$ then $t = 0\ldots010\ldots0 = 2^i$, and

$$\psi_t(\beta(H)) = \psi_{2^i}(\beta(H)) = (-1)^a.$$

We may conclude

$$f(\#\ldots\#a\#\ldots\#) = 2^{-\ell/2}\left(w_0 + (-1)^a w_{2^i}\right).$$

As another example, let us consider the schema $H = \#11$. In this case, $\beta(H) = 011$ and $t \in J(H)$ if t is of the form $0ab$ for some $a, b \in \{0,1\}$. We find

t	$\psi_t(\beta(H))$
$000 = 0$	$+1$
$001 = 1$	-1
$010 = 2$	-1
$011 = 3$	$+1$

So,

$$f(H) = 2^{-3/2} \sum_{t \in J(H)} \psi_t(\beta(H)) w_t = 2^{-3/2}(w_0 - w_1 - w_2 + w_3).$$

Note that the foregoing is a special case of a more general phenomenon:

Proposition IV.11. *For any schema H over Ω, the average $f(H)$ may be computed using only those Walsh coefficients w_t with $u(t) \leq o(H)$.*

(Note: in the above example, $o(H) = 2$ and $u(0) = 0$, $u(1) = u(2) = 1$ and $u(3) = 2$, respectively!)

Proof. Recall that

$$f(H) = \frac{1}{|H|} \sum_{s \in H} f(s) = \sum_{t \in J(H)} v_t \psi_t(\beta(H)).$$

It now suffices to observe that $t \in J(H)$ implies $u(t) \leq o(H)$. Indeed, if $t \in J(H)$, then $h_i = \#$ implies $t_i = 0$. So, $t_i = 1$ implies $h_i \neq \#$, hence

$$u(t) \leq |\{i;\ h_i \neq \#\}| = o(H).$$

□

3 Link with partition coefficients

With any schema H' we want to associate a value $\varepsilon(H')$, the *partition coefficient* of H', such that for every schema H,

$$f(H) = \sum_{H' \supset H} \varepsilon(H'),$$

where $H' \supset H$ indicates H and H' agree on the fixed positions of H' (or otherwise put, $h_i = \# \Rightarrow h'_i = \#$). In particular, this will yield for every $s \in \Omega$ that

$$f(s) = \sum_{s \in H} \varepsilon(H).$$

From the very definition, it is clear that

$$\varepsilon(H) = f(H) - \sum_{H' \supsetneq H} \varepsilon(H'),$$

so, the $\varepsilon(H)$ may be calculated recursively. For example, we obviously have

$$\varepsilon(\#\ldots\#) = f(\Omega)$$

and

$$\begin{aligned}\varepsilon(\#\ldots\#a\#\ldots\#) &= f(\#\ldots\#a\#\ldots\#) - \varepsilon(\#\ldots\#)\\ &= f_{(i,a)} - f(\Omega).\end{aligned}$$

It thus appears that $\varepsilon(\#\ldots\#a\#\ldots\#)$ "corrects" the approximation of $f_{(i,a)}$ by $f(\Omega)$. Let us also point out that

$$\varepsilon(\#\ldots\#) = f(\Omega) = 2^{-\ell/2} w_0$$

and

$$\begin{aligned}\varepsilon(\#\ldots\#a\#\ldots\#) &= f_{(i,a)} - f(\Omega) = 2^{-\ell/2}\left(w_0 + (-1)^a w_{2^i}\right) - 2^{-\ell/2} w_0\\ &= (-1)^a 2^{-\ell/2} w_{2^i}.\end{aligned}$$

One may thus reasonably expect these $\varepsilon(H)$ to be linked to Walsh coefficients.

Let us first give some easy extra examples, before formulating a general result.

Consider $H = \#\ldots\#\overset{j}{\overset{\downarrow}{0}}\#\ldots\#\overset{i}{\overset{\downarrow}{1}}\#\ldots\#$. Clearly

$$\begin{aligned}J(H) &= \{\,0\ldots0a0\ldots0b0\ldots0;\quad a,b \in \{0,1\}\,\}\\ &= \{0, 2^i, 2^j, 2^j + 2^i\}.\end{aligned}$$

Since $\beta(H) = 0\ldots000\ldots010\ldots0 = 2^i$, we find

$$\begin{aligned}\psi_0(\beta(H)) &= \psi_{2^j}(\beta(H)) = 1,\\ \psi_{2^i}(\beta(H)) &= \psi_{2^i+2^j}(\beta(H)) = -1,\end{aligned}$$

and the general formula

$$f(H) = 2^{-\ell/2} \sum_{t \in J(H)} w_t \psi_t(\beta(H))$$

yields

$$f(H) = 2^{-\ell/2}\left(w_0 - w_{2^i} + w_{2^j} - w_{2^i+2^j}\right).$$

It follows that

$$\begin{aligned}\varepsilon(H) = & f(H) - \varepsilon(\#\ldots\#0\#\ldots\#\#\#\ldots\#)\\ & -\varepsilon(\#\ldots\#\#\#\ldots\#1\#\ldots\#)\\ & -\varepsilon(\#\ldots\#)\\ = & 2^{-\ell/2}\left(w_0 - w_{2^i} + w_{2^j} - w_{2^i+2^j}\right) - 2^{-\ell/2} w_{2^j} - \left(-2^{-\ell/2} w_{2^i}\right) - 2^{-\ell/2} w_0.\end{aligned}$$

We thus find

$$\varepsilon(\#\ldots\#0\#\ldots\#1\#\ldots\#) = -2^{-\ell/2} w_{2^i+2^j}.$$

In a similar way, one obtains

$$\begin{aligned}\varepsilon(\#\ldots\#0\#\ldots\#0\#\ldots\#) &= 2^{-\ell/2} w_{2^i+2^j},\\ \varepsilon(\#\ldots\#1\#\ldots\#0\#\ldots\#) &= -2^{-\ell/2} w_{2^i+2^j},\\ \varepsilon(\#\ldots\#1\#\ldots\#1\#\ldots\#) &= 2^{-\ell/2} w_{2^i+2^j}.\end{aligned}$$

Observe that in each of these examples

$$\varepsilon(H) = (-1)^{u(\beta(H))} 2^{-\ell/2} w_{\alpha(H)}.$$

Indeed, for example, with $H = \#\ldots\#0\#\ldots\#1\#\ldots\#$, we have

$$\alpha(H) = 0\ldots010\ldots010\ldots0 = 2^i + 2^j$$

and

$$\beta(H) = 0\ldots000\ldots010\ldots0 = 2^i,$$

so, $u(\beta(H)) = 1$. We will see below that this phenomenon may be generalized. In order to prove this, we need a few rather straightforward lemmas.

Lemma IV.12. *For any $H', H'' \supset H$, we have*

$$H' \neq H'' \Leftrightarrow \alpha(H') \neq \alpha(H'').$$

Proof. Indeed, both H' and H'' arise from H by replacing some fixed positions in H by #. The map α puts a 0 in the corresponding positions, and as there is at least one different position where this occurs for H' and H'' when these are different, this shows that then necessarily $\alpha(H') \neq \alpha(H'')$. The converse is obvious. □

Lemma IV.13. *If $H' \supset H$ then $\alpha(H') \in J(H)$. Conversely, for any $t \in J(H)$, there exists exactly one $H' \supset H$ with $\alpha(H') = t$.*

Proof. That $H' \supset H$ implies $\alpha(H') \in J(H)$ is obvious. Indeed, at every locus i where H has #, so does H', hence $\alpha(H')_i = 0$.

Conversely, for any $t \in J(H)$, there exists exactly one $H' \supset H$ with $\alpha(H') = t$. Obviously, H' is defined by

$$h'_i = \begin{cases} \# & \text{if } t_i = 0 \\ 1 & \text{if } t_i = 1, \end{cases}$$

as one easily checks. □

Lemma IV.14. *For any $H' \supset H$, we have*

$$\psi_{\alpha(H')}(\beta(H)) = (-1)^{u(\beta(H'))}.$$

Proof. Since, obviously, $\alpha(H') \wedge \beta(H) = \beta(H')$, we indeed have

$$\psi_{\alpha(H')}(\beta(H)) = (-1)^{u(\alpha(H') \wedge \beta(H))} = (-1)^{u(\beta(H'))}.$$

□

We may now prove:

Theorem IV.15. *For any schema H over Ω, we have*

$$\varepsilon(H) = (-1)^{u(\beta(H))} 2^{-\ell/2} w_{\alpha(H)}.$$

Proof. First note that the previous lemmas imply

$$f(H) = 2^{-\ell/2} \sum_{t \in J(H)} \psi_t(\beta(H))w_t = 2^{-\ell/2} \sum_{H' \supset H} (-1)^{u(\beta(H'))} w_{\alpha(H')}.$$

Since the result is obviously true for $H = \# \dots \#$, we may now argue by induction, i.e., let us assume that

$$\varepsilon(H') = (-1)^{u(\beta(H'))} 2^{-\ell/2} w_{\alpha(H')},$$

for all $H' \supsetneq H$. We then have

$$\begin{aligned} \varepsilon(H) &= f(H) - \sum_{H' \supsetneq H} \varepsilon(H') \\ &= 2^{-\ell/2} \sum_{H' \supset H} (-1)^{u(\beta(H'))} w_{\alpha(H')} - \sum_{H' \supsetneq H} (-1)^{u(\beta(H'))} 2^{-\ell/2} w_{\alpha(H')} \\ &= (-1)^{u(\beta(H))} 2^{-\ell/2} w_{\alpha(H)}. \end{aligned}$$

This proves the assertion. □

Since for any schema H over Ω we have

$$\sum_{t \in J(H)} \psi_t(\beta(H))w_t = \sum_{H' \supset H} (-1)^{u(\beta(H'))} w_{\alpha(H')},$$

it follows from the previous result that the decompositions

$$f(H) = 2^{-\ell/2} \sum_{t \in J(H)} \psi_t(\beta(H))w_t$$

and

$$f(H) = \sum_{H' \supset H} \varepsilon(H')$$

are essentially the same.

4 Link with epistasis

The aim of this section is to calculate the normalized epistasis of an arbitrary function $f : \Omega \to \mathbb{R}$ in terms of its Walsh coefficients. In order to realize this, let us first introduce the diagonal matrix $\boldsymbol{D}_\ell \in M_{2^\ell}(\mathbb{Z})$, whose only non-zero diagonal entries d_{ii} have value 1 and are situated at $i = 0$ and $i = 2^j$, for $0 \le j < \ell$. So,

$$\boldsymbol{D}_0 = (1) \qquad \boldsymbol{D}_1 = \begin{pmatrix} 10 \\ 01 \end{pmatrix} \qquad \boldsymbol{D}_2 = \begin{pmatrix} 1000 \\ 0100 \\ 0010 \\ 0000 \end{pmatrix}$$

$$\boldsymbol{D}_3 = \begin{pmatrix} 10000000 \\ 01000000 \\ 00100000 \\ 00000000 \\ 00001000 \\ 00000000 \\ 00000000 \\ 00000000 \end{pmatrix}.$$

Lemma IV.16. *With the above notations, and those of chapter II, we have*

$$\boldsymbol{W}_\ell \boldsymbol{E}_\ell \boldsymbol{W}_\ell = \boldsymbol{D}_\ell.$$

Proof. Let us first prove that

$$\boldsymbol{W}_\ell \boldsymbol{U}_\ell \boldsymbol{W}_\ell = 2^\ell \begin{pmatrix} 1 0 \dots 0 \\ 0 0 \dots 0 \\ \vdots \vdots \ddots \vdots \\ 0 0 \dots 0 \end{pmatrix}.$$

The statement is trivial for $\ell = 0$. Let us argue recursively and assume the statement to be true up to $\ell - 1$. Then $\boldsymbol{W}_\ell \boldsymbol{U}_\ell \boldsymbol{W}_\ell$ is equal to

$$2^{-1} \begin{pmatrix} \boldsymbol{W}_{\ell-1} & \boldsymbol{W}_{\ell-1} \\ \boldsymbol{W}_{\ell-1} & -\boldsymbol{W}_{\ell-1} \end{pmatrix} \begin{pmatrix} \boldsymbol{U}_{\ell-1} & \boldsymbol{U}_{\ell-1} \\ \boldsymbol{U}_{\ell-1} & \boldsymbol{U}_{\ell-1} \end{pmatrix} \begin{pmatrix} \boldsymbol{W}_{\ell-1} & \boldsymbol{W}_{\ell-1} \\ \boldsymbol{W}_{\ell-1} & -\boldsymbol{W}_{\ell-1} \end{pmatrix}$$

and this is equal to

$$2 \begin{pmatrix} \boldsymbol{W}_{\ell-1} \boldsymbol{U}_{\ell-1} \boldsymbol{W}_{\ell-1} & \boldsymbol{O}_{\ell-1} \\ \boldsymbol{O}_{\ell-1} & \boldsymbol{O}_{\ell-1} \end{pmatrix},$$

which proves our claim. Finally, using the relation

$$\boldsymbol{G}_\ell = \begin{pmatrix} \boldsymbol{G}_{\ell-1} + \boldsymbol{U}_{\ell-1} & \boldsymbol{G}_{\ell-1} - \boldsymbol{U}_{\ell-1} \\ \boldsymbol{G}_{\ell-1} - \boldsymbol{U}_{\ell-1} & \boldsymbol{G}_{\ell-1} + \boldsymbol{U}_{\ell-1} \end{pmatrix}$$

and the previous remarks, it follows from $\boldsymbol{E}_\ell = 2^{-\ell}\boldsymbol{G}_\ell$ that

$$\boldsymbol{W}_\ell \boldsymbol{E}_\ell \boldsymbol{W}_\ell = \begin{pmatrix} \boldsymbol{W}_{\ell-1}\boldsymbol{E}_{\ell-1}\boldsymbol{W}_{\ell-1} & \boldsymbol{O}_{\ell-1} \\ & \begin{matrix} 1\,0 & \dots & 0 \\ 0\,0 & \dots & 0 \end{matrix} \\ \boldsymbol{O}_{\ell-1} & \begin{matrix} \vdots\,\vdots & \ddots & \vdots \\ 0\,0 & \dots & 0 \end{matrix} \end{pmatrix}.$$

Another straightforward induction argument finishes the proof. □

In particular, the lemma confirms the result of proposition II.7:

$$rk(\boldsymbol{E}_\ell) = rk(\boldsymbol{G}_\ell) = \ell + 1.$$

We may now prove:

Proposition IV.17. *If $w_0, \dots, w_{2^\ell - 1}$ are the Walsh coefficients of $f : \Omega \to \mathbb{R}$, then the normalized epistasis $\varepsilon^*(f)$ of f is given by*

$$\varepsilon^*(f) = 1 - \frac{w_0^2 + \sum_{i=0}^{\ell-1} w_{2^i}^2}{\sum_{j=0}^{2^\ell-1} w_j^2}.$$

Proof. Obviously,

$${}^t\boldsymbol{f}\boldsymbol{f} = {}^t(\boldsymbol{W}\boldsymbol{w})\,(\boldsymbol{W}\boldsymbol{w}) = {}^t\boldsymbol{w}{}^t\boldsymbol{W}\boldsymbol{W}\boldsymbol{w} = {}^t\boldsymbol{w}\boldsymbol{w},$$

as $\boldsymbol{W}$ is symmetric and $\boldsymbol{W}^2 = \boldsymbol{I}$ (we omit the index ℓ in our notations). On the other hand,

$${}^t\boldsymbol{f}\boldsymbol{E}\boldsymbol{f} = {}^t\boldsymbol{w}{}^t\boldsymbol{W}\boldsymbol{E}\boldsymbol{W}\boldsymbol{w} = {}^t\boldsymbol{w}\boldsymbol{W}\boldsymbol{E}\boldsymbol{W}\boldsymbol{w} = {}^t\boldsymbol{w}\boldsymbol{D}\boldsymbol{w}.$$

We thus obtain

$$\varepsilon^*(f) = 1 - \frac{{}^t\boldsymbol{f}\boldsymbol{E}\boldsymbol{f}}{{}^t\boldsymbol{f}\boldsymbol{f}} = 1 - \frac{{}^t\boldsymbol{w}\boldsymbol{D}\boldsymbol{w}}{{}^t\boldsymbol{w}\boldsymbol{w}},$$

and this yields our claim, indeed. □

The previous result sheds new light on the meaning of the notion of epistasis. Indeed, with notations as before, we know that w_0 is the average of f and that the w_{2^i} are the contributions of the schemata whose only non-# entry is on locus i, i.e., exactly the *linear* contributions. The other Walsh coefficients correspond to schemata where more than one locus is non-#, the non-linear contributions.

The above expression for $\varepsilon^*(f)$ may thus be viewed as the ratio between (the sum of the square of the) non-linear Walsh coefficients and the norm of the function. We have seen in chapter II that the eigenspaces $V_0^\ell = \{\boldsymbol{v} \in \mathbb{R}^{2^\ell}; \boldsymbol{E}_\ell \boldsymbol{v} = \boldsymbol{0}\}$ and $V_1^\ell = \{\boldsymbol{v} \in \mathbb{R}^{2^\ell}; \boldsymbol{E}_\ell \boldsymbol{v} = \boldsymbol{v}\}$ corresponding to the eigenvalues 0 and 1 of $\boldsymbol{E}_\ell$ have dimensions $2^\ell - \ell - 1$ and $\ell + 1$, respectively. Therefore, for any $\boldsymbol{v} \in V_0^\ell$ representing a function f with Walsh coefficients $\boldsymbol{w}$, we have that

$$\boldsymbol{W}_\ell \boldsymbol{E}_\ell \boldsymbol{W}_\ell \boldsymbol{W}_\ell \boldsymbol{v} = \boldsymbol{0},$$

and since $\boldsymbol{W}_\ell \boldsymbol{v} = \boldsymbol{w}$ this obviously means that $\boldsymbol{D}_\ell \boldsymbol{w} = \boldsymbol{0}$, or equivalently $w_0 = 0$ and $w_{2^i} = 0$, $0 \leq i < \ell$.

As a direct consequence of the previous result, we have that $\varepsilon^*(f) = 0$. Moreover, the vectors of the columns of the matrix $\boldsymbol{W}_\ell$, which are at positions j, with $j \neq 0, 2^i$ $(0 \leq i < \ell)$, form a basis for V_0^ℓ.

For any vector $\boldsymbol{v} \in V_1^\ell$, obviously $\boldsymbol{D}_\ell \boldsymbol{w} = \boldsymbol{w}$. It follows that $w_j = 0$ for all $j \neq 0, 2^i$ $(0 \leq i < \ell)$ and a basis for V_1^ℓ is given by the vectors of the columns of $\boldsymbol{W}_\ell$ situated at positions $j = 0, 2^i$, with $0 \leq i < \ell$, cf. the remarks following proposition II.7.

Proposition IV.18. *For any "linear" map $f : \Omega \to \mathbb{R}$ we have $\varepsilon^*(f) = 0$.*

Proof. Obviously, f is linear if and only if

$$\boldsymbol{f} = \begin{pmatrix} a \\ a+b \\ \vdots \\ a + (2^\ell - 1)b \end{pmatrix},$$

with $a, b \in \mathbb{R}$. Since $\varepsilon^*(f) = 1 - \frac{w_0^2 + \sum_{i=0}^{\ell-1} w_{2^i}^2}{\sum_{j=0}^{2^\ell - 1} w_j^2}$, it suffices to check that $w_j = 0$ for

$j \neq 0, 2^i$, $0 \leq i < \ell$. Equivalently,

$$\sum_{k=0}^{2^\ell-1} (a+bk)\psi_j(k) = 0$$

for these values. Now,

$$\sum_{k=0}^{2^\ell-1} (a+bk)\psi_j(k) = a\sum_{k=0}^{2^\ell-1} \psi_j(k) + b\sum_{k=0}^{2^\ell-1} k\psi_j(k).$$

To prove the assertion, it suffices to check that $\sum_{k=0}^{2^\ell-1} k\psi_j(k) = 0$, as $\sum_{k=0}^{2^\ell-1} \psi_j(k) = 0$ by corollary IV.2.

Write k as $\hat{k}k_0$, j as $\hat{\jmath}j_0$ with $\hat{k}, \hat{\jmath} \in \Omega_{\ell-1}$. Then

$$\begin{aligned}\sum_{k\in\Omega} k(-1)^{k\cdot j} &= \sum_{\substack{k\in\Omega\\k_0=0}} k(-1)^{k\cdot j} + \sum_{\substack{k\in\Omega\\k_0=1}} k(-1)^{k\cdot j}\\ &= \sum_{\hat{k}\in\Omega_{\ell-1}} 2\hat{k}(-1)^{2\hat{k}\cdot j} + \sum_{\hat{k}\in\Omega_{\ell-1}} (2\hat{k}+1)(-1)^{(2\hat{k}+1)\cdot j}.\end{aligned}$$

1. Case $j_0 = 0$, i.e., $j = 2\hat{\jmath}$.

$$\begin{aligned}\sum_{k\in\Omega} k(-1)^{k\cdot j} &= 2\sum_{\hat{k}\in\Omega_{\ell-1}} \hat{k}(-1)^{2\hat{k}\cdot 2\hat{\jmath}} + 2\sum_{\hat{k}\in\Omega_{\ell-1}} \hat{k}(-1)^{(2\hat{k}+1)\cdot 2\hat{\jmath}} + \sum_{\hat{k}\in\Omega_{\ell-1}} (-1)^{(2\hat{k}+1)\cdot 2\hat{\jmath}}\\ &= 4\sum_{\hat{k}\in\Omega_{\ell-1}} \hat{k}(-1)^{\hat{k}\cdot\hat{\jmath}}.\end{aligned}$$

The result follows as $\hat{\jmath} \neq 0, 2^i$, $0 \leq i < \ell - 1$, by induction.

2. Case $j_0 = 1$, i.e., $j = 2\hat{\jmath} + 1$.

$$\begin{aligned}\sum_{k\in\Omega} k(-1)^{k\cdot j} &= 2\sum_{\hat{k}\in\Omega_{\ell-1}} \hat{k}(-1)^{2\hat{k}\cdot(2\hat{\jmath}+1)} + 2\sum_{\hat{k}\in\Omega_{\ell-1}} \hat{k}(-1)^{(2\hat{k}+1)\cdot(2\hat{\jmath}+1)} +\\ &\quad \sum_{\hat{k}\in\Omega_{\ell-1}} (-1)^{(2\hat{k}+1)\cdot(2\hat{\jmath}+1)}\\ &= 2\sum_{\hat{k}\in\Omega_{\ell-1}} \hat{k}(-1)^{\hat{k}\cdot\hat{\jmath}} - 2\sum_{\hat{k}\in\Omega_{\ell-1}} \hat{k}(-1)^{\hat{k}\cdot\hat{\jmath}} = 0.\end{aligned}$$

□

5 Examples

5.1 Some first, easy examples

The needle-in-a-haystack function

Let us take $f = needle(0)$, i.e., $f(t) = \delta_{t,0}$ as defined in chapter I. In this case the vector representation of f is

$$\boldsymbol{f} = \begin{pmatrix} 1 \\ 0 \\ \vdots \\ 0 \end{pmatrix} .$$

Since $\psi_0(s) = 1$ for all $s \in \Omega$, obviously

$$\boldsymbol{w} = \boldsymbol{W}_\ell \boldsymbol{f} = 2^{-\ell/2} \begin{pmatrix} 1 \\ \vdots \\ 1 \end{pmatrix} ,$$

i.e., $w_s = 2^{-\ell/2}$ for all $s \in \Omega$. From this it trivially follows that

$$\begin{aligned} \varepsilon^*(needle(0)) &= 1 - \frac{w_0^2 + \sum_{i=0}^{\ell-1} w_{2^i}^2}{\|\boldsymbol{w}\|^2} = 1 - \frac{(\ell+1)(2^{-\ell/2})^2}{2^\ell \, (2^{-\ell/2})^2} \\ &= 1 - \frac{\ell+1}{2^\ell} , \end{aligned}$$

as obtained in chapter II.

The camel function

Let us consider the function $f = camel$, defined by $camel(0\dots0) = camel(1\dots1) = 1$ and $camel(t) = 0$ for all other strings $t \in \Omega$. Clearly

$$camel(s) = \delta_{0\dots0,s} + \delta_{1\dots1,s} = \delta_{0,s} + \delta_{2^\ell-1,s} ,$$

and the vector representation of f is

$$\boldsymbol{f} = \begin{pmatrix} 1 \\ 0 \\ \vdots \\ 0 \\ 1 \end{pmatrix}.$$

It follows from $\boldsymbol{w} = \boldsymbol{W}_\ell \boldsymbol{f}$ that, for each $s \in \Omega$, we have

$$w_s = 2^{-\ell/2}(\psi_{0\dots0}(s) + \psi_{1\dots1}(s)).$$

In particular,

$$w_0 = 2^{-\ell/2}(\psi_{0\dots0}(0\dots0) + \psi_{1\dots1}(0\dots0)) = 2 \cdot 2^{-\ell/2}$$

and

$$\begin{aligned} w_{2^i} &= 2^{-\ell/2}(\psi_{0\dots0}(0\dots010\dots0) + \psi_{1\dots1}(0\dots010\dots0)) \\ &= 2^{-\ell/2}\,(1 + (-1)) = 0. \end{aligned}$$

So,

$$\begin{aligned} \varepsilon^*(camel) &= 1 - \frac{w_0^2 + \sum_{i=0}^{\ell-1} w_{2^i}^2}{\|\boldsymbol{w}\|^2} = 1 - \frac{w_0^2}{\|\boldsymbol{f}\|^2} = 1 - \frac{(2 \cdot 2^{-\ell/2})^2}{1^2 + 1^2} \\ &= 1 - \frac{1}{2^{\ell-1}}. \end{aligned}$$

Unitation functions

As a last easy example, let us reconsider unitation functions on Ω, and reconstruct the results from chapter III, section 2.

With the same notations as in chapter III, we start from a function f on Ω such that $f(s) = h(u(s))$, for some real-valued function $h : \{0, \dots, \ell\} \to \mathbb{R}$, where $u(s)$ — the unitation of s — is the Hamming distance between s and the zero-string.

It is intuitively clear that the epistasis of f should then only depend upon the function h, i.e., the components of the vector

$$\boldsymbol{h} = \begin{pmatrix} h(0) \\ \vdots \\ h(\ell) \end{pmatrix}.$$

So, let us consider a unitation function f with associated function h, and let us denote by $w_0, \ldots, w_{2^\ell - 1}$ the Walsh coefficients of f. We know that the coefficient w_0 is, up to a factor, the average of f, i.e.,

$$w_0 = 2^{\ell/2} f(\Omega) = 2^{-\ell/2} \sum_{u=0}^{\ell} \binom{\ell}{u} h(u).$$

On the other hand, using the notation

$$f_{(i,a)} = f(\# \ldots \# \overset{i}{\overset{\downarrow}{a}} \# \ldots \#),$$

we know that the w_{2^i} may be given by

$$\begin{aligned} w_{2^i} &= (-1)^a \left(2^{\ell/2} f_{(i,a)} - w_0\right) \qquad \text{with } a = 0 \\ &= 2^{\ell/2} f_{(i,0)} - 2^{\ell/2} \left(\frac{1}{2} f_{(i,0)} + \frac{1}{2} f_{(i,1)}\right) \\ &= 2^{\ell/2-1} \left(f_{(i,0)} - f_{(i,1)}\right) \\ &= 2^{\ell/2-1} \left(\frac{1}{2^{\ell-1}} \sum_{u=0}^{\ell-1} \left[\binom{\ell-1}{u} h(u)\right] - \frac{1}{2^{\ell-1}} \sum_{u=1}^{\ell} \left[\binom{\ell-1}{u-1} h(u)\right]\right) \\ &= 2^{-\ell/2} \sum_{u=0}^{\ell} \left[\binom{\ell-1}{u} - \binom{\ell-1}{u-1}\right] h(u). \end{aligned}$$

Note that this result is independent of i. We thus find

$$\varepsilon^*(f) = 1 - \frac{w_0^2 + \ell w_1^2}{\|\boldsymbol{w}\|^2},$$

where

$$\|\boldsymbol{w}\|^2 = \|\boldsymbol{f}\|^2 = \sum_{u=0}^{\ell} \binom{\ell}{u} h^2(u).$$

Also note that the previous description of $\varepsilon^*(f)$ coincides with that given in chapter III. Indeed, a straightforward calculation shows that

$$\varepsilon^*(f) = 1 - \frac{1}{2^\ell}\frac{\sum_{p,q=0}^{\ell}\binom{\ell}{p}\binom{\ell}{q}\left(1+\frac{(\ell-2p)(\ell-2q)}{\ell}\right)h(p)h(q)}{||\boldsymbol{f}||^2}$$

(the value of $\varepsilon^*(f)$ given in theorem III.16) is equal to

$$\varepsilon^*(f) = 1 - \frac{1}{2^\ell}\frac{\left(\sum_{u=0}^{\ell}\binom{\ell}{u}h_u\right)^2 + \ell\left(\sum_{u=0}^{\ell}\left(\binom{\ell-1}{u}\binom{\ell-1}{u-1}\right)h_u\right)^2}{||\boldsymbol{f}||^2}$$

(the value of $\varepsilon^*(f)$ obtained in the present section).

If we let

$$h(u) = \begin{cases} 1 & \text{if } u = 0 \\ 0 & \text{if } u \neq 0, \end{cases}$$

then $needle(s) = h(u(s))$. In this case, the above relations yield

$$w_0 = 2^{-\ell/2}\sum_{u=0}^{\ell}\binom{\ell}{u}h(u) = 2^{-\ell/2},$$

$$w_{2^i} = 2^{-\ell/2}\sum_{u=0}^{\ell}\left(\binom{\ell-1}{u} - \binom{\ell-1}{u-1}\right)h(u) = 2^{-\ell/2},$$

and we recover

$$\varepsilon^*(needle) = 1 - \frac{w_0^2 + \ell w_1^2}{\|\boldsymbol{w}\|^2} = 1 - \frac{(\ell+1)(2^{-\ell/2})^2}{1} = 1 - \frac{\ell+1}{2^\ell}.$$

If we let

$$h(u) = \begin{cases} 1 & \text{if } u = 0,\, \ell \\ 0 & \text{if } u \neq 0,\, \ell \end{cases}$$

then $camel(s) = h(u(s))$. In this case, the above relations yield

$$w_0 = 2^{-\ell/2}\sum_{u=0}^{\ell}\binom{\ell}{u}h(u) = 2^{-\ell/2}\left(1\cdot h(0) + 1\cdot h(\ell)\right) = 2\cdot 2^{-\ell/2},$$

$$\begin{aligned} w_{2^i} &= 2^{-\ell/2}\sum_{u=0}^{\ell}\left(\binom{\ell-1}{u} - \binom{\ell-1}{u-1}\right)h(u) \\ &= 2^{-\ell/2}\left((1-0)h(0) + (0-1)h(\ell)\right) = 0, \end{aligned}$$

and we recover

$$\varepsilon^*(camel) = 1 - \frac{w_0^2}{\|\boldsymbol{w}\|^2} = 1 - \frac{4 \cdot 2^{-\ell}}{2} = 1 - \frac{1}{2^{\ell-1}}.$$

5.2 A more complicated example: template functions

In this section we show how the use of Walsh transforms permits an easy calculation of the epistasis of the "template functions". We invite the reader to compare it with the set-up of chapter III.

As we saw in the previous chapter, the template functions T_ℓ^n calculate the fitness of a string of length ℓ, by sliding a fixed string t of length $n \leq \ell$ (the template) over it.

As always, let us denote by $\boldsymbol{T}_\ell^n$ the vector corresponding to the function T_ℓ^n.

In order to calculate the epistasis of this type of functions by applying proposition IV.17, and taking into account the value of $\|\boldsymbol{T}_\ell^n\|^2$ derived in the previous chapter, it only remains to calculate

$$w_0^2 + \sum_{i=0}^{\ell-1} w_{2^i}^2.$$

First of all, let us note that it will be easier to work with $\boldsymbol{v}_\ell^n = \boldsymbol{V}_\ell \boldsymbol{T}_\ell^n$ and $v_{\ell,j}^n = (\boldsymbol{v}_\ell^n)_j$ (for $j = 0, \ldots, 2^\ell - 1$), so it is clear that $\boldsymbol{w}_\ell^n = 2^{-\ell/2} \boldsymbol{v}_\ell^n$.

Let us first consider the case $n = \ell$. To simplify, let us write $\boldsymbol{v}_\ell = \boldsymbol{v}_\ell^\ell$. We have

$$\boldsymbol{v}_\ell = \boldsymbol{V}_\ell \boldsymbol{T}_\ell^\ell = \boldsymbol{V}_\ell \begin{pmatrix} 0 \\ \vdots \\ 0 \\ 1 \end{pmatrix} = \begin{pmatrix} (-1)^{u(0)} \\ \vdots \\ (-1)^{u(2^\ell-1)} \end{pmatrix}$$

where $u(i)$ denotes the unitation of the binary representation of i. So $(\boldsymbol{v}_\ell)_0 = 1$ and $(\boldsymbol{v}_\ell)_{2^i} = -1$, for all $i = 0, \ldots, \ell - 1$. As $\|\boldsymbol{T}_\ell^\ell\|^2 = 1$, we find that

$$\varepsilon^*(\boldsymbol{T}_\ell^\ell) = 1 - \frac{1+\ell}{2^\ell},$$

in accordance with the calculations in chapter III.

More generally, let us now assume $n < \ell$. Recall that

$$\boldsymbol{T}_\ell^n = \begin{pmatrix} T_\ell^n(00\ldots0) \\ \vdots \\ T_\ell^n(11\ldots1) \end{pmatrix} = \begin{pmatrix} \boldsymbol{T}_{\ell-1}^n \\ \boldsymbol{T}_{\ell-1}^n + \boldsymbol{D}_{\ell-1}^n \end{pmatrix} = \begin{pmatrix} \boldsymbol{T}_{\ell-1}^n \\ \boldsymbol{T}_{\ell-1}^n \end{pmatrix} + \boldsymbol{D}_\ell^{n+1}$$

with

$$\boldsymbol{D}_\ell^{n+1} = \begin{pmatrix} \boldsymbol{0}_{\ell-1} \\ \boldsymbol{D}_{\ell-1}^n \end{pmatrix}.$$

Using this, we obtain

$$\begin{aligned} \boldsymbol{v}_\ell^n &= \boldsymbol{V}_\ell \left(\begin{pmatrix} \boldsymbol{T}_{\ell-1}^n \\ \boldsymbol{T}_{\ell-1}^n \end{pmatrix} + \boldsymbol{D}_\ell^{n+1} \right) \\ &= \begin{pmatrix} \boldsymbol{V}_{\ell-1} & \boldsymbol{V}_{\ell-1} \\ \boldsymbol{V}_{\ell-1} & -\boldsymbol{V}_{\ell-1} \end{pmatrix} \left(\begin{pmatrix} \boldsymbol{T}_{\ell-1}^n \\ \boldsymbol{T}_{\ell-1}^n \end{pmatrix} + \boldsymbol{D}_\ell^{n+1} \right) = 2 \begin{pmatrix} \boldsymbol{v}_{\ell-1}^n \\ \boldsymbol{0}_{\ell-1} \end{pmatrix} + \boldsymbol{d}_\ell^{n+1}, \end{aligned}$$

where $\boldsymbol{d}_\ell^n = \boldsymbol{V}_\ell \boldsymbol{D}_\ell^n$.

So,

$$\begin{aligned} \boldsymbol{v}_\ell^n &= 2 \begin{pmatrix} 2 \begin{pmatrix} \boldsymbol{v}_{\ell-2}^n \\ \boldsymbol{0}_{\ell-2} \end{pmatrix} + \boldsymbol{d}_{\ell-1}^{n+1} \\ \boldsymbol{0}_{\ell-1} \end{pmatrix} + \boldsymbol{d}_\ell^{n+1} \\ &= 4 \begin{pmatrix} \boldsymbol{v}_{\ell-2} \\ \boldsymbol{0}_{\ell-2} \\ \boldsymbol{0}_{\ell-1} \end{pmatrix} + 2 \begin{pmatrix} \boldsymbol{d}_{\ell-1}^{n+1} \\ \boldsymbol{0}_{\ell-1} \end{pmatrix} + \boldsymbol{d}_\ell^{n+1} \\ &= 8 \begin{pmatrix} \boldsymbol{v}_{\ell-3} \\ \boldsymbol{0}_{\ell-3} \\ \boldsymbol{0}_{\ell-2} \\ \boldsymbol{0}_{\ell-1} \end{pmatrix} + 4 \begin{pmatrix} \boldsymbol{d}_{\ell-2}^{n+1} \\ \boldsymbol{0}_{\ell-2} \\ \boldsymbol{0}_{\ell-1} \end{pmatrix} + 2 \begin{pmatrix} \boldsymbol{d}_{\ell-1}^{n+1} \\ \boldsymbol{0}_{\ell-1} \end{pmatrix} + \boldsymbol{d}_\ell^{n+1} \\ &\vdots \\ &= 2^{\ell-n} \begin{pmatrix} \boldsymbol{v}_n \\ \boldsymbol{0}_n \\ \vdots \\ \boldsymbol{0}_{\ell-1} \end{pmatrix} + 2^{\ell-n-1} \begin{pmatrix} \boldsymbol{d}_{n+1}^{n+1} \\ \boldsymbol{0}_{n+1} \\ \vdots \\ \boldsymbol{0}_{\ell-1} \end{pmatrix} + \cdots + 2 \begin{pmatrix} \boldsymbol{d}_{\ell-1}^{n+1} \\ \boldsymbol{0}_{\ell-1} \end{pmatrix} + \boldsymbol{d}_\ell^{n+1}. \end{aligned}$$

We may write the above formula in a more elegant way using the Kronecker product (defined in section 1.1 of appendix B). In fact, to calculate $\boldsymbol{d}_i^{n+1}$ for $i = n+1, \dots, \ell$, first note that

$$\begin{aligned}
\boldsymbol{d}_\ell^n = \boldsymbol{V}_\ell \boldsymbol{D}_\ell^n &= \begin{pmatrix} \boldsymbol{V}_{\ell-1} & \boldsymbol{V}_{\ell-1} \\ \boldsymbol{V}_{\ell-1} & -\boldsymbol{V}_{\ell-1} \end{pmatrix} \begin{pmatrix} \boldsymbol{0}_{\ell-1} \\ \boldsymbol{0}_{\ell-2} \\ \vdots \\ \boldsymbol{0}_{\ell-n+1} \\ \boldsymbol{u}_{\ell-n+1} \end{pmatrix} \\
&= \begin{pmatrix} \boldsymbol{V}_{\ell-1} \boldsymbol{D}_{\ell-1}^{n-1} \\ -\boldsymbol{V}_{\ell-1} \boldsymbol{D}_{\ell-1}^{n-1} \end{pmatrix} = \begin{pmatrix} \boldsymbol{d}_{\ell-1}^{n-1} \\ -\boldsymbol{d}_{\ell-1}^{n-1} \end{pmatrix} \\
&= \begin{pmatrix} 1 \\ -1 \end{pmatrix} \otimes \boldsymbol{d}_{\ell-1}^{n-1} = \boldsymbol{v}_1 \otimes \boldsymbol{d}_{\ell-1}^{n-1}.
\end{aligned}$$

Similarly, if we write $\boldsymbol{v}_i = \boldsymbol{v}_1^{\otimes i}$, for any i we have $\boldsymbol{d}_\ell^n = \boldsymbol{v}_i \otimes \boldsymbol{d}_{\ell-i}^{n-i}$. Taking $i = n-1$, we can write

$$\begin{aligned}
\boldsymbol{d}_\ell^{n+1} &= \boldsymbol{v}_{n-1} \otimes \boldsymbol{d}_{\ell-n+1}^2 = \boldsymbol{v}_{n-1} \otimes \boldsymbol{V}_{\ell-n+1} \begin{pmatrix} \boldsymbol{0}_{\ell-n} \\ \boldsymbol{u}_{\ell-n} \end{pmatrix} \\
&= \boldsymbol{v}_{n-1} \otimes \begin{pmatrix} \boldsymbol{V}_{\ell-n} \boldsymbol{u}_{\ell-n} \\ -\boldsymbol{V}_{\ell-n} \boldsymbol{u}_{\ell-n} \end{pmatrix} = \boldsymbol{v}_{n-1} \otimes \boldsymbol{v}_1 \otimes \boldsymbol{V}_{\ell-n} \boldsymbol{u}_{\ell-n} \\
&= \boldsymbol{v}_n \otimes 2^{\ell-n} \boldsymbol{h}_{\ell-n}
\end{aligned}$$

where

$$\boldsymbol{h}_\ell = \begin{pmatrix} 1 \\ 0 \\ \vdots \\ 0 \end{pmatrix} \in \mathbb{R}^{2^\ell}$$

for all $\ell \in \mathbb{N}$. In a similar way,

$$\boldsymbol{d}_i^{n+1} = 2^{i-n} \boldsymbol{v}_n \otimes \boldsymbol{h}_{i-n}$$

for $i = n+1, \ldots, \ell$. We thus obtain

$$\begin{aligned}
\boldsymbol{v}_\ell^n &= 2^{\ell-n}\begin{pmatrix}\boldsymbol{v}_n\\ \boldsymbol{0}_n\\ \vdots\\ \boldsymbol{0}_{\ell-1}\end{pmatrix} + \sum_{i=n+1}^{\ell} 2^{\ell-i}\begin{pmatrix}\boldsymbol{d}_i^{n+1}\\ \boldsymbol{0}_i\\ \vdots\\ \boldsymbol{0}_{\ell-1}\end{pmatrix}\\
&= 2^{\ell-n}\left(\begin{pmatrix}\boldsymbol{v}_n\\ \boldsymbol{0}_n\\ \vdots\\ \boldsymbol{0}_{\ell-1}\end{pmatrix} + \sum_{i=n+1}^{\ell}\begin{pmatrix}\boldsymbol{v}_n\otimes\boldsymbol{h}_{i-n}\\ \boldsymbol{0}_i\\ \vdots\\ \boldsymbol{0}_{\ell-1}\end{pmatrix}\right)\\
&= 2^{\ell-n}\left(\boldsymbol{h}_{\ell-n}\otimes\boldsymbol{v}_n + \sum_{i=n+1}^{\ell}\boldsymbol{h}_{\ell-i}\otimes\boldsymbol{v}_n\otimes\boldsymbol{h}_{i-n}\right)\\
&= 2^{\ell-n}\sum_{i=0}^{\ell-n}\boldsymbol{h}_1^{\otimes i}\otimes\boldsymbol{v}_n\otimes\boldsymbol{h}_1^{\otimes \ell-n-i}.
\end{aligned}$$

(Note that $\boldsymbol{h}_i = \boldsymbol{h}_1^{\otimes i}$ with $\boldsymbol{h}_1^{\otimes 0} = 1$.)

As we have mentioned before, we are only interested in the value of $v_{\ell,0}^n$ and $v_{\ell,2^i}^n$ for $i = 0, \ldots, \ell-1$. For the first one, it is clear that

$$v_{\ell,0}^n = 2^{\ell-n}(\ell-n+1)$$

since $(\boldsymbol{h}_j\otimes\boldsymbol{v}_n\otimes\boldsymbol{h}_{\ell-n-j})_0 = 1$ for all j.

Now, in order to deduce a general formula for the second case, let us first consider two examples, one for the case $n < \ell \le 2n$ and another one for the case $2n \le \ell$.

Example 1. We assume that $\ell = 7$ and $n = 5$.

First, note that

$$\boldsymbol{v}_7^5 = 2^2\sum_{j=0}^{2}\boldsymbol{h}_j\otimes\boldsymbol{v}_5\otimes\boldsymbol{h}_{2-j} = 4(\boldsymbol{h}_0\otimes\boldsymbol{v}_5\otimes\boldsymbol{h}_2 + \boldsymbol{h}_1\otimes\boldsymbol{v}_5\otimes\boldsymbol{h}_1 + \boldsymbol{h}_2\otimes\boldsymbol{v}_5\otimes\boldsymbol{h}_0).$$

It should be clear that

$$\begin{aligned}
v_{7,1}^5 &= v_{7,64}^5 = -2^2\\
v_{7,2}^5 &= v_{7,32}^5 = -2^2 2\\
v_{7,4}^5 &= v_{7,8}^5 = v_{7,16}^5 = -2^2 3.
\end{aligned}$$

We need to distinguish three cases:

1. If $0 \leq i < \ell - n$, the non-zero summands are $\boldsymbol{h}_j \otimes \boldsymbol{v}_n \otimes \boldsymbol{h}_{\ell-n-j}$ with $j = \ell - n - i, \ldots, \ell - n$.
2. If $\ell - n \leq i < n$, the non-zero summands are $\boldsymbol{h}_j \otimes \boldsymbol{v}_n \otimes \boldsymbol{h}_{\ell-n-j}$ with $j = 0, \ldots, \ell - n$.
3. If $n \leq i < \ell$, the non-zero summands are $\boldsymbol{h}_j \otimes \boldsymbol{v}_n \otimes \boldsymbol{h}_{\ell-n-j}$ with $n + \ell - n - j > i$, so $j = 0, \ldots, \ell - i - 1$.

Example 2. We assume that $\ell = 7$ and $n = 2$.

As

$$\boldsymbol{v}_7^2 = 2^5 \sum_{j=0}^{5} \boldsymbol{h}_j \otimes \boldsymbol{v}_2 \otimes \boldsymbol{h}_{5-j} = 32(\boldsymbol{h}_0 \otimes \boldsymbol{v}_2 \otimes \boldsymbol{h}_5 + \boldsymbol{h}_1 \otimes \boldsymbol{v}_2 \otimes \boldsymbol{h}_4$$
$$+ \boldsymbol{h}_2 \otimes \boldsymbol{v}_2 \otimes \boldsymbol{h}_3 + \boldsymbol{h}_3 \otimes \boldsymbol{v}_2 \otimes \boldsymbol{h}_2 + \boldsymbol{h}_4 \otimes \boldsymbol{v}_2 \otimes \boldsymbol{h}_1 + \boldsymbol{h}_5 \otimes \boldsymbol{v}_2 \otimes \boldsymbol{h}_0),$$

it follows that

$$v_{7,1}^2 = v_{7,64}^2 = -2^5$$
$$v_{7,2}^2 = v_{7,4}^2 = v_{7,8}^2 = v_{7,16}^2 = v_{7,32}^2 = -2^5 2.$$

Again, we can distinguish three cases:

1. If $0 \leq i \leq n - 1$, the non-zero summands are $\boldsymbol{h}_j \otimes \boldsymbol{v}_n \otimes \boldsymbol{h}_{\ell-n-j}$ with $j = \ell - n - i, \ldots, \ell - n$.
2. If $n \leq i \leq \ell - n$, the non-zero summands are $\boldsymbol{h}_j \otimes \boldsymbol{v}_n \otimes \boldsymbol{h}_{\ell-n-j}$ with $j = \ell - i - n, \ldots, \ell - i - 1$.
3. If $\ell - n \leq i \leq \ell - 1$, the non-zero summands are $\boldsymbol{h}_j \otimes \boldsymbol{v}_n \otimes \boldsymbol{h}_{\ell-n-j}$ with $n + \ell - n - j > i$, so $j = 0, \ldots, \ell - i - 1$.

The general case works similarly, so we obtain:

1. if $n < \ell \leq 2n$ and $i = 0, \ldots, \ell - 1$,

$$v_{\ell,2^i}^n = \begin{cases} -2^{\ell-n}(i+1) & \text{if } 0 \leq i < \ell - n \\ -2^{\ell-n}(\ell - n + 1) & \text{if } \ell - n \leq i < n \\ -2^{\ell-n}(\ell - i) & \text{if } n \leq i \leq \ell - 1, \end{cases}$$

2. if $2n \leq \ell$ and $i = 0, \ldots, \ell - 1$,

$$v^n_{\ell,2^i} = \begin{cases} -2^{\ell-n}(i+1) & \text{if } 0 \leq i \leq n-1 \\ -2^{\ell-n}n & \text{if } n \leq i \leq \ell-n \\ -2^{\ell-n}(\ell-i) & \text{if } \ell-n \leq i \leq \ell-1. \end{cases}$$

In the first case, we have

$$\begin{aligned}
&(v^n_{\ell,0})^2 + \textstyle\sum_{i=0}^{\ell-1}(v^n_{\ell,2^i})^2 = 4^{\ell-n}\{(\ell-n+1)^2 + 2\sum_{i=0}^{\ell-n} i^2 + (2n-\ell)(\ell-n+1)^2\} = \\
&= 4^{\ell-n}\{(\ell-n+1)^2(2n-\ell+1) + \tfrac{1}{3}(\ell-n)(\ell-n+1)(2\ell-2n+1)\} \\
&= 4^{\ell-n}\{1 + n(\ell-n+1)^2 + \tfrac{\ell-n}{3}(4-(\ell-n)^2)\}.
\end{aligned}$$

Similarly, in the second case,

$$\begin{aligned}
&(v^n_{\ell,0})^2 + \textstyle\sum_{i=0}^{\ell-1}(v^n_{\ell,2^i})^2 = 4^{\ell-n}\{(\ell-n+1)^2 + 2\sum_{i=0}^{n} i^2 + (\ell-2n)n^2\} \\
&= 4^{\ell-n}\{(\ell-n+1)^2 + \tfrac{1}{3}n(n+1)(2n+1) + (\ell-2n)n^2\} \\
&= 4^{\ell-n}\{(\ell-2n)(n^2+\ell+2) + \tfrac{n}{3}(2n^2+7) + 2n^2+1\}.
\end{aligned}$$

Finally, the combination of all of the previous results with the value of the norm of $\boldsymbol{T}^n_\ell$ and proposition IV.17 yields:

Theorem IV.19. *The epistasis of the template function T^n_ℓ is:*

$$\varepsilon^*(T^n_\ell) = \begin{cases} 1 - \dfrac{1+n(\ell-n+1)^2+\frac{(\ell-n)}{3}\left(4-(\ell-n)^2\right)}{2^n\left(3(\ell-n)-1+2^{n-\ell+1}\right)} & \textit{if } n \leq \ell \leq 2n \\ \\ 1 - \dfrac{(\ell-2n)(n^2+\ell+2)+\frac{n}{3}(2n^2+7)+2n^2+1}{2^n(3(\ell-n)-1)+(\ell-2n)^2+2(n+1)-\ell} & \textit{if } 2n \leq \ell. \end{cases}$$

Proof. Note that

$$w_0^2 + \sum_{i=0}^{\ell-1} w_{2^i}^2 = 2^{-\ell}\left((v^n_{\ell,0})^2 + \sum_{i=0}^{\ell-1}(v^n_{\ell,2^i})^2\right)$$

and $\displaystyle\sum_{j=0}^{2^\ell-1} w_j^2 = \|\boldsymbol{T}^n_\ell\|^2$. □

6 Minimal epistasis and Walsh coefficients

Let us use notations as before. In view of the fact that

$$\varepsilon^*(f) = 1 - \frac{w_0^2 + \sum_{i=0}^{\ell-1} w_{2^i}^2}{\sum_{j=0}^{2^\ell-1} w_j^2},$$

it is clear that the minimal value $\varepsilon^*(f) = 0$ will be reached exactly by functions $f : \Omega \to \mathbb{R}$ with

$$w_0^2 + \sum_{i=0}^{\ell-1} w_{2^i}^2 = \sum_{j=0}^{2^\ell-1} w_j^2,$$

i.e., whose only non-zero Walsh coefficients are w_0 and w_{2^i} for $0 \leq i < \ell$. It appears that these are exactly the first order functions ((I.3), page 46), i.e., those $f : \Omega \to \mathbb{R}$ which have the property that there exist

$$g_i : \{0,1\} \to \mathbb{R}$$

such that

$$f(s) = \sum_{i=0}^{\ell-1} g_i(s_i)$$

for all $s = s_{\ell-1} \ldots s_0 \in \Omega$. To prove this, we will need some preparations first. Let us denote $g_i(a)$ by $g_{i,a}$. It is easy to see that the average of f is given by

$$v_0 = \frac{1}{2^\ell} \sum_{s \in \Omega} f(s) = \frac{1}{2^\ell} \sum_{s \in \Omega} \sum_{i=0}^{\ell-1} g_{i,s_i} = \frac{1}{2} \sum_{i=0}^{\ell-1} (g_{i,0} + g_{i,1}).$$

On the other hand, for any $0 \leq j < \ell$, we have

$$\begin{aligned} v_{2^j} &= f(\# \ldots \# \overset{\overset{j}{\downarrow}}{0} \# \ldots \#) - v_0 \\ &= \frac{1}{2} \sum_{\substack{0 \leq i < \ell \\ i \neq j}} (g_{i,0} + g_{i,1}) + g_{j,0} - \frac{1}{2} \sum_{0 \leq i < \ell} (g_{i,0} + g_{i,1}) \\ &= g_{j,0} - \frac{1}{2} (g_{j,0} + g_{j,1}) \\ &= \frac{1}{2} g_{j,0} - \frac{1}{2} g_{j,1}. \end{aligned}$$

As we pointed out before, v_j corresponds to the schema H, with 0 in the loci where a 1 occurs in the binary representation of j and a # in the other loci. We will denote this correspondence by $j \sim H$.

Example IV.20. If $\ell = 5$, then v_{13} corresponds to $H = \#00\#0$, as the binary representation of 13 is 01101. So, $13 \sim \#00\#0$.

In particular, the coefficients v_{2^i} correspond to the schemata $\#\ldots\#0\#\ldots\#$, with 0 in the i-th locus. We may now prove:

Lemma IV.21. *If f is a first order function, then $v_t = 0$ for all $t \neq 0, 2^j$, $0 \leq j < \ell$.*

Proof. Let $t \neq 0, 2^j$ and let H be the schema corresponding to t, i.e., $t \sim H$ with 0 at all defined loci. Let $k = o(H)$.
Let us first assume $k = 2$. If the non-# entries are situated at the loci j_1 and j_2, then

$$\begin{aligned} v_t =& f(H) - v_{2^{j_1}} - v_{2^{j_2}} - v_0 \\ =& \frac{1}{2} \sum_{\substack{0 \leq i \leq \ell-1 \\ i \neq j_1, j_2}} (g_{i,0} + g_{i,1}) + g_{j_1,0} + g_{j_2,0} \\ & - \frac{1}{2} g_{j_1,0} + \frac{1}{2} g_{j_1,1} - \frac{1}{2} g_{j_2,0} + \frac{1}{2} g_{j_2,1} - \frac{1}{2} \sum_{0 \leq i < \ell} (g_{i,0} + g_{i,1}) \\ =& 0. \end{aligned}$$

In order to apply induction, let us now assume the result to be correct for all H with $o(H) < k$. Then, denoting by $A(H)$ the set of non-# loci of H,

$$v_t = f(H) - v_0 - \sum_{j \in A(H)} v_{2^j},$$

as the other components vanish in view of the induction hypothesis. So,

$$\begin{aligned} v_t =& \frac{1}{2} \sum_{\substack{0 \leq i < \ell \\ i \notin A(H)}} (g_{i,0} + g_{i,1}) + \sum_{j \in A(H)} g_{j,0} \\ & - \frac{1}{2} \sum_{j \in A(H)} (g_{i,0} - g_{i,1}) - \frac{1}{2} \sum_{0 \leq i < \ell} (g_{i,0} + g_{i,1}) \\ =& 0. \end{aligned}$$

This proves the assertion. □

Since the Walsh coefficients w_t are multiples of the corresponding v_t, this already proves that every first order function f has the property that $\varepsilon^*(f) = 0$.

The converse is also true. Indeed, if $\varepsilon^*(f) = 0$ for some $f : \Omega \to \mathbb{R}$, then we have already seen that all Walsh coefficients w_t vanish for $t \neq 0, 2^j$. We then find

$$\begin{aligned} f(s) &= (\boldsymbol{W}_\ell \boldsymbol{w})_s = 2^{-\frac{\ell}{2}} (\boldsymbol{V}_\ell \boldsymbol{w})_s = 2^{-\frac{\ell}{2}} \sum_{r \in \Omega} (-1)^{r \cdot s} w_r \\ &= 2^{-\frac{\ell}{2}} w_0 + 2^{-\frac{\ell}{2}} \sum_{i=0}^{\ell-1} (-1)^{s_i} w_{2^i} \\ &= \sum_{i=0}^{\ell-1} g_i(s_i), \end{aligned}$$

where g_i is given as

$$g_i : \{0,1\} \to \mathbb{R} : a \mapsto 2^{-\frac{\ell}{2}} \left(\frac{1}{\ell} w_0 + (-1)^a w_{2^i} \right).$$

It follows that f is a first order function, indeed. We thus have proved:

Theorem IV.22. *For any $f : \Omega \to \mathbb{R}$, the following assertions are equivalent:*

1. $\varepsilon^*(f) = 0$;
2. *f is a first order function.*

In particular, it appears that proposition IV.18 is a straightforward consequence of this result.

Chapter V

Multary epistasis

This chapter extends the epistasis theory of the previous chapters to multary representations — fixed-length string encodings where the alphabet contains more than two symbols. Our motivation is twofold. From a mathematical point of view, the extension is natural and can be carried out without too many complications. GA practitioners, on the other hand, are used to multary encodings since the majority of real world search problems are best encoded in a non-binary way.

We briefly discuss the implications of abandoning $\Omega = \{0,1\}^\ell$ in the first section of this chapter. We extend the definition and formulation of normalized epistasis to multary strings in the second section, and compute its extreme values in the third section. We finish the chapter by writing out, as an example, the epistasis of (generalized) unitation functions.

1 Multary representations

Let us use the well known *traveling salesman problem* (TSP) as a first example of a search problem where a non-binary representation is much more natural than any binary one. The problem is defined as follows: given a fixed number of cities, conveniently labeled 1, 2, 3, ..., n and distances between each pair of cities, find the shortest tour allowing the traveling salesman to visit each city exactly once.

The obvious representation for this problem is the space of permutations of the set

$\{1, \ldots, n\}$, as each permutation accords to exactly one possible tour for the salesman. We write the permutation as a string of symbols, where each symbol occurs exactly once. It is not too difficult to think of a way of encoding a permutation as a binary string; suppose we choose the following one: we set $\ell = n\lceil \log n \rceil$[1], and use the first $\lceil \log n \rceil$ bits as an indication for the first city visited, the next $\lceil \log n \rceil$ bits for the second city, and so on. Note that this is a valid representation, as each possible tour (individual) can be represented as a binary string of length ℓ. It is also similar to the representation that will be implicitly used by the computer when we store or manipulate a permutation, the only difference is that the $\lceil \log n \rceil$ will probably be replaced by a multiple of 8.

The difference between the two encodings does not lie in the actual way of writing or storing the individuals, but in the way the genetic operators, mutation and crossover, manipulate them. A mutation in permutation space is a small modification to a permutation which yields another permutation (think of swapping the order of two adjacent cities, for example). A mutation in binary string space typically amounts to the flipping of one or a few bits; the result is, of course, another bit string. But in the case of our binary representation of the TSP problem, this new string may not be a valid permutation anymore. Two constraints may be violated. If n is not a power of 2, then $2^{\lceil \log n \rceil} > n$, and the mutation may have resulted in a string with a non-existing city at a given position. Or, when the result is a different city indeed, this city may occur at a different position already. It is easy to see that none of the classical crossover operators can guarantee that these two constraints are never violated.

A second example illustrating a different issue is that of encoding a real or integer valued domain in the space of bit strings. The interval $[0, 1]$, for example, can be represented by length ℓ binary strings if we accept a finite precision. In the standard way of encoding numbers to bit strings, a mutation of the most significant bit implies a much larger jump in $[0, 1]$ than a mutation of the least significant bit does. As these mutations typically occur with equal probability, the mutation operator is very different from the more natural Gaussian noise which would normally be applied to

[1] As before, we use the notation $\lceil x \rceil$ for the smallest integer greater than or equal to x.

make a small modification to a real number. The difference between the encodings is again thinking of the number as a real number, and not as a sequence of bits.

Let us give another example of this second "compatibility" issue between representation and operators. If a variable takes 4 values, is it best to use two bits to encode it, or should we better use a multary encoding with an alphabet of size 4 for this variable? In the former case, using the classical operators, $01 = 1$ and $10 = 3$ are one mutation away from $00 = 0$, while $11 = 4$ is further away from $00 = 0$. In the latter case, again with the traditional "change to another value, all choices being equal", this is obviously not the case. It depends on the application to decide which choice is best, but the two choices will result in different GA behavior.

There are, of course, many other issues that make a successful combination of representation and genetic operators, but they do not relate directly to the difference between binary and multary encodings. In practice, multary encodings often arise as a more natural choice than binary ones. The best known example of a multary encoding is of course the DNA encoding, which consists of (variable length) strings over the 4-symbol alphabet $\{A, C, G, T\}$. Three of these nucleotides encode for one amino acid. Because the number of different amino acids (20 frequently occurring, and a few rare ones) is smaller than $4^3 = 64$, there is room for redundancy: some amino acids are represented by more than one triplet. Searching in amino acid sequence space (i.e., protein space), however, is best done with a 20-valued alphabet, with a very specific mutation operator that, for example, takes the frequency of occurring of amino acids into account.

2 Epistasis in the multary case

In view of the previous considerations, it is necessary to reconsider some topics dealt with in the literature and in the previous chapters. In particular, in chapter II, we introduced the matrices $\boldsymbol{E}_\ell$ and $\boldsymbol{G}_\ell$ and applied them to calculate the normalized epistasis of a fitness function f on $\Omega = \{0,1\}^\ell$ as

$$\varepsilon^*(f) = 1 - \frac{{}^t\boldsymbol{f}\boldsymbol{E}_\ell\boldsymbol{f}}{{}^t\boldsymbol{f}\boldsymbol{f}}.$$

In order to study a similar notion in the multary case, throughout this chapter, we will fix positive integers n and ℓ, and we will work over a fixed alphabet Σ of cardinality n, which we will usually identify with the set of integers $\{0, 1, \ldots, n-1\}$. The set Σ^ℓ of length ℓ strings $s = s_{\ell-1} \ldots s_0$ over Σ will be denoted by Ω_n.

2.1 The epistasis value of a function

We start by introducing the notion of epistasis value $\varepsilon^2(f)$ for a fitness function f acting on strings over a not necessarily binary alphabet.

Following ideas used in the binary case, we will only be working with the full search space Ω_n, so $|\Omega_n| = n^\ell$. Arguing as in the binary case, the global epistasis of $s \in \Omega_n$ may be given by

$$\varepsilon(s) \equiv \varepsilon_{\Omega_n}(s) = f(s) - \frac{1}{n^{\ell-1}} \sum_{i=0}^{\ell-1} \sum_{t \in \Omega_n(i,s_i)} f(t) + \frac{\ell-1}{n^\ell} \sum_{t \in \Omega_n} f(t),$$

where $\Omega_n(i, s_i)$ consists of all strings in Ω_n which have value s_i at position i. The epistasis value of f is then defined to be

$$\varepsilon^2(f) \equiv \varepsilon^2_{\Omega_n}(f) = \sum_{s \in \Omega_n} \varepsilon^2(s).$$

2.2 Matrix representation

Mimicking the binary case, we will show that the previous definition may be rewritten more elegantly in matrix form. In order to realize this, we first introduce analogs of the matrices $\boldsymbol{E}_\ell$ and $\boldsymbol{G}_\ell$, defined for the binary case in section 3.1 of chapter II.

The matrix $G_{n,\ell}$ and its basic properties

For any $0 \leq i, j < n^\ell$, we will denote by d_{ij} (or $\mathrm{d}_{ij}^{n,\ell}$, if ambiguity may arise) the Hamming distance between the (length ℓ) n-ary representation of i and j. For example, since the ternary representation of 23 and 47 are 0212 and 1202, respectively, we have $\mathrm{d}_{23,47}^{3,4} = 2$.

We define the matrix $\boldsymbol{G}_{n,\ell} = (g_{ij}^{n,\ell})$ by letting $g_{ij}^{n,\ell} = (n-1)\ell + 1 - n\mathrm{d}_{ij}^{n,\ell}$ for every $0 \leq i, j < n^{\ell}$. If no ambiguity arises, we will use the same notation as in the binary case, i.e., we will just write $\boldsymbol{G}_{\ell}$ and g_{ij}^{ℓ} for $\boldsymbol{G}_{n,\ell}$ and $g_{ij}^{n,\ell}$, respectively. We will also use the matrix $\boldsymbol{E}_{n,\ell} = n^{-\ell}\boldsymbol{G}_{n,\ell}$, with rational entries.

Let us consider the vectors $\boldsymbol{\varepsilon} = {}^t(\varepsilon(0), \ldots, \varepsilon(n^{\ell}-1))$ and $\boldsymbol{f} = {}^t(f_0, \ldots, f_{n^{\ell}-1})$, where we write $\varepsilon(n^{\ell}-1)$ for $\varepsilon((n-1)\ldots(n-1))$ and $f_{n^{\ell}-1}$ for $f((n-1)\ldots(n-1))$. A similar reasoning to the one used in the binary representation easily shows that $\boldsymbol{\varepsilon} = \boldsymbol{f} - \boldsymbol{E}_{n,\ell}\boldsymbol{f}$. We thus obtain that the epistasis value of f is given by

$$\varepsilon^2(f) = ||\boldsymbol{\varepsilon}||^2 = ||\boldsymbol{f} - \boldsymbol{E}_{n,\ell}\boldsymbol{f}||^2.$$

The following result allows $\boldsymbol{G}_{\ell}$ to be calculated recursively:

Lemma V.1. *For any positive integer ℓ, we have:*

$$\boldsymbol{G}_{\ell} = \begin{pmatrix} \boldsymbol{G}_{\ell-1} + (n-1)\boldsymbol{U}_{\ell-1} & \boldsymbol{G}_{\ell-1} - \boldsymbol{U}_{\ell-1} & \ldots & \boldsymbol{G}_{\ell-1} - \boldsymbol{U}_{\ell-1} \\ \boldsymbol{G}_{\ell-1} - \boldsymbol{U}_{\ell-1} & \boldsymbol{G}_{\ell-1} + (n-1)\boldsymbol{U}_{\ell-1} \ldots & & \boldsymbol{G}_{\ell-1} - \boldsymbol{U}_{\ell-1} \\ \vdots & \vdots & \ddots & \vdots \\ \boldsymbol{G}_{\ell-1} - \boldsymbol{U}_{\ell-1} & \boldsymbol{G}_{\ell-1} - \boldsymbol{U}_{\ell-1} & \ldots & \boldsymbol{G}_{\ell-1} + (n-1)\boldsymbol{U}_{\ell-1} \end{pmatrix},$$

where, for any positive integer k, the n^k-dimensional matrix $\boldsymbol{U}_k$ is given by

$$\boldsymbol{U}_k = \begin{pmatrix} 1 \ldots 1 \\ \vdots \ddots \vdots \\ 1 \ldots 1 \end{pmatrix}.$$

Proof. The length ℓ words over the alphabet $\Sigma = \{0, 1, \ldots, n-1\}$, with cardinality n, may be subdivided into n subclasses, each of these determined by the value at position ℓ. This subdivision allows us to view the matrix $\boldsymbol{G}_{\ell}$ as composed of n^2 submatrices $\boldsymbol{G}_{pq}$, say

$$\boldsymbol{G}_{\ell} = \begin{pmatrix} \boldsymbol{G}_{0,0} & \ldots & \boldsymbol{G}_{0,n-1} \\ \vdots & \ddots & \vdots \\ \boldsymbol{G}_{n-1,0} & \ldots & \boldsymbol{G}_{n-1,n-1} \end{pmatrix},$$

with $\boldsymbol{G}_{pq} = (g_{ij}^{n,\ell})$, where i and j vary through the elements in $\Omega_n = \Sigma^{\ell}$ with values p and q at bit-position ℓ, respectively.

For any pair of strings i and j of length ℓ, we will denote by $\mathrm{d}_{ij}^{n,\ell-1}$ (or just $\mathrm{d}_{ij}^{\ell-1}$ if no ambiguity arises) the Hamming distance between the (length $\ell-1$) strings obtained from the previous ones by eliminating the ℓ-th position. For example, since the ternary representation of 23 and 47 are 0212 and 1202, respectively, we have $d_{23,47}^{3,4} = 2$, while $d_{23,47}^{3,3} = 1$.

For every $0 \leq p < n$, we have $\boldsymbol{G}_{pp} = \boldsymbol{G}_{\ell-1} + (n-1)\boldsymbol{U}_{\ell-1}$. Indeed,

$$\begin{aligned}\boldsymbol{G}_{pp} &= (g_{ij}^{\ell}) = ((n-1)\ell + 1 - n\mathrm{d}_{ij}^{\ell})\\ &= ((n-1)(\ell-1) + 1 - n\mathrm{d}_{ij}^{\ell} + (n-1))\\ &= ((n-1)(\ell-1) + 1 - n\mathrm{d}_{ij}^{\ell-1} + (n-1))\\ &= \boldsymbol{G}_{\ell-1} + (n-1)\boldsymbol{U}_{\ell-1},\end{aligned}$$

since, in this case, we always have $\mathrm{d}_{ij}^{\ell} = \mathrm{d}_{ij}^{\ell-1}$.

Outside of the diagonal, i.e., with $0 \leq p \neq q < n$, we have

$$\begin{aligned}\boldsymbol{G}_{p,q} &= (g_{ij}^{\ell}) = ((n-1)\ell + 1 - n\mathrm{d}_{ij}^{\ell})\\ &= ((n-1)(\ell-1) + 1 - n\mathrm{d}_{ij}^{\ell} + (n-1))\\ &= ((n-1)(\ell-1) + 1 - n(\mathrm{d}_{ij}^{\ell-1} + 1) + (n-1))\\ &= \boldsymbol{G}_{\ell-1} - \boldsymbol{U}_{\ell-1},\end{aligned}$$

since, in this case, we always have $\mathrm{d}_{ij}^{\ell} = \mathrm{d}_{ij}^{\ell-1} + 1$. This finishes the proof. □

Using this result, it is easy to check that the recursion relation

$$\boldsymbol{G}_{\ell} = \boldsymbol{U}_{\ell-m} \otimes \boldsymbol{G}_{m} + (\boldsymbol{G}_{\ell-m} - \boldsymbol{U}_{\ell-m}) \otimes \boldsymbol{U}_{m},$$

valid for any pair of positive integers $m \leq n$ by lemma II.3, remains valid in the multary case.

As a consequence, let us mention:

Corollary V.2. *For any positive integer ℓ, we have $\boldsymbol{G}_{\ell}^{2} = n^{\ell}\boldsymbol{G}_{\ell}$.*

Proof. The statement obviously holds true for $\ell = 0$ and $\ell = 1$, where $\boldsymbol{G}_0 = (1)$ and $\boldsymbol{G}_1 = n\boldsymbol{I}_n$, respectively ($\boldsymbol{I}_n$ denoting the n-dimensional identity matrix). Using the previous result, the general case follows from a straightforward induction argument. □

Eigenvalues and eigenspaces

It is easy to see that the last result implies that the eigenvalues of $\boldsymbol{G}_\ell$ are 0 and n^ℓ. Indeed, as $\boldsymbol{G}_\ell$ is a symmetric real matrix, its eigenvalues are real (proposition B.38). On the other hand, if $\boldsymbol{v}$ is an eigenvector of $\boldsymbol{G}_\ell$ with eigenvalue λ, i.e., $\boldsymbol{G}_\ell \boldsymbol{v} = \lambda \boldsymbol{v}$, then

$$n^\ell \lambda \boldsymbol{v} = n^\ell \boldsymbol{G}_\ell \boldsymbol{v} = \boldsymbol{G}_\ell^2 \boldsymbol{v} = \lambda^2 \boldsymbol{v}.$$

So, $\lambda = 0$ or $\lambda = n^\ell$, as we claimed.

From the previous corollary and the identity $\boldsymbol{G}_\ell = n^\ell \boldsymbol{E}_\ell$, it clearly follows:

Corollary V.3. *For any positive integer ℓ, the matrix $\boldsymbol{E}_\ell$ is idempotent. In particular, $\boldsymbol{E}_\ell$ has eigenvalues 0 and 1.*

Just as in the binary case, we now define the normalized epistasis $\varepsilon^*(f) \equiv \varepsilon^*_{n,\ell}(f)$ of a given fitness function f over Ω_n as

$$\varepsilon^*(f) = \varepsilon^2\left(\frac{f}{\|\boldsymbol{f}\|}\right) = \frac{\varepsilon^2(f)}{\|\boldsymbol{f}\|^2} = \frac{{}^t\boldsymbol{f}\,(\boldsymbol{I}_\ell - \boldsymbol{E}_{n,\ell})\,\boldsymbol{f}}{{}^t\boldsymbol{f}\boldsymbol{f}} = \cos^2\left(\boldsymbol{f}, \boldsymbol{F}_{n,\ell}\boldsymbol{f}\right),$$

where $\boldsymbol{F}_{n,\ell} = \boldsymbol{I}_\ell - \boldsymbol{E}_{n,\ell}$ is an orthogonal projection (being idempotent and symmetric). In particular, it follows that $0 \leq \varepsilon^*(f) \leq 1$ for any function f.

In order to characterize the functions with maximal and minimal normalized epistasis, we will determine the eigenspaces of $\boldsymbol{G}_\ell$ (and $\boldsymbol{E}_\ell$). As a first step, let us calculate the rank of $\boldsymbol{G}_\ell$.

Lemma V.4. *For any positive integer ℓ, we have*

$$\operatorname{rk}(\boldsymbol{G}_\ell) = (n-1)\ell + 1.$$

Proof. Let us argue by induction on ℓ. The assertion holds true for $\ell = 1$. On the other hand, applying lemma V.1, elementary row and column operations reduce $\boldsymbol{G}_\ell$ to the form

$$\begin{pmatrix} n\boldsymbol{G}_{\ell-1} & 0 & \dots & 0 \\ 0 & \boldsymbol{U}_{\ell-1} & \dots & 0 \\ \vdots & \vdots & \ddots & \vdots \\ 0 & 0 & \dots & \boldsymbol{U}_{\ell-1} \end{pmatrix}.$$

This yields that

$$\begin{aligned}\operatorname{rk}(\boldsymbol{G}_\ell) &= \operatorname{rk}(\boldsymbol{G}_{\ell-1}) + (n-1)\operatorname{rk}(\boldsymbol{U}_{\ell-1})\\ &= ((n-1)(\ell-1)+1) + (n-1) = (n-1)\ell + 1,\end{aligned}$$

which proves the assertion. □

With the same notations as chapter II, let us now denote by V_0^ℓ and V_1^ℓ the eigenspaces in $\mathbb{R}^{n^\ell}$ corresponding to the eigenvalues 0 and n^ℓ of $\boldsymbol{G}_\ell$, respectively (or, equivalently, the eigenvalues 0 and 1 of $\boldsymbol{E}_\ell$). Then $\mathbb{R}^{n^\ell} = V_0^\ell \oplus V_1^\ell$ and, as $\boldsymbol{E}_\ell$ is idempotent, $\mathbb{R}^{n^\ell} = \operatorname{Ker}(\boldsymbol{E}_\ell) \oplus \operatorname{Im}(\boldsymbol{E}_\ell)$ (proposition B.25) with $\operatorname{Ker}(\boldsymbol{E}_\ell) = V_0^\ell$. On the other hand, $\operatorname{Im}(\boldsymbol{E}_\ell) \subseteq V_1^\ell$. Indeed, if $\boldsymbol{x} \in \operatorname{Im}(\boldsymbol{E}_\ell)$, then there exists some $\boldsymbol{y} \in \mathbb{R}^{n^\ell}$ with $\boldsymbol{x} = \boldsymbol{E}_\ell \boldsymbol{y}$ and so, as

$$\boldsymbol{E}_\ell \boldsymbol{x} = \boldsymbol{E}_\ell^2 \boldsymbol{y} = \boldsymbol{E}_\ell \boldsymbol{y} = \boldsymbol{x},$$

obviously, $\boldsymbol{x} \in V_1^\ell$.
Finally, as $\dim(\operatorname{Im}(\boldsymbol{E}_\ell)) = \dim(V_1^\ell)$, we obtain that $\operatorname{Im}(\boldsymbol{E}_\ell) = V_1^\ell$. Now, the previous result yields that

$$\dim(V_0^\ell) = n^\ell - (n-1)\ell - 1$$

and

$$\dim(V_1^\ell) = (n-1)\ell + 1.$$

An explicit orthogonal basis for V_1^ℓ may be constructed as follows. Start from $\boldsymbol{v}_0^0 = 1$ and suppose we already constructed a subset

$$\{\boldsymbol{v}_0^{\ell-1}, \ldots, \boldsymbol{v}_{(n-1)(\ell-1)}^{\ell-1}\} \subseteq \mathbb{R}^{n^{\ell-1}}.$$

We construct a new subset

$$\{\boldsymbol{v}_0^\ell, \ldots, \boldsymbol{v}_{(n-1)\ell}^\ell\} \subseteq \mathbb{R}^{n^\ell},$$

where

$$\boldsymbol{v}_k^\ell = \begin{pmatrix} \boldsymbol{v}_k^{\ell-1} \\ \vdots \\ \boldsymbol{v}_k^{\ell-1} \end{pmatrix}$$

for all $0 \leq k \leq (n-1)(\ell-1)$ and where $\boldsymbol{v}^{\ell}_{(n-1)(\ell-1)+1}, \boldsymbol{v}^{\ell}_{(n-1)(\ell-1)+2}, \ldots, \boldsymbol{v}^{\ell}_{(n-1)\ell}$ are given by

$$\begin{pmatrix} \boldsymbol{u}_{\ell-1} \\ -\boldsymbol{u}_{\ell-1} \\ \boldsymbol{0}_{\ell-1} \\ \boldsymbol{0}_{\ell-1} \\ \vdots \\ \boldsymbol{0}_{\ell-1} \end{pmatrix}, \begin{pmatrix} \boldsymbol{u}_{\ell-1} \\ \boldsymbol{u}_{\ell-1} \\ -2\boldsymbol{u}_{\ell-1} \\ \boldsymbol{0}_{\ell-1} \\ \vdots \\ \boldsymbol{0}_{\ell-1} \end{pmatrix}, \ldots, \begin{pmatrix} \boldsymbol{u}_{\ell-1} \\ \boldsymbol{u}_{\ell-1} \\ \vdots \\ \boldsymbol{u}_{\ell-1} \\ \boldsymbol{u}_{\ell-1} \\ -(n-1)\boldsymbol{u}_{\ell-1} \end{pmatrix},$$

with, as before,

$$\boldsymbol{u}_{\ell-1} = \begin{pmatrix} 1 \\ \vdots \\ 1 \end{pmatrix}, \qquad \boldsymbol{0}_{\ell-1} = \begin{pmatrix} 0 \\ \vdots \\ 0 \end{pmatrix}$$

within $\mathbb{R}^{n^{\ell-1}}$.

As an example, if $n = 3$ and $\ell = 1$, then

$$\boldsymbol{v}^1_0 = \begin{pmatrix} 1 \\ 1 \\ 1 \end{pmatrix}, \boldsymbol{v}^1_1 = \begin{pmatrix} 1 \\ -1 \\ 0 \end{pmatrix}, \boldsymbol{v}^1_2 = \begin{pmatrix} 1 \\ 1 \\ -2 \end{pmatrix},$$

so, for $n = 3$ and $\ell = 2$, we obtain

$$\boldsymbol{v}^2_0 = \begin{pmatrix} 1 \\ 1 \\ 1 \\ 1 \\ 1 \\ 1 \\ 1 \\ 1 \\ 1 \end{pmatrix}, \boldsymbol{v}^2_1 = \begin{pmatrix} 1 \\ -1 \\ 0 \\ 1 \\ -1 \\ 0 \\ 1 \\ -1 \\ 0 \end{pmatrix}, \boldsymbol{v}^2_2 = \begin{pmatrix} 1 \\ 1 \\ -2 \\ 1 \\ 1 \\ -2 \\ 1 \\ 1 \\ -2 \end{pmatrix}, \boldsymbol{v}^2_3 = \begin{pmatrix} 1 \\ 1 \\ 1 \\ -1 \\ -1 \\ -1 \\ 0 \\ 0 \\ 0 \end{pmatrix}, \boldsymbol{v}^2_4 = \begin{pmatrix} 1 \\ 1 \\ 1 \\ 1 \\ 1 \\ 1 \\ -2 \\ -2 \\ -2 \end{pmatrix}.$$

We may now prove:

Proposition V.5. *With the previous notations, for every positive integer ℓ, the set*

$$\{\boldsymbol{v}_0^\ell, \ldots, \boldsymbol{v}_{(n-1)\ell}^\ell\}$$

is an orthogonal basis for V_1^ℓ.

Proof. For $\ell = 0$, the statement is obvious. Suppose the assertion holds true for strings of length $0, \ldots, \ell-1$ and let us prove it for strings of length ℓ. In this case, if $0 \leq k \neq k' \leq (n-1)(\ell-1)$, then the induction hypothesis implies that ${}^t\boldsymbol{v}_k^\ell \boldsymbol{v}_{k'}^\ell = 0$. On the other hand, if

$$0 \leq k \leq (n-1)(\ell-1) < k' \leq (n-1)\ell,$$

then

$${}^t\boldsymbol{v}_k^\ell \boldsymbol{v}_{k'}^\ell = ({}^t\boldsymbol{v}_k^{\ell-1}, \ldots, {}^t\boldsymbol{v}_k^{\ell-1}) \begin{pmatrix} \boldsymbol{u}_{\ell-1} \\ \vdots \\ \boldsymbol{u}_{\ell-1} \\ -i\boldsymbol{u}_{\ell-1} \\ \boldsymbol{0}_{\ell-1} \\ \vdots \\ \boldsymbol{0}_{\ell-1} \end{pmatrix} = i\,{}^t\boldsymbol{v}_k^{\ell-1}\boldsymbol{u}_{\ell-1} - i\,{}^t\boldsymbol{v}_k^{\ell-1}\boldsymbol{u}_{\ell-1} = 0,$$

with $i = k' - (n-1)(\ell-1)$. Finally, if $(n-1)(\ell-1)+1 \leq k \neq k' \leq (n-1)\ell$,

$${}^t\boldsymbol{v}_k^\ell \boldsymbol{v}_{k'}^\ell = ({}^t\boldsymbol{u}_{\ell-1}, \ldots, {}^t\boldsymbol{u}_{\ell-1}, -i\,{}^t\boldsymbol{u}_{\ell-1}, {}^t\boldsymbol{0}_{\ell-1}, \ldots, {}^t\boldsymbol{0}_{\ell-1}) \begin{pmatrix} \boldsymbol{u}_{\ell-1} \\ \vdots \\ \boldsymbol{u}_{\ell-1} \\ -j\boldsymbol{u}_{\ell-1} \\ \boldsymbol{0}_{\ell-1} \\ \vdots \\ \boldsymbol{0}_{\ell-1} \end{pmatrix} = 0,$$

with $i = k - (n-1)(\ell-1)$ and $j = k' - (n-1)(\ell-1)$.

Since the vectors $\boldsymbol{v}_0^\ell, \ldots, \boldsymbol{v}_{(n-1)\ell}^\ell$ are obviously linearly independent, it thus suffices to verify that they belong to V_1^ℓ, as we have seen that $\dim(V_1^\ell) = (n-1)\ell + 1$.

Let us again argue by induction on ℓ. For $\ell = 0$, the statement is obvious, so let us assume it to hold true for length $0, \ldots, \ell - 1$ and prove it for length ℓ. First, if $0 \le k \le (n-1)(\ell - 1)$, then

$$
\begin{aligned}
\boldsymbol{G}_\ell \boldsymbol{v}_k^\ell &= \begin{pmatrix} \boldsymbol{G}_{\ell-1} + (n-1)\boldsymbol{U}_{\ell-1} \ldots & \boldsymbol{G}_{\ell-1} - \boldsymbol{U}_{\ell-1} \\ \vdots \quad\quad \ddots & \vdots \\ \boldsymbol{G}_{\ell-1} - \boldsymbol{U}_{\ell-1} & \ldots \boldsymbol{G}_{\ell-1} + (n-1)\boldsymbol{U}_{\ell-1} \end{pmatrix} \begin{pmatrix} \boldsymbol{v}_k^{\ell-1} \\ \vdots \\ \boldsymbol{v}_k^{\ell-1} \end{pmatrix} \\
&= \begin{pmatrix} n\boldsymbol{G}_{\ell-1}\boldsymbol{v}_k^{\ell-1} \\ \vdots \\ n\boldsymbol{G}_{\ell-1}\boldsymbol{v}_k^{\ell-1} \end{pmatrix} \\
&= n.n^{\ell-1} \begin{pmatrix} \boldsymbol{v}_k^{\ell-1} \\ \vdots \\ \boldsymbol{v}_k^{\ell-1} \end{pmatrix} \\
&= n^\ell \boldsymbol{v}_k^\ell.
\end{aligned}
$$

On the other hand, if $k = (n-1)(\ell - 1) + i$ with $1 \le i \le n - 1$, then

$$
\boldsymbol{G}_\ell \boldsymbol{v}_k^\ell = \begin{pmatrix} \boldsymbol{G}_{\ell-1} + (n-1)\boldsymbol{U}_{\ell-1} \ldots & \boldsymbol{G}_{\ell-1} - \boldsymbol{U}_{\ell-1} \\ \vdots \quad\quad \ddots & \vdots \\ \boldsymbol{G}_{\ell-1} - \boldsymbol{U}_{\ell-1} & \ldots \boldsymbol{G}_{\ell-1} + (n-1)\boldsymbol{U}_{\ell-1} \end{pmatrix} \begin{pmatrix} \boldsymbol{u}_{\ell-1} \\ \vdots \\ \boldsymbol{u}_{\ell-1} \\ -i\boldsymbol{u}_{\ell-1} \\ \boldsymbol{0}_{\ell-1} \\ \vdots \\ \boldsymbol{0}_{\ell-1} \end{pmatrix}
$$

$$
= n \begin{pmatrix} \boldsymbol{U}_{\ell-1}\boldsymbol{u}_{\ell-1} \\ \vdots \\ \boldsymbol{U}_{\ell-1}\boldsymbol{u}_{\ell-1} \\ -i\boldsymbol{U}_{\ell-1}\boldsymbol{u}_{\ell-1} \\ \boldsymbol{0}_{\ell-1} \\ \vdots \\ \boldsymbol{0}_{\ell-1} \end{pmatrix} = n.n^{\ell-1} \begin{pmatrix} \boldsymbol{u}_{\ell-1} \\ \vdots \\ \boldsymbol{u}_{\ell-1} \\ -i\boldsymbol{u}_{\ell-1} \\ \boldsymbol{0}_{\ell-1} \\ \vdots \\ \boldsymbol{0}_{\ell-1} \end{pmatrix} = n^\ell \boldsymbol{v}_k^\ell.
$$

This finishes the proof. □

We already pointed out that $0 \leq \varepsilon^*_{n,\ell}(f) \leq 1$ for any fitness function f. From the previous remarks, it is now clear that $\varepsilon^*(f) = 0$ and $\varepsilon^*(f) = 1$ exactly when $\boldsymbol{f} \in V_1^\ell$ and $\boldsymbol{f} \in V_0^\ell$, respectively. As an example, if $n = 3$ and $\ell = 2$, it thus follows that $\boldsymbol{f} \in V_1^\ell$ if and only if it belongs to the vector space generated by the vectors

$$\begin{pmatrix}1\\1\\1\\1\\1\\1\\1\\1\\1\end{pmatrix}, \begin{pmatrix}1\\-1\\0\\1\\-1\\0\\1\\-1\\0\end{pmatrix}, \begin{pmatrix}1\\1\\-2\\1\\1\\-2\\1\\1\\-2\end{pmatrix}, \begin{pmatrix}1\\1\\1\\-1\\-1\\-1\\0\\0\\0\end{pmatrix}, \begin{pmatrix}1\\1\\1\\1\\1\\1\\-2\\-2\\-2\end{pmatrix}.$$

This is easily seen to be equivalent to

$$f_{01} + f_{02} + f_{10} + f_{12} + f_{20} + f_{21} = 2(f_{00} + f_{11} + f_{22}).$$

2.3 Comparing epistasis

In this section we present two simple fitness functions and encode them on both binary and multary strings. We show that the normalized epistasis of the multary version is lower than that of the binary one.

The function that we consider first is given by

$$f(s) = \begin{cases} 1 - \frac{1}{2^\ell} & u(s) = 0 \\ 1 - \frac{1+u(s)}{\ell} & 0 < u(s) < \ell \\ 1 & u(s) = \ell, \end{cases}$$

where $s \in \Omega_2 = \{0,1\}^\ell$ and $u(s)$ denotes, as usual, the number of bits in s with value 1. Fixing the length $\ell = 4$, it easily follows that the associated vector $\boldsymbol{f}$ is

$${}^t\boldsymbol{f} = \left(\frac{15}{16}, \frac{1}{2}, \frac{1}{2}, \frac{1}{4}, \frac{1}{2}, \frac{1}{4}, \frac{1}{4}, 0, \frac{1}{2}, \frac{1}{4}, \frac{1}{4}, 0, \frac{1}{4}, 0, 0, 1\right).$$

As f is a unitation function on Ω_2, it is completely determined by the vector

$$h = \begin{pmatrix} \frac{15}{16} \\ \frac{1}{2} \\ \frac{1}{4} \\ 0 \\ 1 \end{pmatrix}.$$

The associated matrix $\boldsymbol{B}_4$ (see chapter III, section 2.4) is

$$\boldsymbol{B}_4 = \begin{pmatrix} 5 & 12 & 6 & -4 & -3 \\ 12 & 32 & 24 & 0 & -4 \\ 6 & 24 & 36 & 24 & 6 \\ -4 & 0 & 24 & 32 & 12 \\ -3 & -4 & 6 & 12 & 5 \end{pmatrix}.$$

Hence, the normalized epistasis of f is

$$\varepsilon^*_{2,4}(f) = 1 - \frac{{}^t\boldsymbol{f}\boldsymbol{B}_4\boldsymbol{f}}{2^4 \|\boldsymbol{f}\|^2} = 0.483.$$

Since the cardinality of the search space is 16, we can also encode this function using a quaternary alphabet, i.e., it can be defined on $\Omega_4 = \{0, 1, 2, 3\}^2$. The normalized epistasis is then given by

$$\varepsilon^*_{4,2}(f) = 1 - \frac{{}^t\boldsymbol{f}\boldsymbol{G}_{4,2}\boldsymbol{f}}{4^2 \|\boldsymbol{f}\|^2} = 0.3.$$

Note that the normalized epistasis has a smaller value in the multary case.

As a second example we consider the schemata $H_1 = 11\#\#$ and $H_2 = \#\#11$ of $\Omega_2 = \{0, 1\}^4$, and define the function

$$f(s) = |\{i; s \in H_i\}|.$$

Its normalized epistasis is

$$\varepsilon^*_{2,4}(f) = 1 - \frac{{}^t\boldsymbol{f}\boldsymbol{G}_{2,4}\boldsymbol{f}}{2^4 \|\boldsymbol{f}\|^2} = 0.2.$$

Note that $\boldsymbol{G}_{2,4}$ is the matrix given by the recursion relation:

$$\boldsymbol{G}_{2,4} = \begin{pmatrix} \boldsymbol{G}_{2,3} + \boldsymbol{U}_{2,3} & \boldsymbol{G}_{2,3} - \boldsymbol{U}_{2,3} \\ \boldsymbol{G}_{2,3} - \boldsymbol{U}_{2,3} & \boldsymbol{G}_{2,3} + \boldsymbol{U}_{2,3} \end{pmatrix},$$

where

$$\boldsymbol{G}_{2,3} = \begin{pmatrix} 4 & 2 & 2 & 0 & 2 & 0 & 0 & -2 \\ 2 & 4 & 0 & 2 & 0 & 2 & -2 & 0 \\ 2 & 0 & 4 & 2 & 0 & -2 & 2 & 0 \\ 0 & 2 & 2 & 4 & -2 & 0 & 0 & 2 \\ 2 & 0 & 0 & -2 & 4 & 2 & 2 & 0 \\ 0 & 2 & -2 & 0 & 2 & 4 & 0 & 2 \\ 0 & -2 & 2 & 0 & 2 & 0 & 4 & 2 \\ -2 & 0 & 0 & 2 & 0 & 2 & 2 & 4 \end{pmatrix}.$$

If we use a quaternary representation, i.e., $\Omega_4 = \{0,1,2,3\}^2$, and define f in terms of $H_1' = 3\#$ and $H_2' = \#3$, the normalized epistasis becomes

$$\varepsilon_{4,2}^*(f) = 1 - \frac{{}^t\boldsymbol{f}\boldsymbol{G}_{4,2}\boldsymbol{f}}{4^2 \|\boldsymbol{f}\|^2} = 1 - \frac{16\|\boldsymbol{f}\|^2}{4^2\|\boldsymbol{f}\|^2} = 0,$$

where, the recursion relation for $\boldsymbol{G}_{4,2}$ is now

$$\boldsymbol{G}_{4,2} = \begin{pmatrix} \boldsymbol{G}_{4,1} + 3\boldsymbol{U}_{4,1} & \boldsymbol{G}_{4,1} - \boldsymbol{U}_{4,1} & \boldsymbol{G}_{4,1} - \boldsymbol{U}_{4,1} & \boldsymbol{G}_{4,1} - \boldsymbol{U}_{4,1} \\ \boldsymbol{G}_{4,1} - \boldsymbol{U}_{4,1} & \boldsymbol{G}_{4,1} + 3\boldsymbol{U}_{4,1} & \boldsymbol{G}_{4,1} - \boldsymbol{U}_{4,1} & \boldsymbol{G}_{4,1} - \boldsymbol{U}_{4,1} \\ \boldsymbol{G}_{4,1} - \boldsymbol{U}_{4,1} & \boldsymbol{G}_{4,1} - \boldsymbol{U}_{4,1} & \boldsymbol{G}_{4,1} + 3\boldsymbol{U}_{4,1} & \boldsymbol{G}_{4,1} - \boldsymbol{U}_{4,1} \\ \boldsymbol{G}_{4,1} - \boldsymbol{U}_{4,1} & \boldsymbol{G}_{4,1} - \boldsymbol{U}_{4,1} & \boldsymbol{G}_{4,1} - \boldsymbol{U}_{4,1} & \boldsymbol{G}_{4,1} + 3\boldsymbol{U}_{4,1} \end{pmatrix}$$

with $\boldsymbol{G}_{4,1} = 4\boldsymbol{I}_4$.

3 Extreme values

In this section we take a closer look at the extreme values of the normalized epistasis of a fitness functions on multary encodings. We observe, as before, that the minimal and maximal values of $\varepsilon^*(f) \equiv \varepsilon_{n,\ell}^*(f)$ correspond to the maximal and minimal values of

$$\gamma(f) \equiv \gamma_{n,\ell}(f) = {}^t\boldsymbol{f}\boldsymbol{G}_{n,\ell}\boldsymbol{f}, \qquad (\|\boldsymbol{f}\| = 1)$$

respectively. In particular, $0 \leq \gamma(f) \leq n^\ell$.

3.1 Minimal epistasis

Let us first point out that the theoretical minimal value $\varepsilon^*(f) = 0$ (or, equivalently, the maximal value $\gamma(f) = n^\ell$) may actually be reached. Indeed, if $\ell = 1$, then $\dim(V_1^1) = n$, so $V_1^1 = \mathbb{R}^n$, and any $f : \{0, \dots, n-1\} \to \mathbb{R}$ (with $||\boldsymbol{f}|| = 1$, $\boldsymbol{f} \in \mathbb{R}^n$) satisfies

$$\begin{aligned}
\gamma(f) &= {}^t\boldsymbol{f}\boldsymbol{G}_{n,1}\boldsymbol{f} = (f_0 \dots f_{n-1}) \begin{pmatrix} n \dots 0 \\ \vdots \ddots \vdots \\ 0 \dots n \end{pmatrix} \begin{pmatrix} f_0 \\ \vdots \\ f_{n-1} \end{pmatrix} \\
&= n \sum_{i=0}^{n-1} f_i^2 = n.
\end{aligned}$$

In the general case, i.e., when $\ell > 1$, we will need the following result:

Lemma V.6. *For any positive integer ℓ, we have*

$$\sum_{0 \le i,j < n^\ell} g_{ij}^{n,\ell} = n^{2\ell}.$$

Proof. Let us apply induction on ℓ. For $\ell = 1$, we have $\boldsymbol{G}_{n,1} = n\boldsymbol{I}_n$, so the result is obviously correct. Assume it holds true for length $1, \dots, \ell - 1$ and let us prove it for length ℓ. It suffices to apply lemma V.1, which easily yields that

$$\sum_{0 \le i,j < n^\ell} g_{ij}^{\ell,n} = n^2 \sum_{0 \le i,j < n^\ell} g_{ij}^{\ell-1,n} = n^2 n^{2(\ell-1)} = n^{2\ell}.$$

This proves the assertion. □

Consider the vector

$$\boldsymbol{f}' = \begin{pmatrix} \boldsymbol{u}_{\ell-1} \\ \boldsymbol{0}_{\ell-1} \\ \vdots \\ \boldsymbol{0}_{\ell-1} \\ \boldsymbol{u}_{\ell-1} \end{pmatrix}$$

and put $\boldsymbol{f} = \boldsymbol{f}'/||\boldsymbol{f}'||$. Then we claim that $\varepsilon^*(f) = 0$, which proves that minimal normalized epistasis may always be realized. Indeed,

$$\begin{aligned}
\gamma(f) &= {}^t\boldsymbol{f}\boldsymbol{G}_\ell\boldsymbol{f} = \frac{1}{||\boldsymbol{f}'||^2}{}^t\boldsymbol{f}'\mathbf{G}_\ell\boldsymbol{f}' \\
&= \frac{1}{2n^{\ell-1}}{}^t\boldsymbol{f}'\begin{pmatrix} \boldsymbol{G}_{\ell-1}+(n-1)\boldsymbol{U}_{\ell-1}\ldots & \boldsymbol{G}_{\ell-1}-\boldsymbol{U}_{\ell-1} \\ \vdots \qquad \ddots & \vdots \\ \boldsymbol{G}_{\ell-1}-\boldsymbol{U}_{\ell-1} & \ldots\ \boldsymbol{G}_{\ell-1}+(n-1)\boldsymbol{U}_{\ell-1} \end{pmatrix}\boldsymbol{f}' \\
&= 2\frac{1}{2n^{\ell-1}}{}^t\boldsymbol{u}_{\ell-1}(2\boldsymbol{G}_{\ell-1}+(n-2)\boldsymbol{U}_{\ell-1})\boldsymbol{u}_{\ell-1} \\
&= \frac{1}{n^{\ell-1}}(2\sum_{0\le i,j<n^\ell} g_{ij}^{\ell-1}+(n-2)\sum_{0\le i,j<n^\ell} u_{ij}^{\ell-1}) \\
&= \frac{1}{n^{\ell-1}}(2n^{2(\ell-1)}+(n-2)n^{2(\ell-1)}) = n^\ell.
\end{aligned}$$

In chapter II we saw that for any fitness function f defined on binary strings, we have $\varepsilon^*(f) = 0$ if and only if f is a first order function, i.e., f is of the form $f(s) = \sum_{i=0}^{\ell-1} g_i(s_i)$, for some functions g_i which only depend upon the i-th bit s_i of $s = s_{\ell-1}\ldots s_0$. In order to extend this result to multary encodings, let us define for any $0 \le i < \ell$ and $0 \le j < n$ the map

$$h_{i,j}^{n,\ell} : \Omega_n \longrightarrow \mathbb{R} : s = s_{\ell-1}\ldots s_0 \mapsto \begin{cases} 1 & \text{if } s_i = j \\ 0 & \text{if } s_i \ne j. \end{cases}$$

Denote by $\boldsymbol{h}_{i,j}^{n,\ell}$ (or just $\boldsymbol{h}_{i,j}^{\ell}$) the corresponding vector in $\mathbb{R}^{n^\ell}$. Clearly, for any $0 \le i < \ell - 1$, we have

$$\boldsymbol{h}_{i,j}^{\ell} = \begin{pmatrix} \boldsymbol{h}_{i,j}^{\ell-1} \\ \vdots \\ \boldsymbol{h}_{i,j}^{\ell-1} \end{pmatrix},$$

with

$$\boldsymbol{h}_{\ell-1,1}^{\ell} = \begin{pmatrix} \mathbf{0}_{\ell-1} \\ \boldsymbol{u}_{\ell-1} \\ \mathbf{0}_{\ell-1} \\ \mathbf{0}_{\ell-1} \\ \vdots \\ \mathbf{0}_{\ell-1} \end{pmatrix}, \boldsymbol{h}_{\ell-1,2}^{\ell} = \begin{pmatrix} \mathbf{0}_{\ell-1} \\ \mathbf{0}_{\ell-1} \\ \boldsymbol{u}_{\ell-1} \\ \mathbf{0}_{\ell-1} \\ \vdots \\ \mathbf{0}_{\ell-1} \end{pmatrix}, \ldots, \boldsymbol{h}_{\ell-1,n-1}^{\ell} = \begin{pmatrix} \mathbf{0}_{\ell-1} \\ \mathbf{0}_{\ell-1} \\ \vdots \\ \mathbf{0}_{\ell-1} \\ \mathbf{0}_{\ell-1} \\ \boldsymbol{u}_{\ell-1} \end{pmatrix} \in \mathbb{R}^{n^\ell}.$$

Lemma V.7. *The set*

$$\{\boldsymbol{u}_\ell, \boldsymbol{h}^\ell_{i,j};\ 0 \le i < \ell,\ 1 \le j < n\}$$

is linearly independent.

Proof. Suppose that

$$\sum_{i=0}^{\ell-1} \alpha_{i1} \boldsymbol{h}^\ell_{i,1} + \cdots + \sum_{i=0}^{\ell-1} \alpha_{i,n-1} \boldsymbol{h}^\ell_{i,n-1} + \beta \boldsymbol{u}_\ell = \boldsymbol{0}_\ell,$$

and denote by g the corresponding real-valued function

$$\sum_{i=1}^{\ell-1} \alpha_{i1} h^\ell_{i,1} + \cdots + \sum_{i=1}^{\ell-1} \alpha_{i,n-1} h^\ell_{i,n-1} + \beta u_\ell$$

on Ω_n, where u_ℓ denotes the constant function with value equal to 1. We then clearly have $\beta = g(0) = 0$. Moreover, for every $1 \le j < n-1$ and $0 \le i < \ell$, we have

$$0 = g(jn^i) = \sum_{k=0}^{\ell-1} \alpha_{kj} h^\ell_{k,j}(jn^i) + \beta = \alpha_{ij}.$$

This proves the assertion. □

Arguing as in proposition V.5 shows that the vectors $\boldsymbol{h}^\ell_{i,j}$ and $\boldsymbol{u}_\ell$ belong to V^ℓ_1, so it follows that they actually form a basis for V^ℓ_1. We are now ready to prove:

Theorem V.8. *For any fitness function f on Ω_n, the following assertions are equivalent:*

1. *f is a (generalized) first order function, i.e., $f = \sum_{i=0}^{\ell-1} g_i$ for some functions g_i on Ω_n which only depend on the i-th position,*
2. *$\varepsilon^*(f) = 0$.*

Proof. Clearly, if f is of the form $\sum_{i=0}^{\ell-1} g_i$, where g_i only depends upon the i-th position, then $\boldsymbol{f} \in V^\ell_1$, hence $\varepsilon^*(f) = 0$. Indeed, it suffices to verify this for each of the $\boldsymbol{g}_i$ corresponding to the functions g_i. Now, if we let a_{ij} denote the common value of all $g_i(s)$ with value j on the i-th position (with $1 \leq j < n$), then

$$g_i = \sum_{j=1}^{n-1} a_{ij} h^\ell_{i,j} + a_{i0}(u_\ell - h^\ell_{i,1} - \cdots - h^\ell_{i,n-1}),$$

so $\boldsymbol{g}_i \in < \boldsymbol{h}^\ell_{i,j}, \boldsymbol{u}_\ell > = V_1^\ell$. Conversely, if $\boldsymbol{f} \in V_1^\ell$, then

$$\boldsymbol{f} = \sum_{i,j} \alpha_{ij} \boldsymbol{h}^\ell_{i,j} + \beta \boldsymbol{u}_\ell = \sum_{i=0}^{\ell-1} (\sum_{j=1}^{n-1} \alpha_{ij} \boldsymbol{h}^\ell_{i,j}) + \beta \boldsymbol{u}_\ell = \sum_{i=0}^{\ell-1} \boldsymbol{g}_i,$$

where

$$\boldsymbol{g}_0 = \sum_{j=1}^{n-1} \alpha_{0j} \boldsymbol{h}^\ell_{0,j} + \beta \boldsymbol{u}_\ell$$

and

$$\boldsymbol{g}_i = \sum_{j=1}^{n-1} \alpha_{ij} \boldsymbol{h}^\ell_{i,j}$$

with $1 \leq i < \ell$.

□

3.2 Maximal epistasis

We have already pointed out that $\varepsilon^*(f) \leq 1$. Moreover, for any $\boldsymbol{f} \in V_0^\ell$ this maximum value may actually be reached. However, if we add the extra restriction that all coordinates of $\boldsymbol{f}$ be positive, i.e., that $\boldsymbol{f}$ corresponds to a positive valued function on Ω_n, then it appears that the maximal value of $\varepsilon^*(f)$ is $1 - \frac{1}{n^{\ell-1}}$. Put differently: the minimal value of $\gamma(f)$ with $||\boldsymbol{f}|| = 1$ is n.

Let us first point out that the extreme value $\gamma(f) = n$ may actually be reached. Indeed, consider the vector $\boldsymbol{f} \in \mathbb{R}^{n^\ell}$, given by

$${}^t\boldsymbol{f} = (\alpha, 0, \ldots, 0, \alpha, 0, \ldots, 0, \alpha),$$

where $\alpha = \sqrt{n}/n$ appears at the $i\frac{n^\ell-1}{n-1}$-th position, for $0 \leq i < n$. Obviously

$||\boldsymbol{f}|| = 1$. Moreover, with $m = \frac{n^\ell - 1}{n-1}$, we have

$$\begin{aligned}\gamma(f) &= {}^t\boldsymbol{f}\boldsymbol{G}_\ell\boldsymbol{f}\\ &= \alpha^2((g_{0,0} + g_{0,m} + \cdots + g_{0,n^\ell-1}) + (g_{m,0} + \cdots + g_{m,n^\ell-1})\\ &\quad + \cdots + (g_{n^\ell-1,0} + \cdots + g_{n^\ell-1,n^\ell-1}))\\ &= \frac{1}{n}((g_{0,0} + g_{m,m} + \cdots + g_{n^\ell-1,n^\ell-1}) + 2(g_{0,m} + \cdots + g_{(n-2)m,m}))\\ &= \frac{1}{n}(ng_{0,0} + 2\sum_{i<j} g_{im,jm}).\end{aligned}$$

Since each of the $g_{im,jm}$ has the same value $(n-1)\ell + 1 - n\ell$, it thus easily follows that $\gamma(f) = n$, as claimed.

Theorem V.9. *For any positive integer ℓ and any positive valued fitness function f with $||\boldsymbol{f}|| = 1$, we have*

$$\varepsilon^*(f) \leq 1 - \frac{1}{n^{\ell-1}}.$$

Proof. Let us argue as in proposition II.13. As the matrix $\boldsymbol{G}_\ell$ is symmetric, we may find an orthogonal matrix $\boldsymbol{S}$ which diagonalizes it, i.e., with the property that ${}^t\boldsymbol{S}\boldsymbol{G}_\ell\boldsymbol{S} = \boldsymbol{D}$ is a diagonal matrix, whose diagonal entries are then, of course, the eigenvalues of $\boldsymbol{G}_\ell$ (taking into account multiplicities). We may thus assume

$$\boldsymbol{D} = \begin{pmatrix} n^\ell \boldsymbol{I}_{(n-1)\ell+1} & \mathbf{0}' \\ \mathbf{0}' & \mathbf{0}_{n^\ell-(n-1)\ell-1} \end{pmatrix},$$

where $\mathbf{0}_{n^\ell-(n-1)\ell-1}$ is the square zero-matrix of dimension $n^\ell - (n-1)\ell - 1$ and $\mathbf{0}'$ the zero-matrix of dimensions $(n-1)\ell + 1 \times n^\ell - (n-1)\ell - 1$.

Put $\boldsymbol{g} = {}^t\boldsymbol{S}\boldsymbol{f}$. Then, obviously, $\gamma(f) = n^\ell \sum_{i=0}^{(n-1)\ell} g_i^2$. The columns of the matrix $\boldsymbol{S}$ consist of (normalized) eigenvectors of $\boldsymbol{G}_\ell$. In particular, its first $(n-1)\ell+1$ columns may be chosen to be the normalizations of the vectors $\boldsymbol{v}_0^\ell, \ldots, \boldsymbol{v}_{(n-1)\ell}^\ell$ constructed before. So, let us consider the orthonormal basis

$$\{\boldsymbol{w}_0^\ell, \ldots, \boldsymbol{w}_{(n-1)(\ell-1)}^\ell, \boldsymbol{z}_1^\ell, \ldots, \boldsymbol{z}_{n-1}^\ell\}$$

of V_1^ℓ, where $\boldsymbol{w}_k^\ell = n^{-\ell/2}\, \boldsymbol{v}_k^\ell$ for $0 \leq k \leq (n-1)(\ell-1)$ and where

$$\boldsymbol{z}_i^\ell = (i^2+i)^{-1/2}\, n^{(1-\ell)/2}\, \boldsymbol{v}_{(n-1)(\ell-1)+i}^\ell$$

for $1 \leq i < n$. We then obtain

$$\begin{aligned}\gamma(f) &= \gamma(f_0, \dots, f_{n^\ell - 1}) \\ &= n^\ell \sum_{k=0}^{(n-1)(\ell-1)} ({}^t\boldsymbol{w}_k^\ell \boldsymbol{f})^2 + n^\ell \sum_{i=1}^{n-1} ({}^t\boldsymbol{z}_i^\ell \boldsymbol{f})^2 \\ &= n^\ell (n^{-\ell/2})^2 \sum_{k=0}^{(n-1)(\ell-1)} ({}^t\boldsymbol{v}_k^\ell \boldsymbol{f})^2 + n^\ell (n^{(1-\ell)/2})^2 \sum_{i=1}^{n-1} \frac{1}{i(i+1)} ({}^t\boldsymbol{v}_{(n-1)(\ell-1)+i} \boldsymbol{f})^2.\end{aligned}$$

By construction, we thus obtain that $\gamma(f_0, \dots, f_{n^\ell-1})$ is equal to

$$\begin{aligned}&\gamma_{\ell-1}(f_0 + f_{n^{\ell-1}} + \dots + f_{(n-1)n^{\ell-1}}, \dots, f_{n^{\ell-1}-1} + \dots + f_{n^\ell-1}) \\ &+ n(\frac{1}{2}((f_0 + \dots + f_{n^{\ell-1}-1}) - (f_{n^{\ell-1}} + \dots + f_{2n^{\ell-1}-1}))^2 \\ &+ \frac{1}{2.3}((f_0 + \dots + f_{n^{\ell-1}-1}) + (f_{n^{\ell-1}} + \dots + f_{2n^{\ell-1}-1}) - 2(f_{2n^{\ell-1}} + \dots + f_{3n^{\ell-1}-1}))^2 \\ &+ \dots \\ &+ \frac{1}{n(n-1)}((f_0 + \dots + f_{n^{\ell-1}-1}) + \dots - (n-1)(f_{(n-1)n^{\ell-1}} + \dots + f_{n^\ell-1}))^2).\end{aligned}$$

Let us write

$$\hat{\boldsymbol{f}} = \begin{pmatrix} f_0 + f_{n^{\ell-1}} + \dots + f_{(n-1)n^{\ell-1}} \\ \vdots \\ f_{n^{\ell-1}-1} + \dots + f_{n^\ell-1} \end{pmatrix} \in \mathbb{R}^{n^{\ell-1}},$$

then

$$\begin{aligned}||\hat{\boldsymbol{f}}||^2 &= (f_0 + \dots + f_{(n-1)n^{\ell-1}})^2 + \dots + (f_{n^{\ell-1}-1} + \dots + f_{n^\ell-1})^2 \\ &= f_0^2 + \dots + f_{n^\ell-1}^2 + 2(f_0 f_{n^{\ell-1}} + \dots + f_{n^{\ell-1}-1} f_{n^\ell-1}) = a^2,\end{aligned}$$

for some $a \geq 1$.

Let $\boldsymbol{f}' = \frac{1}{a}\hat{\boldsymbol{f}}$, then $||\boldsymbol{f}'|| = 1$ and

$$\gamma(f') = \gamma(\frac{1}{a}\hat{f}) = \frac{1}{a^2}\gamma(\hat{f}).$$

Let us now assume that, for some positive integer ℓ, we have $\gamma(f) < n$, for some fitness function $f : \Omega_n \to \mathbb{R}$ with $||\boldsymbol{f}|| = 1$. Then

$$\begin{aligned}\gamma(\hat{f}) &\leq \gamma(\hat{f}) + \sum_{i=1}^{n-1} \frac{n}{i(i+1)}((f_0 + \cdots + f_{n^{\ell-1}-1}) + \dots \\ &\quad - (i-1)(f_{(i-1)n^{\ell-1}} + \cdots + f_{in^{\ell-1}-1}))^2 \\ &= \gamma(f) < n.\end{aligned}$$

It follows that we also have

$$\gamma(f') = \frac{1}{a^2}\gamma(\hat{f}) < \frac{n}{a^2} \leq n.$$

Iterating this process, we would thus find some fitness function $f : \{0, \dots, n-1\} \to \mathbb{R}$ with $||\boldsymbol{f}|| = 1$, and $\gamma(f) < n$. However, this is impossible, as γ is easily seen to have constant value n on normalized fitness functions of one variable only. This contradiction proves our assertion. □

We just pointed out that the minimal value $\gamma(f) = n$, corresponding to maximal normalized epistasis, may actually be reached. Let us now conclude this section by solving the problem of completely describing the class of all fitness functions f for which $\gamma(f) = n$.

Fix a positive integer $\ell \geq 2$ and consider mutually distinct indices $0 \leq i_0, \dots, i_{n-1} < n^{\ell-1}$ with the property that

1. $\sum_{r=0}^{n-1} i_r = \frac{n}{2}(n^{\ell-1} - 1)$,

2. $\mathrm{d}_{i_r i_s} = \ell - 1$ for any $0 \leq r \neq s < n$.

For each such family of indices $i_0, \dots, i_{n-1}$, we define the vectors

$$\boldsymbol{q}^{\ell}_{i_0,\dots,i_{n-1}} = \frac{\sqrt{n}}{n}\begin{pmatrix} \boldsymbol{e}_{i_0} \\ \vdots \\ \boldsymbol{e}_{i_{n-1}} \end{pmatrix} \in \mathbb{R}^{n^\ell},$$

where $\{\boldsymbol{e}_0, \dots, \boldsymbol{e}_{n^{\ell-1}-1}\}$ is the canonical basis of $\mathbb{R}^{n^{\ell-1}}$.

For example, if $n = \ell = 2$, then we necessarily have $\{i_0, i_1\} = \{0, 1\}$ as suitable indices, and this corresponds to

$$\boldsymbol{q}^2_{0,1} = \frac{\sqrt{2}}{2}\begin{pmatrix}1\\0\\0\\1\end{pmatrix}, \quad \boldsymbol{q}^2_{1,0} = \frac{\sqrt{2}}{2}\begin{pmatrix}0\\1\\1\\0\end{pmatrix}.$$

In general, still with $n = 2$, suitable indices are given by couples $0 \leq i_0, i_1 < 2^{\ell-1}$, with $i_0 + i_1 = 2^{\ell-1} - 1$ (which automatically implies that $\mathrm{d}_{i_0 i_1} = \ell - 1$), and this yields vectors of the form

$${}^t\boldsymbol{q}^\ell_{k,2^{\ell-1}-k} = \frac{\sqrt{2}}{2}(0, \ldots, 0, 1, 0, \ldots, 0, 1, 0, \ldots, 0) \in \mathbb{R}^{2^\ell}$$

with entry 1 at positions k and $2^\ell - k - 1$.

As another example, with $n = 3$ and $\ell = 2$, we necessarily have $\{i_0, i_1, i_2\} = \{0, 1, 2\}$ with, e.g.,

$$\begin{aligned}{}^t\boldsymbol{q}^2_{0,1,2} &= \frac{\sqrt{3}}{3}(1, 0, 0, 0, 1, 0, 0, 0, 1)\\ {}^t\boldsymbol{q}^2_{2,1,0} &= \frac{\sqrt{3}}{3}(0, 0, 1, 0, 1, 0, 1, 0, 0).\end{aligned}$$

One should view the corresponding fitness functions $q^\ell_{i_0,\ldots,i_{n-1}}$, which we may refer to as *generalized camel functions*, as having n peaks lying as far apart as possible. Note also that suitable sets of indices $i_0, \ldots, i_{n-1}$ may always be found. For example, putting

$$i_r = r\frac{n^\ell - 1}{n - 1}$$

with $0 \leq r < n$ obviously does the trick.

Let us now prove the following result which completely describes the positive functions with maximal normalized epistasis:

Theorem V.10. *For any $\ell \geq 2$ and any positive $\boldsymbol{f} \in \mathbb{R}^{n^\ell}$ with $||\boldsymbol{f}|| = 1$, the following assertions are equivalent:*

1. $\varepsilon^*(f) = 1 - \dfrac{1}{n^{\ell-1}}$,

2. $\boldsymbol{f} = \boldsymbol{q}^{\ell}_{i_0,\dots,i_{n-1}}$ *for suitable indices* $0 \leq i_0, \dots, i_{n-1} < n^{\ell-1}$.

Proof. Let us start by proving that the second assertion implies the first one. For any choice of suitable indices $i_0, \dots, i_{n-1}$ we have

$$\begin{aligned}
\gamma(f) &= {}^t\boldsymbol{f}\boldsymbol{G}_\ell\boldsymbol{f} = (\frac{\sqrt{n}}{n})^2({}^t\boldsymbol{e}_{i_0}, \dots, {}^t\boldsymbol{e}_{i_{n-1}})\boldsymbol{G}_\ell \begin{pmatrix} \boldsymbol{e}_{i_0} \\ \vdots \\ \boldsymbol{e}_{i_{n-1}} \end{pmatrix} \\
&= \frac{1}{n}({}^t\boldsymbol{e}_{i_0}, \dots, {}^t\boldsymbol{e}_{i_{n-1}}) \begin{pmatrix} g_{0,i_0} + g_{0,n^{\ell-1}+i_1} + \dots + g_{0,(n-1)n^{\ell-1}+i_{n-1}} \\ \vdots \\ g_{n^\ell-1,i_0} + \dots + g_{n^\ell-1,(n-1)n^{\ell-1}+i_{n-1}} \end{pmatrix} \\
&= \frac{1}{n}\{(g_{i_0,i_0} + \dots + g_{i_0,(n-1)n^{\ell-1}+i_{n-1}}) + \dots \\
&\quad \dots + (g_{(n-1)n^{\ell-1}+i_{n-1},i_0} + \dots + g_{(n-1)n^{\ell-1}+i_{n-1},(n-1)n^{\ell-1}+i_{n-1}})\} \\
&= \frac{1}{n}\{ng_{0,0} + 2((g_{i_0,n^{\ell-1}+i_1} + \dots + g_{i_0,(n-1)n^{\ell-1}+i_{n-1}}) + \dots \\
&\quad \dots + (g_{(n-2)n^{\ell-1}+i_{n-2},(n-1)n^{\ell-1}+i_{n-1}}))\} \\
&= \frac{1}{n}\{n((n-1)\ell+1) + 2((n-1)((n-1)\ell+1) \\
&\quad - n(\mathrm{d}_{i_0,n^{\ell-1}+i_1} + \dots + \mathrm{d}_{i_0,(n-1)n^{\ell-1}+i_{n-1}}) + (n-2)((n-1)\ell+1) \\
&\quad - n(\mathrm{d}_{n^{\ell-1}+i_1,2n^{\ell-1}+i_2} + \dots + \mathrm{d}_{n^{\ell-1}+i_1,(n-1)n^{\ell-1}+i_{n-1}}) + \dots \\
&\quad \dots + 1((n-1)\ell+1) - n\mathrm{d}_{(n-2)n^{\ell-1}+i_{n-2},(n-1)n^{\ell-1}+i_{n-1}})\} \\
&= \frac{1}{n}\{((n-1)\ell+1)n^2 - 2n(\mathrm{d}_{i_0,n^{\ell-1}+i_1} + \dots + \mathrm{d}_{(n-2)n^{\ell-1}+i_{n-2},(n-1)n^{\ell-1}+i_{n-1}})\} \\
&= n((n-1)\ell+1) - 2\binom{n}{2}\ell = n,
\end{aligned}$$

which proves our claim.

To prove the converse, we will use induction on ℓ. Consider a fitness function f whose corresponding vector $\boldsymbol{f} \in \mathbb{R}^{n^\ell}$ is normalized and has the property that $\gamma(f) = n$.

With notations as in theorem V.9, this means that

$$\begin{aligned} n &= \gamma(f_0, \ldots, f_{n^\ell - 1}) \\ &= \gamma(\hat{f}) + n\{\frac{1}{2}((f_0 + \cdots + f_{n^{\ell-1}-1}) - (f_{n^{\ell-1}} + \cdots + f_{2n^{\ell-1}-1}))^2 + \ldots \\ &+ \frac{1}{n(n-1)}((f_0 + \cdots + f_{n^{\ell-1}-1}) + \cdots - (n-1)(f_{(n-1)n^{\ell-1}} + \cdots + f_{n^\ell - 1}))^2\}, \end{aligned}$$

hence $\gamma(\hat{f}) \leq \gamma(f)$. Moreover, if we consider again $f' = \hat{f}/||\hat{\boldsymbol{f}}||$ then we have

$$\gamma(f') = \frac{1}{||\hat{\boldsymbol{f}}||^2}\gamma(\hat{f}) \leq \frac{1}{||\hat{\boldsymbol{f}}||^2}\gamma(f) \leq \frac{n}{||\hat{\boldsymbol{f}}||^2} \leq n,$$

so we necessarily have $\gamma(f') = n$, in view of the lower bound obtained on the value of γ, and, of course, this yields $||\hat{\boldsymbol{f}}|| = 1$. We thus obtain $\hat{\boldsymbol{f}} = \boldsymbol{f}'$ and $\gamma(\hat{f}) = n$, whence the following identities:

$$\begin{aligned} f_0 + \cdots + f_{n^{\ell-1}-1} &= f_{n^{\ell-1}} + \cdots + f_{2n^{\ell-1}-1} \\ &\vdots \\ &= f_{(n-1)n^{\ell-1}} + \cdots + f_{n^\ell - 1}. \end{aligned}$$

On the other hand, as $||\hat{\boldsymbol{f}}|| = ||\boldsymbol{f}|| = 1$, we also have

$$\begin{array}{lllll} f_0 f_{n^{\ell-1}} & = \ldots = & f_0 f_{(n-1)n^{\ell-1}} & = \ldots = & f_{(n-2)n^{\ell-1}} f_{(n-1)n^{\ell-1}} = 0 \\ \vdots & & & & \\ f_i f_{n^{\ell-1}+i} & = \ldots = & f_i f_{(n-1)n^{\ell-1}+i} & = \ldots = & f_{(n-2)n^{\ell-1}+i} f_{(n-1)n^{\ell-1}+i} = 0 \\ \vdots & & & & \\ f_{n^{\ell-1}-1} f_{2n^{\ell-1}-1} & = \ldots = & f_{n^{\ell-1}-1} f_{n^\ell - 1} & = \ldots = & f_{(n-1)n^{\ell-1}-1} f_{n^\ell-1} = 0. \end{array}$$

In particular, if $\ell = 2$ and $\gamma(f_0, \ldots, f_{n^2-1}) = n = \gamma(\hat{f})$, then the previous equations reduce to

$$\begin{aligned} f_0 + \cdots + f_{n-1} &= f_n + \cdots + f_{2n-1} \\ &\vdots \\ &= f_{(n-1)n} + \cdots + f_{n^2-1} \end{aligned}$$

and

$$f_0 f_n \quad = \ldots = f_{(n-2)n} f_{(n-1)n} = 0$$
$$\vdots$$
$$f_{n-1} f_{2n-1} = \ldots = f_{(n-1)n-1} f_{n^2-1} = 0.$$

Solving this system of equations easily yields that

$$\boldsymbol{f} = \frac{\sqrt{n}}{n} \begin{pmatrix} \boldsymbol{e}_{\sigma(0)} \\ \vdots \\ \boldsymbol{e}_{\sigma(n-1)} \end{pmatrix} = \boldsymbol{q}^2_{\sigma(0),\ldots,\sigma(n-1)} \in \mathbb{R}^{n^2},$$

where σ is a permutation of $\{0, \ldots, n-1\}$ and $\{\boldsymbol{e}_0, \ldots, \boldsymbol{e}_{n-1}\}$ is the canonical basis of $\mathbb{R}^n$. Of course, if $0 \leq r \neq s < n$, then $\sigma(r) \neq \sigma(s)$ and

$$\sum_{r=0}^{n-1} \sigma(r) = \sum_{r=0}^{n-1} r = \frac{n}{2}(n-1),$$

so the indices $\sigma(0), \ldots, \sigma(r-1)$ satisfy the necessary requirements.

Let us now assume our assertion to hold true for strings of length $2, 3, \ldots, \ell - 1$ and let us prove it for length ℓ. Consider a normalized fitness function f defined on strings of length ℓ and suppose that $\gamma(f) = n$. Then, by induction,

$$\hat{\boldsymbol{f}} = \boldsymbol{f}' = \boldsymbol{q}^{\ell-1}_{i_0,\ldots,i_{n-1}} \in \mathbb{R}^{n^{\ell-1}},$$

for indices $0 \leq i_0, \ldots, i_{n-1} < n^{\ell-2}$, with the property that $\mathrm{d}_{i_r,i_s} = \ell - 1$ for $r \neq s$ and that $\sum_{r=0}^{n-1} i_r = \frac{n}{2}(n^{\ell-2} - 1)$. From the very definition of $\hat{\boldsymbol{f}} = \boldsymbol{q}^{\ell-1}_{i_0,\ldots,i_{n-1}}$, it follows that its non-zero components may be found in the rows $kn^{\ell-2} + i_k$ $(0 \leq k < n)$, whose expression, for any k, is

$$f_{kn^{\ell-2}+i_k} + f_{n^{\ell-1}+(kn^{\ell-2}+i_k)} + \cdots + f_{(n-1)n^{\ell-1}+(kn^{\ell-2}+i_k)} = \frac{\sqrt{n}}{n}.$$

On the other hand, the above systems of equations applied to $\hat{\boldsymbol{f}} = \boldsymbol{q}^{\ell-1}_{i_0,\ldots,i_{n-1}}$ reduce to

$$\begin{aligned} & f_{i_0} + f_{n^{\ell-2}+i_1} + \cdots + f_{(n-1)n^{\ell-2}+i_{n-1}} = \ldots \\ & = f_{(n-1)n^{\ell-1}+i_0} + f_{(n-1)n^{\ell-1}+(n^{\ell-2}+i_1)} + \cdots + f_{(n-1)n^{\ell-1}+((n-1)n^{\ell-2}+i_{n-1})} \end{aligned}$$

and

$$f_{i_0} f_{n^{\ell-1}+i_0} = \cdots = f_{i_0} f_{(n-1)n^{\ell-1}+i_0} = \cdots = f_{(n-2)n^{\ell-1}+i_0} f_{(n-1)n^{\ell-1}+i_0} = 0$$

$$\vdots$$

$$\begin{aligned} & f_{(n-1)n^{\ell-2}+i_{n-1}} f_{n^{\ell-1}+((n-1)n^{\ell-2}+i_{n-1})} = \dots \\ & = f_{(n-1)n^{\ell-2}+i_{n-1}} f_{(n-1)n^{\ell-1}+((n-1)n^{\ell-2}+i_{n-1})} = \dots \\ & = f_{(n-2)n^{\ell-1}+((n-1)n^{\ell-2}+i_{n-1})} f_{(n-1)n^{\ell-1}+((n-1)n^{\ell-2}+i_{n-1})} = 0. \end{aligned}$$

Let us put $x_j^k = f_{jn^{\ell-1}+kn^{\ell-2}+i_k}$ for any $0 \leq j, k < n$. The above systems of equations are then shown to be equivalent to

$$(a) \begin{cases} x_0^0 + x_1^0 + \cdots + x_{n-1}^0 = \frac{\sqrt{n}}{n} \\ \qquad \vdots \\ x_0^{n-1} + \cdots + x_{n-1}^{n-1} = \frac{\sqrt{n}}{n} \end{cases}$$

$$(b) \begin{cases} x_0^0 + x_0^1 + \cdots + x_0^{n-1} = \cdots = \\ x_{n-1}^0 + x_{n-1}^1 + \cdots + x_{n-1}^{n-1} \end{cases}$$

$$(c_1) \begin{cases} x_0^0 x_1^0 = \dots = x_0^0 x_{n-1}^0 \\ \qquad \ddots \\ \qquad = x_{n-2}^0 x_{n-1}^0 = 0 \end{cases}$$

$$\vdots$$

$$(c_{n-1}) \begin{cases} x_0^{n-1} x_1^{n-1} = \dots = x_0^{n-1} x_{n-1}^{n-1} \\ \qquad \ddots \\ \qquad = x_{n-2}^{n-1} x_{n-1}^{n-1} = 0. \end{cases}$$

In view of the fact that $x_j^k \geq 0$ for all indices j, k, it follows that in each of the equations in (a) at least one of the summands has to be non-zero. The equations (c) imply the uniqueness of this summand. It thus follows that the system of equations (a) reduces to

$$x_{r_0}^0 = x_{r_1}^1 = \cdots = x_{r_{n-1}}^{n-1} = \frac{\sqrt{n}}{n}$$

for certain $0 \leq r_i < n$. Moreover, analyzing the equations (b), it follows that in each of the composing equations there should be the same number of non-zero terms.

A tedious, but essentially straightforward verification, shows that in each of them there is actually exactly just one non-zero component. The solutions are thus of the form

$$x^0_{r_0} = x^1_{r_1} = \cdots = x^{n-1}_{r_{n-1}} = \frac{\sqrt{n}}{n},$$

with $r_i \neq r_j$ if $i \neq j$. In other words,

$$\boldsymbol{f} = \frac{\sqrt{n}}{n} \begin{pmatrix} \boldsymbol{e}_{\hat{\imath}_0} \\ \vdots \\ \boldsymbol{e}_{\hat{\imath}_{n-1}} \end{pmatrix} \in \mathbb{R}^{n^\ell}$$

where $\hat{\imath}_j = pn^{\ell-2} + i_p$ for some suitable indices i_p, such that the $0 \leq \hat{\imath}_0, \ldots, \hat{\imath}_{n-1} < n^{\ell-1}$ are mutually distinct, and such that

$$\begin{aligned} \sum_{r=0}^{n-1} \hat{\imath}_r &= \sum_{r=0}^{n-1} i_r + \sum_{r=0}^{n-1} rn^{\ell-2} \\ &= \frac{n}{2}(n^{\ell-2} - 1) + n^{\ell-2}\frac{n(n-1)}{2} \\ &= \frac{n}{2}(n^{\ell-1} - 1) \end{aligned}$$

and

$$\mathrm{d}_{\hat{\imath}_j,\hat{\imath}_k} = \mathrm{d}_{pn^{\ell-2}+i_p, qn^{\ell-2}+i_q} = 1 + \mathrm{d}_{i_p,i_q} = 1 + (\ell - 1) = \ell.$$

This finishes the proof. □

4 Example: Generalized unitation functions

The main purpose of this section is to take a detailed look at the epistasis of a particular class of functions which we may call *generalized unitation functions*.

A generalized unitation function f on $\Omega_n = \{0, 1, \ldots, n-1\}^\ell$ is characterized by the fact that its value on any $s \in \Omega_n$ is identical to that on any of the strings obtained from s by permutation of its components. In other words, for any $s \in \Omega_n$,

$$f(s) = f(s_{\ell-1} \ldots s_0) = f(s_{\sigma(\ell-1)} \ldots s_{\sigma(0)}),$$

for any permutation σ of $\{0, 1, \ldots, \ell - 1\}$. Note that, just as in the binary case, generalized unitation functions only take a maximum of $[n, \ell] \equiv \binom{n+\ell-1}{\ell}$ different values.

4.1 Normalized epistasis

In order not to complicate notations and calculations unnecessarily, we restrict ourselves to the case $n = 3$. As in the binary case, we consider the vector

$$\boldsymbol{h} = \begin{pmatrix} f(0\ldots000) \\ f(0\ldots001) \\ f(0\ldots002) \\ f(0\ldots011) \\ f(0\ldots012) \\ \vdots \\ f(2\ldots222) \end{pmatrix} \in \mathbb{R}^{\binom{\ell+2}{\ell}},$$

whose components will be denoted by $h_0, h_1, \ldots, h_m$, with $m = \binom{\ell+2}{\ell} - 1$. To obtain the matrix $\boldsymbol{B}_{3,\ell}$ with the property ${}^t\boldsymbol{f}\boldsymbol{G}_{3,\ell}\boldsymbol{f} = {}^t\boldsymbol{h}\boldsymbol{B}_{3,\ell}\boldsymbol{h}$, we argue as follows. First, let us inductively define (for any positive integer ℓ) the $3^\ell \times (m+1)$-dimensional matrix $\boldsymbol{A}_\ell$ given by

$$\boldsymbol{A}_\ell = \begin{pmatrix} \boldsymbol{A}_{\ell-1} & \boldsymbol{O}_{\ell-1} & \boldsymbol{0}_{\ell-1} \\ \boldsymbol{0}_{\ell-1} & \boldsymbol{A}'_{\ell-1} & \boldsymbol{0}_{\ell-1} \\ \boldsymbol{0}_{\ell-1} & \boldsymbol{0}_{\ell-1} & \boldsymbol{A}'_{\ell-1} \end{pmatrix}, \qquad \boldsymbol{A}'_{\ell-1} = \begin{pmatrix} \boldsymbol{A}'_{\ell-2} & \boldsymbol{O}_{\ell-2} & \boldsymbol{0}_{\ell-2} \\ \boldsymbol{O}''_{\ell-2} & \boldsymbol{A}''_{\ell-2} & \boldsymbol{0}_{\ell-2} \\ \boldsymbol{O}''_{\ell-2} & \boldsymbol{0}_{\ell-2} & \boldsymbol{A}''_{\ell-2} \end{pmatrix},$$

$$\boldsymbol{A}''_{\ell-2} = \begin{pmatrix} \boldsymbol{A}''_{\ell-3} & \boldsymbol{O}_{\ell-3} & \boldsymbol{0}_{\ell-3} \\ \boldsymbol{O}^{(3)}_{\ell-3} & \boldsymbol{A}^{(3)}_{\ell-3} & \boldsymbol{0}_{\ell-3} \\ \boldsymbol{O}^{(3)}_{\ell-3} & \boldsymbol{0}_{\ell-3} & \boldsymbol{A}^{(3)}_{\ell-3} \end{pmatrix}, \ldots, \quad \boldsymbol{A}^{(\ell-1)}_1 = \begin{pmatrix} 1\overbrace{0\ldots0}^{\ell-1}00 \\ 00\ldots010 \\ 00\ldots001 \end{pmatrix}$$

where $\cdot^{(i)}$ indicates an object with i primes, and

— $\boldsymbol{A}_1$ is the 3-dimensional identity matrix and $\boldsymbol{A}^{(i)}_{\ell-i}$ is $3^{\ell-i} \times \left(m+1-\frac{i(i+3)}{2}\right)$-dimensional. Moreover, for all $1 \le j < \ell$,

$$\boldsymbol{A}^{(j)}_1 = \begin{pmatrix} 1\overbrace{0\ldots0}^{j}00 \\ 00\ldots010 \\ 00\ldots001 \end{pmatrix};$$

— $\boldsymbol{O}_{\ell-i}$ denotes the zero-matrix of dimension $3^{\ell-i} \times \ell$, $\mathbf{0}_{\ell-i}$ is the zero-vector of $\mathbb{R}^{3^{\ell-i}}$, and $\boldsymbol{O}^{(i)}_{\ell-i}$ is the $3^{\ell-i} \times i$-dimensional zero-matrix.

A tedious but essentially straightforward verification shows that $\boldsymbol{f} = \boldsymbol{A}_\ell \boldsymbol{h}$. So, we have that $\boldsymbol{B}_{3,\ell} = {}^t\boldsymbol{A}_\ell \boldsymbol{G}_{3,\ell} \boldsymbol{A}_\ell$ which, in particular, implies that $\boldsymbol{B}_{3,\ell}$ is a square matrix of dimension $[3,\ell\,] = \binom{\ell+2}{\ell}$.

In order to describe explicitly the components $(b_{3,\ell})_{pq}$ of $\boldsymbol{B}_{3,\ell}$, denote by $\widetilde{\boldsymbol{O}}_\ell, \widehat{\boldsymbol{O}}_\ell$ and $\widehat{\widehat{\boldsymbol{O}}}_{\ell-1}$ the zero-matrices of dimensions $[3,\ell\,] \times (\ell+1)$, $3^{\ell-1} \times [3,\ell\,] - 2$ and $3^{\ell-1} \times [3,\ell-1]$, respectively. Denote by $\widetilde{\mathbf{0}}_{\ell-1}$ and $\widetilde{\widetilde{\mathbf{0}}}_\ell$ the zero-vectors in $\mathbb{R}^{[3,\ell-1]}$ and $\mathbb{R}^{[3,\ell]-2}$, respectively. We may then rewrite $\boldsymbol{A}_\ell$ as

$$\boldsymbol{A}_\ell = \begin{pmatrix} \boldsymbol{A}_{\ell-1} & \widehat{\boldsymbol{O}}_\ell & \widehat{\boldsymbol{O}}_\ell \\ \widehat{\widehat{\boldsymbol{O}}}_{\ell-1} & \boldsymbol{A}'_{\ell-1} & \widehat{\boldsymbol{O}}_\ell \\ \widehat{\widehat{\boldsymbol{O}}}_{\ell-1} & \widehat{\boldsymbol{O}}_\ell & \boldsymbol{A}'_{\ell-1} \end{pmatrix} \begin{pmatrix} \boldsymbol{I}_{[3,\ell-1]} & \widetilde{\boldsymbol{O}}_{\ell-1} & \widetilde{\mathbf{0}}_{\ell-1} \\ \widetilde{\widetilde{\mathbf{0}}}_\ell & \boldsymbol{I}_{[3,\ell]-2} & \widetilde{\widetilde{\mathbf{0}}}_\ell \\ \widetilde{\widetilde{\mathbf{0}}}_\ell & \widetilde{\widetilde{\mathbf{0}}}_\ell & \boldsymbol{I}_{[3,\ell]-2} \end{pmatrix}$$

$$= \begin{pmatrix} \boldsymbol{A}_{\ell-1} & \widehat{\boldsymbol{O}}_\ell & \widehat{\boldsymbol{O}}_\ell \\ \widehat{\widehat{\boldsymbol{O}}}_{\ell-1} & \boldsymbol{A}'_{\ell-1} & \widehat{\boldsymbol{O}}_\ell \\ \widehat{\widehat{\boldsymbol{O}}}_{\ell-1} & \widehat{\boldsymbol{O}}_\ell & \boldsymbol{A}'_{\ell-1} \end{pmatrix} \boldsymbol{Y}_\ell,$$

where the submatrices situated on the diagonal of $\boldsymbol{Y}_\ell$ are the identity matrices of dimensions $[3,\ell-1]$ and $[3,\ell\,]-2$.

Using the induction formula given by lemma V.1 for $n = 3$, it then follows that

$$\boldsymbol{B}_{3,\ell} = \boldsymbol{B}'_{3,\ell} + \boldsymbol{B}''_{3,\ell}$$

with

$$\boldsymbol{B}'_{3,\ell} = {}^t\boldsymbol{Y}_\ell \begin{pmatrix} 2\,{}^t\boldsymbol{A}_{\ell-1}\boldsymbol{U}_{3,\ell-1}\boldsymbol{A}_{\ell-1} & -{}^t\boldsymbol{A}_{\ell-1}\boldsymbol{U}_{3,\ell-1}\boldsymbol{A}'_{\ell-1} & -{}^t\boldsymbol{A}_{\ell-1}\boldsymbol{U}_{3,\ell-1}\boldsymbol{A}'_{\ell-1} \\ -{}^t\boldsymbol{A}'_{\ell-1}\boldsymbol{U}_{3,\ell-1}\boldsymbol{A}_{\ell-1} & 2\,{}^t\boldsymbol{A}'_{\ell-1}\boldsymbol{U}_{3,\ell-1}\boldsymbol{A}'_{\ell-1} & -{}^t\boldsymbol{A}'_{\ell-1}\boldsymbol{U}_{3,\ell-1}\boldsymbol{A}'_{\ell-1} \\ {}^t\boldsymbol{A}'_{\ell-1}\boldsymbol{U}_{3,\ell-1}\boldsymbol{A}_{\ell-1} & -{}^t\boldsymbol{A}'_{\ell-1}\boldsymbol{U}_{3,\ell-1}\boldsymbol{A}'_{\ell-1} & 2\,{}^t\boldsymbol{A}'_{\ell-1}\boldsymbol{U}_{3,\ell-1}\boldsymbol{A}'_{\ell-1} \end{pmatrix} \boldsymbol{Y}_\ell,$$

and

$$\boldsymbol{B}''_{3,\ell} = {}^t\boldsymbol{Y}_\ell \begin{pmatrix} {}^t\boldsymbol{A}_{\ell-1}\boldsymbol{G}_{3,\ell-1}\boldsymbol{A}_{\ell-1} & {}^t\boldsymbol{A}_{\ell-1}\boldsymbol{G}_{3,\ell-1}\boldsymbol{A}'_{\ell-1} & {}^t\boldsymbol{A}_{\ell-1}\boldsymbol{G}_{3,\ell-1}\boldsymbol{A}'_{\ell-1} \\ {}^t\boldsymbol{A}'_{\ell-1}\boldsymbol{G}_{3,\ell-1}\boldsymbol{A}_{\ell-1} & {}^t\boldsymbol{A}'_{\ell-1}\boldsymbol{G}_{3,\ell-1}\boldsymbol{A}'_{\ell-1} & {}^t\boldsymbol{A}'_{\ell-1}\boldsymbol{G}_{3,\ell-1}\boldsymbol{A}'_{\ell-1} \\ {}^t\boldsymbol{A}'_{\ell-1}\boldsymbol{G}_{3,\ell-1}\boldsymbol{A}_{\ell-1} & {}^t\boldsymbol{A}'_{\ell-1}\boldsymbol{G}_{3,\ell-1}\boldsymbol{A}'_{\ell-1} & {}^t\boldsymbol{A}'_{\ell-1}\boldsymbol{G}_{3,\ell-1}\boldsymbol{A}'_{\ell-1} \end{pmatrix} \boldsymbol{Y}_\ell.$$

Some more words about notations. From here to the end of this section we will denote by k, r, s, t the positive integers with the two following properties:

1. $0 \leq k, r, s, t \leq \ell$,

2. for any $0 \leq p, q \leq [3, \ell] - 1$, we have $\frac{k(k+1)}{2} + r = p$ and $\frac{s(s+1)}{2} + t = q$, where k and s are the largest positive integers that satisfy $\frac{k(k+1)}{2} \leq p$ and $\frac{s(s+1)}{2} \leq q$, respectively.

In order to describe the matrix $\boldsymbol{B}'_{3,\ell}$, we need the following result:

Lemma V.11. *For any positive integer ℓ, consider the matrix*

$$\boldsymbol{C}_\ell = (c_{pq}) = {}^t\boldsymbol{A}_\ell \boldsymbol{U}_{3,\ell} \boldsymbol{A}_\ell.$$

Then, for any $0 \leq p, q < [3, \ell]$, we have

$$c_{pq} = \binom{\ell}{k}\binom{k}{r}\binom{\ell}{s}\binom{s}{t}.$$

Proof. Let us argue inductively on ℓ. For $\ell = 1$, the matrix $\boldsymbol{A}_1$ is the three-dimensional identity matrix, hence the assertion holds true. Assume the statement to be correct for $1, 2, \ldots, \ell - 1$ and let us prove it in dimension ℓ. Let us consider the vectors $\boldsymbol{v}_\ell \in \mathbb{R}^{[3,\ell]}$ and $\boldsymbol{w}_\ell \in \mathbb{R}^{[3,\ell+1]-2}$ defined by

$${}^t\boldsymbol{v}_\ell = \left(\binom{\ell}{0}\binom{0}{0}, \binom{\ell}{1}\binom{1}{0}, \binom{\ell}{1}\binom{1}{1}, \binom{\ell}{2}\binom{2}{0}, \ldots, \binom{\ell}{\ell}\binom{\ell}{0}, \ldots, \binom{\ell}{\ell}\binom{\ell}{\ell}\right)$$

and

$${}^t\boldsymbol{w}_\ell = \left(\binom{\ell}{0}\binom{0}{0}, 0, \binom{\ell}{1}\binom{1}{0}, \binom{\ell}{1}\binom{1}{1}, 0, \binom{\ell}{2}\binom{2}{0}, \ldots, 0, \binom{\ell}{\ell}\binom{\ell}{0}, \ldots, \binom{\ell}{\ell}\binom{\ell}{\ell}\right),$$

respectively. It immediately follows that the matrix $\boldsymbol{C}_\ell = {}^t\boldsymbol{A}_\ell \boldsymbol{U}_{3,\ell} \boldsymbol{A}_\ell$ can be rewritten as

$$
\begin{aligned}
&{}^t\boldsymbol{Y}_\ell \begin{pmatrix} {}^t\boldsymbol{A}_{\ell-1} & {}^t\widehat{\widehat{\boldsymbol{O}}}_{\ell-1} & {}^t\widehat{\widehat{\boldsymbol{O}}}_{\ell-1} \\ {}^t\widehat{\boldsymbol{O}}_\ell & {}^t\boldsymbol{A}'_{\ell-1} & {}^t\widehat{\boldsymbol{O}}_\ell \\ {}^t\widehat{\boldsymbol{O}}_\ell & {}^t\widehat{\boldsymbol{O}}_\ell & {}^t\boldsymbol{A}'_{\ell-1} \end{pmatrix} \begin{pmatrix} \boldsymbol{U}_{3,\ell-1} & \boldsymbol{U}_{3,\ell-1} & \boldsymbol{U}_{3,\ell-1} \\ \boldsymbol{U}_{3,\ell-1} & \boldsymbol{U}_{3,\ell-1} & \boldsymbol{U}_{3,\ell-1} \\ \boldsymbol{U}_{3,\ell-1} & \boldsymbol{U}_{3,\ell-1} & \boldsymbol{U}_{3,\ell-1} \end{pmatrix} \begin{pmatrix} \boldsymbol{A}_{\ell-1} & \widehat{\boldsymbol{O}}_\ell & \widehat{\boldsymbol{O}}_\ell \\ \widehat{\widehat{\boldsymbol{O}}}_{\ell-1} & \boldsymbol{A}'_{\ell-1} & \widehat{\boldsymbol{O}}_\ell \\ \widehat{\widehat{\boldsymbol{O}}}_{\ell-1} & \widehat{\boldsymbol{O}}_\ell & \boldsymbol{A}'_{\ell-1} \end{pmatrix} \boldsymbol{Y}_\ell \\
&= {}^t\boldsymbol{Y}_\ell \begin{pmatrix} {}^t\boldsymbol{A}_{\ell-1}\boldsymbol{U}_{3,\ell-1}\boldsymbol{A}_{\ell-1} & {}^t\boldsymbol{A}_{\ell-1}\boldsymbol{U}_{3,\ell-1}\boldsymbol{A}'_{\ell-1} & {}^t\boldsymbol{A}_{\ell-1}\boldsymbol{U}_{3,\ell-1}\boldsymbol{A}'_{\ell-1} \\ {}^t\boldsymbol{A}'_{\ell-1}\boldsymbol{U}_{3,\ell-1}\boldsymbol{A}_{\ell-1} & {}^t\boldsymbol{A}'_{\ell-1}\boldsymbol{U}_{3,\ell-1}\boldsymbol{A}'_{\ell-1} & {}^t\boldsymbol{A}'_{\ell-1}\boldsymbol{U}_{3,\ell-1}\boldsymbol{A}'_{\ell-1} \\ {}^t\boldsymbol{A}'_{\ell-1}\boldsymbol{U}_{3,\ell-1}\boldsymbol{A}_{\ell-1} & {}^t\boldsymbol{A}'_{\ell-1}\boldsymbol{U}_{3,\ell-1}\boldsymbol{A}'_{\ell-1} & {}^t\boldsymbol{A}'_{\ell-1}\boldsymbol{U}_{3,\ell-1}\boldsymbol{A}'_{\ell-1} \end{pmatrix} \boldsymbol{Y}_\ell \\
&= {}^t\boldsymbol{Y}_\ell \begin{pmatrix} \boldsymbol{v}_{\ell-1} \\ \boldsymbol{w}_{\ell-1} \\ \boldsymbol{w}_{\ell-1} \end{pmatrix} \begin{pmatrix} {}^t\boldsymbol{v}_{\ell-1} & {}^t\boldsymbol{w}_{\ell-1} & {}^t\boldsymbol{w}_{\ell-1} \end{pmatrix} \boldsymbol{Y}_\ell,
\end{aligned}
$$

and, as

$$
\begin{aligned}
{}^t\boldsymbol{Y}_\ell \begin{pmatrix} \boldsymbol{v}_{\ell-1} \\ \boldsymbol{w}_{\ell-1} \\ \boldsymbol{w}_{\ell-1} \end{pmatrix} &= \begin{pmatrix} \boldsymbol{I}_{[3,\ell-1]} & {}^t\widetilde{\widetilde{\mathbf{0}}}_\ell & {}^t\widetilde{\widetilde{\mathbf{0}}}_\ell \\ {}^t\widetilde{\boldsymbol{O}}_{\ell-1} & \boldsymbol{I}_{[3,\ell]-2} & {}^t\widetilde{\widetilde{\mathbf{0}}}_\ell \\ {}^t\widetilde{\mathbf{0}}_{\ell-1} & {}^t\widetilde{\widetilde{\mathbf{0}}}_\ell & \boldsymbol{I}_{[3,\ell]-2} \end{pmatrix} \begin{pmatrix} \boldsymbol{v}_{\ell-1} \\ \boldsymbol{w}_{\ell-1} \\ \boldsymbol{w}_{\ell-1} \end{pmatrix} \\
&= \begin{pmatrix} \binom{\ell}{0}\binom{0}{0} \\ \binom{\ell-1}{1}\binom{1}{0} + \binom{\ell-1}{0}\binom{0}{0} \\ \binom{\ell-1}{1}\binom{1}{1} + \binom{\ell-1}{0}\binom{0}{0} \\ \binom{\ell-1}{2}\binom{2}{0} + \binom{\ell-1}{1}\binom{1}{0} \\ \binom{\ell-1}{2}\binom{2}{1} + \binom{\ell-1}{1}\binom{1}{1} + \binom{\ell-1}{1}\binom{1}{0} \\ \vdots \\ \binom{\ell-1}{\ell-1}\binom{\ell-1}{\ell-1} \end{pmatrix} = \boldsymbol{v}_\ell,
\end{aligned}
$$

we find that $\boldsymbol{C}_\ell = \boldsymbol{v}_\ell\, {}^t\boldsymbol{v}_\ell$, which proves the assertion. □

Let us put $\widetilde{\boldsymbol{C}}_{\ell-1} = {}^t\boldsymbol{A}_{\ell-1}\boldsymbol{U}_{3,\ell-1}\boldsymbol{A}'_{\ell-1}$ and $\widetilde{\widetilde{\boldsymbol{C}}}_{\ell-1} = {}^t\boldsymbol{A}'_{\ell-1}\boldsymbol{U}_{3,\ell-1}\boldsymbol{A}'_{\ell-1}$. Then, using the previous lemma and the fact that

$$
\begin{aligned}
\widetilde{\boldsymbol{C}}_{\ell-1} &= \boldsymbol{v}_{\ell-1}\, {}^t\boldsymbol{w}_{\ell-1} \\
\widetilde{\widetilde{\boldsymbol{C}}}_{\ell-1} &= \boldsymbol{w}_{\ell-1}\, {}^t\boldsymbol{w}_{\ell-1},
\end{aligned}
$$

we can rewrite

$$
\begin{aligned}
\boldsymbol{B}'_{3,\ell} &= {}^t\boldsymbol{Y}_\ell \begin{pmatrix} 2\boldsymbol{C}_{\ell-1} & -\widetilde{\boldsymbol{C}}_{\ell-1} & -\widetilde{\boldsymbol{C}}_{\ell-1} \\ -{}^t\widetilde{\boldsymbol{C}}_{\ell-1} & 2\widetilde{\widetilde{\boldsymbol{C}}}_{\ell-1} & -\widetilde{\widetilde{\boldsymbol{C}}}_{\ell-1} \\ -{}^t\widetilde{\boldsymbol{C}}_{\ell-1} & -\widetilde{\widetilde{\boldsymbol{C}}}_{\ell-1} & 2\widetilde{\widetilde{\boldsymbol{C}}}_{\ell-1} \end{pmatrix} \boldsymbol{Y}_\ell \\
&= {}^t\boldsymbol{Y}_\ell \begin{pmatrix} 2\boldsymbol{v}_{\ell-1}\,{}^t\boldsymbol{v}_{\ell-1} & -\boldsymbol{v}_{\ell-1}\,{}^t\boldsymbol{w}_{\ell-1} & -\boldsymbol{v}_{\ell-1}\,{}^t\boldsymbol{w}_{\ell-1} \\ -\boldsymbol{w}_{\ell-1}\,{}^t\boldsymbol{v}_{\ell-1} & 2\boldsymbol{w}_{\ell-1}\,{}^t\boldsymbol{w}_{\ell-1} & -\boldsymbol{w}_{\ell-1}\,{}^t\boldsymbol{w}_{\ell-1} \\ -\boldsymbol{w}_{\ell-1}\,{}^t\boldsymbol{v}_{\ell-1} & -\boldsymbol{w}_{\ell-1}\,{}^t\boldsymbol{w}_{\ell-1} & 2\boldsymbol{w}_{\ell-1}\,{}^t\boldsymbol{w}_{\ell-1} \end{pmatrix} \boldsymbol{Y}_\ell
\end{aligned}
$$

as

$$
\begin{aligned}
\boldsymbol{B}'_{3,\ell} &= {}^t\boldsymbol{Y}_\ell \begin{pmatrix} \boldsymbol{v}_{\ell-1} \\ -\boldsymbol{w}_{\ell-1} \\ \widetilde{\widetilde{\mathbf{0}}}_\ell \end{pmatrix} \left({}^t\boldsymbol{v}_{\ell-1}, -{}^t\boldsymbol{w}_{\ell-1}, {}^t\widetilde{\widetilde{\mathbf{0}}}_\ell\right) \boldsymbol{Y}_\ell \\
&\quad + {}^t\boldsymbol{Y}_\ell \begin{pmatrix} \boldsymbol{v}_{\ell-1} \\ \widetilde{\widetilde{\mathbf{0}}}_\ell \\ -\boldsymbol{w}_{\ell-1} \end{pmatrix} \left({}^t\boldsymbol{v}_{\ell-1}, {}^t\widetilde{\widetilde{\mathbf{0}}}_\ell, -{}^t\boldsymbol{w}_{\ell-1}\right) \boldsymbol{Y}_\ell \\
&\quad + {}^t\boldsymbol{Y}_\ell \begin{pmatrix} \widetilde{\widetilde{\mathbf{0}}}_\ell \\ -\boldsymbol{w}_{\ell-1} \\ \boldsymbol{w}_{\ell-1} \end{pmatrix} \left({}^t\widetilde{\widetilde{\mathbf{0}}}_\ell, -{}^t\boldsymbol{w}_{\ell-1}, {}^t\boldsymbol{w}_{\ell-1}\right) \boldsymbol{Y}_\ell \\
&= (\boldsymbol{B}'_{3,\ell})_1 + (\boldsymbol{B}'_{3,\ell})_2 + (\boldsymbol{B}'_{3,\ell})_3.
\end{aligned}
$$

Since

$$
{}^t\boldsymbol{Y}_\ell \begin{pmatrix} \boldsymbol{v}_{\ell-1} \\ \boldsymbol{w}_{\ell-1} \\ \widetilde{\widetilde{\mathbf{0}}}_\ell \end{pmatrix} = \begin{pmatrix} \boldsymbol{I}_{[3,\ell-1]} & {}^t\widetilde{\widetilde{\mathbf{0}}}_\ell & {}^t\widetilde{\widetilde{\mathbf{0}}}_\ell \\ {}^t\widetilde{\boldsymbol{O}}_{\ell-1} & \boldsymbol{I}_{[3,\ell]-2} & {}^t\widetilde{\widetilde{\mathbf{0}}}_\ell \\ {}^t\widetilde{\mathbf{0}}_{\ell-1} & {}^t\widetilde{\widetilde{\mathbf{0}}}_\ell & \boldsymbol{I}_{[3,\ell]-2} \end{pmatrix} \begin{pmatrix} \boldsymbol{v}_{\ell-1} \\ \boldsymbol{w}_{\ell-1} \\ \widetilde{\widetilde{\mathbf{0}}}_\ell \end{pmatrix}
= \begin{pmatrix} \binom{\ell-1}{0}\binom{0}{0} \\ \binom{\ell-1}{1}\binom{1}{0} - \binom{\ell-1}{0}\binom{0}{0} \\ \binom{\ell-1}{1}\binom{1}{1} \\ \binom{\ell-1}{2}\binom{2}{0} - \binom{\ell-1}{1}\binom{1}{0} \\ \vdots \\ -\binom{\ell-1}{\ell-1}\binom{\ell-1}{0} \\ \vdots \\ -\binom{\ell-1}{\ell-1}\binom{\ell-1}{\ell-1} \\ 0 \end{pmatrix}
$$

and since

$$
\binom{\ell-1}{i}\binom{i}{j} - \binom{\ell-1}{i-1}\binom{i-1}{j} = \binom{\ell}{i}\binom{i}{j}\frac{\ell-2i+j}{\ell},
$$

$$
\binom{\ell-1}{i}\binom{i}{i} = \binom{\ell}{i}\binom{i}{i}\frac{\ell-i}{\ell}
$$

and

$$
-\binom{\ell-1}{j}\binom{\ell-1}{\ell-1} = \binom{\ell}{\ell}\binom{\ell}{j}\frac{j-\ell}{\ell},
$$

we obtain that

$$
{}^t\boldsymbol{Y}_\ell \begin{pmatrix} \boldsymbol{v}_{\ell-1} \\ \boldsymbol{w}_{\ell-1} \\ \widetilde{\widetilde{\mathbf{0}}}_\ell \end{pmatrix} = \begin{pmatrix} \binom{\ell}{0} \\ \binom{\ell}{1}\binom{1}{0}\frac{\ell-2}{\ell} \\ \binom{\ell}{1}\binom{1}{1}\frac{\ell-1}{\ell} \\ \vdots \\ \binom{\ell}{k}\binom{k}{r}\frac{\ell-2k+r}{\ell} \\ \vdots \\ 0 \end{pmatrix}.
$$

In a similar way, using

$$\binom{\ell-1}{i}\binom{i}{j}-\binom{\ell-1}{i-1}\binom{i-1}{j-1}=\binom{\ell}{i}\binom{i}{j}\frac{\ell-i-j}{\ell}$$
$$\binom{\ell-1}{i}\binom{i}{0}=\binom{\ell}{i}\binom{i}{0}\frac{\ell-i}{\ell}$$
$$-\binom{\ell-1}{\ell-1}\binom{\ell-1}{j-1}=\binom{\ell}{\ell}\binom{\ell}{j}\frac{(-j)}{\ell}$$

and

$$\binom{\ell-1}{j}\binom{j}{j-1}-\binom{\ell-1}{j}\binom{j}{j}=\binom{\ell}{j+1}\binom{j+1}{j}\frac{j-1}{\ell}$$
$$-\binom{\ell-1}{j}\binom{j}{0}=\binom{\ell}{j+1}\binom{j+1}{0}\frac{-(j+1)}{\ell}$$
$$\binom{\ell-1}{j}\binom{j}{j}=\binom{\ell}{j+1}\binom{j+1}{j+1}\frac{j+1}{\ell},$$

one can prove that

$$^t\boldsymbol{Y}_\ell\begin{pmatrix}\boldsymbol{v}_{\ell-1}\\ \widetilde{\widetilde{\mathbf{0}}}_\ell\\ -\boldsymbol{w}_{\ell-1}\end{pmatrix}=\begin{pmatrix}1\\ \binom{\ell}{1}\binom{1}{0}\frac{\ell-1}{\ell}\\ \vdots\\ \binom{\ell}{k}\binom{k}{r}\frac{\ell-k-r}{\ell}\\ \vdots\\ -1\end{pmatrix}$$

and

$$^t\boldsymbol{Y}_\ell\begin{pmatrix}\widetilde{\widetilde{\mathbf{0}}}_\ell\\ -\boldsymbol{w}_{\ell-1}\\ \boldsymbol{w}_{\ell-1}\end{pmatrix}=\begin{pmatrix}0\\ \binom{\ell}{1}\binom{1}{0}\frac{(-1)}{\ell}\\ \binom{\ell}{1}\binom{1}{1}\frac{1}{\ell}\\ \vdots\\ \binom{\ell}{k}\binom{k}{r}\frac{(2r-k)}{\ell}\\ \vdots\\ 1\end{pmatrix},$$

respectively. As a consequence, it is easy to check that

$$(\boldsymbol{B}'_{3,\ell})_1 = (b'_{pq,1}) = \binom{\ell}{k}\binom{k}{r}\binom{\ell}{s}\binom{s}{t}\frac{(\ell-2k+r)(\ell-2s+t)}{\ell^2},$$
$$(\boldsymbol{B}'_{3,\ell})_2 = (b'_{pq,2}) = \binom{\ell}{k}\binom{k}{r}\binom{\ell}{s}\binom{s}{t}\frac{(\ell-k-r)(\ell-s-t)}{\ell^2},$$
$$(\boldsymbol{B}'_{3,\ell})_3 = (b'_{pq,3}) = \binom{\ell}{k}\binom{k}{r}\binom{\ell}{s}\binom{s}{t}\frac{(k-2r)(s-2t)}{\ell^2}.$$

With notations as before, one now easily proves:

Proposition V.12. *For any $0 \leq p, q < [3, \ell]$, the component $(b'_{3,\ell})_{pq}$ of the matrix $\boldsymbol{B}'_{3,\ell}$ is given by*

$$(b'_{3,\ell})_{pq} = \binom{\ell}{k}\binom{k}{r}\binom{\ell}{s}\binom{s}{t}\times$$
$$\frac{1}{\ell^2}\left((\ell-2k+r)(\ell-2s+t)+(\ell-k-r)(\ell-s-t)+(k-2r)(s-2t)\right).$$

Proof. This trivially follows from the previous expressions of $(\boldsymbol{B}'_{3,\ell})_1$, $(\boldsymbol{B}'_{3,\ell})_2$ and $(\boldsymbol{B}'_{3,\ell})_3$. □

To obtain an explicit expression of the matrix $\boldsymbol{B}''_{3,\ell}$, we now consider the matrices

$$\widetilde{\boldsymbol{B}}_{3,\ell-1} = \left((\widetilde{b}_{3,\ell-1})_{pq}\right) = {}^t\boldsymbol{A}_{\ell-1}\boldsymbol{G}_{3,\ell-1}\boldsymbol{A}'_{\ell-1}$$
$$\widetilde{\widetilde{\boldsymbol{B}}}_{3,\ell-1} = \left((\widetilde{\widetilde{b}}_{3,\ell-1})_{pq}\right) = {}^t\boldsymbol{A}'_{\ell-1}\boldsymbol{G}_{3,\ell-1}\boldsymbol{A}'_{\ell-1},$$

and write μ and η for $[3, \ell-1]-1$ and $[3, \ell]-3$, respectively. Then, the next result calculates the matrix $\boldsymbol{B}''_{3,\ell} = (b''_{3,\ell})_{pq}$:

Proposition V.13. *The components $(b''_{3,\ell})_{pq}$ of the matrix $\boldsymbol{B}''_{3,\ell}$ are determined by:*

1.

$$\begin{aligned}
(b''_{3,\ell})_{00} &= (b_{3,\ell-1})_{00} \\
(b''_{3,\ell})_{01} &= (b_{3,\ell-1})_{01} + (\widetilde{b}_{3,\ell-1})_{00} \\
(b''_{3,\ell})_{0,\eta+2} &= (\widetilde{b}_{3,\ell-1})_{0\eta} \\
(b''_{3,\ell})_{11} &= (b_{3,\ell-1})_{11} + 2(\widetilde{b}_{3,\ell-1})_{10} + (\widetilde{\widetilde{b}}_{3,\ell-1})_{00} \\
(b''_{3,\ell})_{1,\eta+2} &= (\widetilde{b}_{3,\ell-1})_{1\eta} + (\widetilde{\widetilde{b}}_{3,\ell-1})_{0\eta} \\
(b''_{3,\ell})_{\eta+2,\eta+2} &= (\widetilde{\widetilde{b}}_{3,\ell-1})_{\eta\eta},
\end{aligned}$$

2. (a) *if* $2 \leq q \leq \mu$, *then*

$$\begin{aligned}
(b''_{3,\ell})_{0q} &= (b_{3,\ell-1})_{0,q} + (\widetilde{b}_{3,\ell-1})_{0,q-1} + (\widetilde{b}_{3,\ell-1})_{0,q-2} \\
(b''_{3,\ell})_{1,q} &= (b_{3,\ell-1})_{1,q} + (\widetilde{b}_{3,\ell-1})_{1,q-1} + (\widetilde{b}_{3,\ell-1})_{1,q-2} \\
&\quad + (\widetilde{b}_{3,\ell-1})_{q,0} + (\widetilde{\widetilde{b}}_{3,\ell-1})_{0,q-1} + (\widetilde{\widetilde{b}}_{3,\ell-1})_{0,q-2},
\end{aligned}$$

(b) *if* $\mu < q \leq \eta + 1$, *then*

$$\begin{aligned}
(b''_{3,\ell})_{0q} &= (\widetilde{b}_{3,\ell-1})_{0,q-1} + (\widetilde{b}_{3,\ell-1})_{0,q-2} \\
(b''_{3,\ell})_{1q} &= (\widetilde{b}_{3,\ell-1})_{1,q-1} + (\widetilde{b}_{3,\ell-1})_{1,q-2} + (\widetilde{\widetilde{b}}_{3,\ell-1})_{0,q-1} + (\widetilde{\widetilde{b}}_{3,\ell-1})_{0,q-2},
\end{aligned}$$

3. *for any* $2 \leq p \leq \mu$, *we have*

(a) *if* $p \leq q \leq \mu$, *then*

$$\begin{aligned}
(b''_{3,\ell})_{pq} &= (b_{3,\ell-1})_{pq} + (\widetilde{b}_{3,\ell-1})_{p,q-1} + (\widetilde{b}_{3,\ell-1})_{p,q-2} \\
&\quad + (\widetilde{b}_{3,\ell-1})_{q,p-1} + (\widetilde{b}_{3,\ell-1})_{q,p-2} + (\widetilde{\widetilde{b}}_{3,\ell-1})_{p-1,q-1} \\
&\quad + (\widetilde{\widetilde{b}}_{3,\ell-1})_{p-1,q-2} + (\widetilde{\widetilde{b}}_{3,\ell-1})_{p-2,q-1} + (\widetilde{\widetilde{b}}_{3,\ell-1})_{p-2,q-2},
\end{aligned}$$

(b) *if* $\mu < q \leq \eta + 1$, *then*

$$\begin{aligned}
(b''_{3,\ell})_{pq} &= (\widetilde{b}_{3,\ell-1})_{p,q-1} + (\widetilde{b}_{3,\ell-1})_{p,q-2} + (\widetilde{\widetilde{b}}_{3,\ell-1})_{p-1,q-1} \\
&\quad + (\widetilde{\widetilde{b}}_{3,\ell-1})_{p-1,q-2} + (\widetilde{\widetilde{b}}_{3,\ell-1})_{p-2,q-1} + (\widetilde{\widetilde{b}}_{3,\ell-1})_{p-2,q-2},
\end{aligned}$$

(c)

$$(b''_{3,\ell})_{p,\eta+2} = (\widetilde{b}_{3,\ell-1})_{p,\eta} + (\widetilde{\widetilde{b}}_{3,\ell-1})_{p-1,\eta} + (\widetilde{\widetilde{b}}_{3,\ell-1})_{p-2,\eta},$$

4. (a) *if* $\mu < p \leq q \leq \eta + 1$, *then*

$$\begin{aligned}(b''_{3,\ell})_{pq} = {} & (\widetilde{\widetilde{b}}_{3,\ell-1})_{p-1,q-1} + (\widetilde{\widetilde{b}}_{3,\ell-1})_{p-1,q-2} \\ & + (\widetilde{\widetilde{b}}_{3,\ell-1})_{p-2,q-1} + (\widetilde{\widetilde{b}}_{3,\ell-1})_{p-2,q-2},\end{aligned}$$

(b) *if* $\mu < p \leq \eta + 1$, *then*

$$(b''_{3,\ell})_{p,\eta+2} = (\widetilde{\widetilde{b}}_{3,\ell-1})_{p-1,\eta} + (\widetilde{\widetilde{b}}_{3,\ell-1}).$$

Proof. In order to prove the validity of the previous relations, it suffices to use the fact that $\boldsymbol{B}''_{3,\ell}$ can be written as

$$\begin{aligned}\boldsymbol{B}''_{3,\ell} &= {}^t\boldsymbol{Y}_\ell \begin{pmatrix} \boldsymbol{B}_{3,\ell-1} & \widetilde{\boldsymbol{B}}_{3,\ell-1} & \widetilde{\boldsymbol{B}}_{3,\ell-1} \\ {}^t\widetilde{\boldsymbol{B}}_{3,\ell-1} & \widetilde{\widetilde{\boldsymbol{B}}}_{3,\ell-1} & \widetilde{\widetilde{\boldsymbol{B}}}_{3,\ell-1} \\ {}^t\widetilde{\boldsymbol{B}}_{3,\ell-1} & \widetilde{\widetilde{\boldsymbol{B}}}_{3,\ell-1} & \widetilde{\widetilde{\boldsymbol{B}}}_{3,\ell-1} \end{pmatrix} \boldsymbol{Y}_\ell \\ &= {}^t\boldsymbol{Y}_\ell \begin{pmatrix} \boldsymbol{B}_{3,\ell-1} & \widetilde{\boldsymbol{B}}_{3,\ell-1} & \widetilde{\boldsymbol{B}}_{3,\ell-1} \\ {}^t\widetilde{\boldsymbol{B}}_{3,\ell-1} & \boldsymbol{O}_\beta & \boldsymbol{O}_\beta \\ {}^t\widetilde{\boldsymbol{B}}_{3,\ell-1} & \boldsymbol{O}_\beta & \boldsymbol{O}_\beta \end{pmatrix} \boldsymbol{Y}_\ell + {}^t\boldsymbol{Y}_\ell \begin{pmatrix} \boldsymbol{O}_\alpha & \boldsymbol{O}_{\alpha\beta} & \boldsymbol{O}_{\alpha\beta} \\ {}^t\boldsymbol{O}_{\alpha\beta} & \widetilde{\widetilde{\boldsymbol{B}}}_{3,\ell-1} & \widetilde{\widetilde{\boldsymbol{B}}}_{3,\ell-1} \\ {}^t\boldsymbol{O}_{\alpha\beta} & \widetilde{\widetilde{\boldsymbol{B}}}_{3,\ell-1} & \widetilde{\widetilde{\boldsymbol{B}}}_{3,\ell-1} \end{pmatrix} \boldsymbol{Y}_\ell,\end{aligned}$$

where $\boldsymbol{O}_{\alpha\beta}$ is the zero-matrix of dimensions $[3, \ell-1] \times [3, \ell] - 2$ and $\boldsymbol{O}_\alpha$ and $\boldsymbol{O}_\beta$ are the square zero-matrices of dimension $[3, \ell-1]$ and $[3, \ell] - 2$, respectively.

□

From the previous results and the fact that $\boldsymbol{B}_{3,\ell} = \boldsymbol{B}'_{3,\ell} + \boldsymbol{B}''_{3,\ell}$, it trivially follows:

Corollary V.14. *The components* $(b_{3,\ell})_{pq}$ *of the matrix* $\boldsymbol{B}_{3,\ell}$ *are determined by*

1.

$$\begin{aligned}(b_{3,\ell})_{00} &= (b_{3,\ell})_{\eta+2,\eta+2} = 1 + 2\ell \\ (b_{3,\ell})_{01} &= 2\ell(\ell-1) \\ (b_{3,\ell})_{0,\eta+2} &= 1 - \ell \\ (b_{3,\ell})_{11} &= (b_{3,\ell-1})_{11} + 2(b_{3,\ell-1})_{01} + (b_{3,\ell-1})_{00} + (\ell-2)^2 + (\ell-1)^2 + 1 \\ (b_{3,\ell})_{1,\eta+2} &= (b_{3,\ell-1})_{1\mu} + (b_{3,\ell-1})_{0\mu} - \ell,\end{aligned}$$

2. (a) *if* $2 \leq q \leq \mu$, *then*

$$
\begin{aligned}
(b_{3,\ell})_{0q} &= (b_{3,\ell-1})_{0q} + (\widetilde{b}_{3,\ell-1})_{0,q-1} + (\widetilde{b}_{3,\ell-1})_{0,q-2} + \tbinom{\ell}{s}\tbinom{s}{t}\tfrac{(2\ell-3s)}{\ell} \\
(b_{3,\ell})_{1q} &= (b_{3,\ell-1})_{1q} + (\widetilde{b}_{3,\ell-1})_{1,q-1} + (\widetilde{b}_{3,\ell-1})_{1,q-2} \\
&\quad + (\widetilde{b}_{3,\ell-1})_{q,0} + (\widetilde{\widetilde{b}}_{3,\ell-1})_{0,q-1}(\widetilde{\widetilde{b}}_{3,\ell-1})_{0,q-2} \\
&\quad + \tbinom{\ell}{s}\tbinom{s}{t}\frac{1}{\ell}\left((\ell-2)(\ell-2s+t) + (\ell-1)(\ell-s-t) + (s-2t)\right),
\end{aligned}
$$

(b) *if* $\mu < q \leq \eta + 1$, *then*

$$
\begin{aligned}
(b_{3,\ell})_{0q} &= (\widetilde{b}_{3,\ell-1})_{0,q-1} + (\widetilde{b}_{3,\ell-1})_{0,q-2} + \tbinom{\ell}{s}\tbinom{s}{t}\tfrac{(2\ell-3s)}{\ell} \\
(b_{3,\ell})_{1q} &= (\widetilde{b}_{3,\ell-1})_{1,q-1} + (\widetilde{b}_{3,\ell-1})_{1,q-2} + (\widetilde{\widetilde{b}}_{3,\ell-1})_{0,q-1} + (\widetilde{\widetilde{b}}_{3,\ell-1})_{0,q-2} \\
&\quad + \tbinom{\ell}{s}\tbinom{s}{t}\frac{1}{\ell}\left((\ell-2)(\ell-2s+t) + (\ell-1)(\ell-s-t) + (s-2t)\right),
\end{aligned}
$$

3. *for any* $2 \leq p \leq \mu$, *we have*

(a) *if* $p \leq q \leq \mu$, *then*

$$
\begin{aligned}
(b_{3,\ell})_{pq} &= (b_{3,\ell-1})_{pq} + (\widetilde{b}_{3,\ell-1})_{p,q-1} + (\widetilde{b}_{3,\ell-1})_{p,q-2} + (\widetilde{b}_{3,\ell-1})_{q,p-1} \\
&\quad + (\widetilde{b}_{3,\ell-1})_{q,p-2} + (\widetilde{\widetilde{b}}_{3,\ell-1})_{p-1,q-1} + (\widetilde{\widetilde{b}}_{3,\ell-1})_{p-1,q-2} \\
&\quad + (\widetilde{\widetilde{b}}_{3,\ell-1})_{p-2,q-1} + (\widetilde{\widetilde{b}}_{3,\ell-1})_{p-2,q-2} \\
&\quad + \tbinom{\ell}{k}\tbinom{k}{r}\tbinom{\ell}{s}\tbinom{s}{t}\tfrac{(\ell-2k+r)(\ell-2s+t)+(\ell-k-r)(\ell-s-t)+(k-2r)(s-2t)}{\ell^2},
\end{aligned}
$$

(b) *if* $\mu < q \leq \eta + 1$, *then*

$$
\begin{aligned}
(b_{3,\ell})_{pq} &= (\widetilde{b}_{3,\ell-1})_{p,q-1} + (\widetilde{b}_{3,\ell-1})_{p,q-2} + (\widetilde{\widetilde{b}}_{3,\ell-1})_{p-1,q-1} \\
&\quad + (\widetilde{\widetilde{b}}_{3,\ell-1})_{p-1,q-2} + (\widetilde{\widetilde{b}}_{3,\ell-1})_{p-2,q-1} + (\widetilde{\widetilde{b}}_{3,\ell-1})_{p-2,q-2} \\
&\quad + \tbinom{\ell}{k}\tbinom{k}{r}\tbinom{\ell}{s}\tbinom{s}{t}\tfrac{(\ell-2k+r)(\ell-2s+t)+(\ell-k-r)(\ell-s-t)+(k-2r)(s-2t)}{\ell^2},
\end{aligned}
$$

(c)

$$
(b_{3,\ell})_{p,\eta+2} = (\widetilde{b}_{3,\ell-1})_{p,\eta} + (\widetilde{\widetilde{b}}_{3,\ell-1})_{p-1,\eta} + (\widetilde{\widetilde{b}}_{3,\ell-1})_{p-2,\eta} + \binom{\ell}{k}\binom{k}{r}\frac{(3r-\ell)}{\ell},
$$

4. *if* $\mu < p \leq \eta + 1$, *then we have*

(a) if $p \leq q \leq \eta + 1$*, then*

$$(b_{3,\ell})_{pq} = (\tilde{\tilde{b}}_{3,\ell-1})_{p-1,q-1} + (\tilde{\tilde{b}}_{3,\ell-1})_{p-1,q-2} + (\tilde{\tilde{b}}_{3,\ell-1})_{p-2,q-1} + (\tilde{\tilde{b}}_{3,\ell-1})_{p-2,q-2} + \binom{\ell}{k}\binom{k}{r}\binom{\ell}{s}\binom{s}{t}\frac{(\ell-2k+r)(\ell-2s+t)+(\ell-k-r)(\ell-s-t)+(k-2r)(s-2t)}{\ell^2},$$

(b)

$$(b_{3,\ell})_{p,\eta+2} = (\tilde{\tilde{b}}_{3,\ell-1})_{p-1,\eta} + (\tilde{\tilde{b}}_{3,\ell-1})_{p-2,\eta} + \binom{\ell}{k}\binom{k}{r}\frac{(3r-\ell)}{\ell}.$$

We now finally may prove:

Theorem V.15. *For any* $0 \leq p, q < [3, \ell]$*, the component* $(b_{3,\ell})_{p,q}$ *of the matrix* $\boldsymbol{B}_{3,\ell}$ *is given by*

$$(b_{3,\ell})_{p,q} = \binom{\ell}{k}\binom{k}{r}\binom{\ell}{s}\binom{s}{t}\left(1 + \frac{(\ell-2k+r)(\ell-2s+t)+(\ell-k-r)(\ell-s-t)+(k-2r)(s-2t)}{\ell}\right).$$

Proof. The previous result directly yields that our claim is true for $(b_{3,\ell})_{00}$, $(b_{3,\ell})_{01}$, $(b_{3,\ell})_{0,\eta+2}$ and $(b_{3,\ell})_{\eta+2,\eta+2}$. Using an induction argument on ℓ, one easily verifies the expression for $(b_{3,\ell})_{11}$ and $(b_{3,\ell})_{1,\eta+2}$.

For the other elements of the matrix, one also proceeds by induction on ℓ, for each of the different cases. As the technique is long but rather straightforward, we will only detail the involved calculations for the remaining elements of the first row of $\boldsymbol{B}_{3,\ell}$, i.e., those $(b_{3,\ell})_{0q}$ with $q \geq 2$. The other components may be calculated in a similar way.

Observe that for the components $(b_{3,\ell})_{0q}$ we have $k = r = 0$.

Let us first assume that $q \leq \mu$. Hence for $\ell = 1$, we have $(b_{3,1})_{02} = 0$, implying the statement to hold true. Assume the statement to be correct for $1, 2, \ldots, \ell - 1$ and let us prove it in dimension ℓ.

Fixing $q = \frac{s(s+1)}{2} + t$, we will distinguish three cases, according to the values of t ($t = s$, $t = 0$ and $t \neq s, 0$).

i) If $t = s$ then $q - 1 = \frac{s(s+1)}{2} + (s-1)$ and $(\tilde{b}_{3,\ell-1})_{0,q-1} = 0$. Moreover, $(\tilde{b}_{3,\ell-1})_{0,q-2} = (b_{3,\ell-1})_{0,\widehat{q}}$, with $\widehat{q} = (q-2) - (s-1) = \frac{(s-1)s}{2} + (s-1)$. (Note that the columns of $\widetilde{\mathbf{B}}_{3,\ell-1}$ which are located at positions $\frac{(s+1)(s+2)}{2} - 1$, with $s \geq 1$, are zero-columns.) Then, from the previous corollary, we have

$$(b_{3,\ell})_{0q} = (b_{3,\ell-1})_{0,q} + (b_{3,\ell-1})_{0,\widehat{q}} + \binom{\ell}{s}\binom{s}{t}\frac{(2\ell-3s)}{\ell}.$$

Now, using the induction hypothesis on ℓ, it follows that

$$\begin{aligned}(b_{3,\ell})_{0q} &= \tbinom{\ell-1}{s}(2\ell-3s-1)+\tbinom{\ell-1}{s-1}(2\ell-3s+2)+\tbinom{\ell}{s}\tfrac{(2\ell-3s)}{\ell}\\ &= \tbinom{\ell}{s}(2\ell-3s+1)\\ &= \tbinom{\ell}{0}\tbinom{0}{0}\tbinom{\ell}{s}\tbinom{s}{s}\left(1+\tfrac{\ell(\ell-s)+\ell(\ell-2s)}{\ell}\right).\end{aligned}$$

ii) If $t=0$, then $(\widetilde{b}_{3,\ell-1})_{0,q-1}=(b_{3,\ell-1})_{0,\widehat{q}}$, with $\widehat{q}=\frac{(s-1)s}{2}$ and $(\widetilde{b}_{3,\ell-1})_{0,q-2}=0$. From the previous result and the induction hypothesis on ℓ, we again obtain that

$$(b_{3,\ell})_{0q}=\binom{\ell}{s}(2\ell-3s+1).$$

iii) If $t\neq s,0$ then, $(\widetilde{b}_{3,\ell-1})_{0,q-1}=(b_{3,\ell-1})_{0,\widehat{q}}$, with $\widehat{q}=\frac{(s-1)s}{2}+t$ and $(\widetilde{b}_{3,\ell-1})_{0,q-2}=(b_{3,\ell-1})_{0,\widehat{\widehat{q}}}$ where $\widehat{\widehat{q}}=\frac{(s-1)s}{2}+(t-1)=\widehat{q}-1$. Then the same reasoning as before shows that

$$\begin{aligned}(b_{3,\ell})_{0q} &= (b_{3,\ell-1})_{0,q-1}+(b_{3,\ell-1})_{0,\widehat{q}}+(b_{3,\ell-1})_{0,\widehat{q}-1}+\tbinom{\ell}{s}\tbinom{s}{t}\tfrac{(2\ell-3s)}{\ell}\\ &= \tbinom{\ell-1}{s}\tbinom{s}{t}(2\ell-3s-1)+\tbinom{\ell-1}{s-1}\tbinom{s-1}{t}(2\ell-3s+2)\\ &\quad+\tbinom{\ell-1}{s-1}\tbinom{s-1}{t-1}(2\ell-3s+2)+\tbinom{\ell}{s}\tbinom{s}{t}\tfrac{(2\ell-3s)}{\ell}\\ &= \tbinom{\ell}{s}\tbinom{s}{t}(2\ell-3s+1),\end{aligned}$$

as wanted.

Let us now assume that $\mu<q\leq\eta+1$ (hence $\ell\geq 2$, and $q=\frac{\ell(\ell+1)}{2}+t$, with $0\leq t<\ell$). We again argue by induction on ℓ. If $\ell=2$, then (as $\mu=2$ and $\eta=3$), we have $(b_{3,2})_{03}=-1$ and $(b_{3,2})_{04}=-2$, which satisfy the required expression.

Let us suppose that the assertion is true for length $2,3,\dots,\ell-1$ and let us prove it for ℓ. Applying corollary V.14 then yields

$$(b_{3,\ell})_{0,q}=(\widetilde{b}_{3,\ell-1})_{0,q-1}+(\widetilde{b}_{3,\ell-1})_{0,q-2}-\tbinom{\ell}{t}.$$

As $(\widetilde{b}_{3,\ell-1})_{0,q-1}=(b_{3,\ell})_{0,\widehat{q}}$, with $\widehat{q}=\frac{(\ell-1)\ell}{2}+t$, we have $(\widetilde{b}_{3,\ell-1})_{0,q-2}=(b_{3,\ell})_{0,\widehat{q}-1}$. Then

$$\begin{aligned}(b_{3,\ell})_{0,q} &= (b_{3,\ell})_{0,\widehat{q}}+(b_{3,\ell})_{0,\widehat{q}-1}-\tbinom{\ell}{t}\\ &= \left(\tbinom{\ell-1}{t}+\tbinom{\ell-1}{t-1}\right)(2-\ell)-\tbinom{\ell}{t}\\ &= \tbinom{\ell}{t}(1-\ell),\end{aligned}$$

which finishes the proof for the components of the first row of $\boldsymbol{B}_{3,\ell}$. □

With the above matrix $\boldsymbol{B}_{3,\ell}$ and the fact that $\boldsymbol{f}$ is completely determined by the vector $\boldsymbol{h}$, the definition of normalized epistasis will then simplify to

$$\varepsilon^*_{3,\ell}(f) = 1 - \frac{1}{3^\ell}\frac{{}^t\boldsymbol{h}\,\boldsymbol{B}_{3,\ell}\,\boldsymbol{h}}{\|\boldsymbol{f}\|^2},$$

with

$$\|\boldsymbol{f}\|^2 = \sum_{p=0}^{m} \binom{\ell}{k}\binom{k}{r} h_p^2,$$

where, as above, $k, r \in \mathbb{N}$ with $0 \leq k, r \leq \ell$ and k is the greatest non-negative integer such that $p = \frac{k(k+1)}{2} + r$.

To conclude this section, we now include two easy examples for which we will obtain the normalized epistasis using the above formula.

Let us start by considering the needle-in-a-haystack function, located at $t = 0$, as an example of a generalized unitation function on $\Omega_3 = \{0, 1, 2\}^\ell$. Then, as $needle(t) = \delta_{t,0}$, for all t, we have that the vector $\boldsymbol{h}$ is given by

$$\boldsymbol{h} = {}^t(1, 0, \ldots, 0) \in \mathbb{R}^{[3,\ell]}$$

and so, from theorem V.15, it follows that

$$\varepsilon^*_{3,\ell}(needle) = 1 - \frac{(b_3^\ell)_{00}}{3^\ell} = 1 - \frac{2\ell + 1}{3^\ell}.$$

Let us now consider generalized camel functions, and let us denote by i_0, i_1, i_2 the indices such that $f(c_{i_0}) = f(c_{i_1}) = f(c_{i_2}) = \frac{\sqrt{3}}{3}$, (they are the only components of $\boldsymbol{f}$ whose values are different from zero, and they are at a pairwise Hamming distance of ℓ). So, the associated vector $\boldsymbol{f}$ of the function f is the sum of three (suitable) vectors of the canonical basis of $\mathbb{R}^{3^\ell}$. To keep the calculations of the normalized epistasis simple, we may assume that $i_0 = 0$, $i_1 = \frac{3^\ell - 1}{2}$ and $i_2 = 3^\ell - 1$, and therefore $\boldsymbol{f} = \sum_{k=0}^{2} \boldsymbol{e}_{k\frac{3^\ell - 1}{2}}$, i.e., the non-zero components of $\boldsymbol{f}$ corresponding to the images of the strings $00\ldots0$, $11\ldots1$ and $22\ldots2$. This is equivalent to saying that the components of the $[3,\ell]$-dimensional vector $\boldsymbol{h}$ associated to $\boldsymbol{f}$ are given by

$$h_j = \begin{cases} 1 & \text{if } j = 0, \nu, \eta + 2 \\ 0 & \text{otherwise,} \end{cases}$$

for $0 \leq j < \binom{\ell+2}{2}$ and where $\nu = \binom{\ell+1}{2}$ and $\eta = \binom{\ell+2}{2} - 2$ (as in theorem V.15), and

$$\begin{aligned}\varepsilon^*_{3,\ell}(f) &= 1 - \frac{1}{3^\ell}\frac{{}^t\boldsymbol{h}\,\boldsymbol{B}_{3,\ell}\,\boldsymbol{h}}{\|\boldsymbol{f}\|^2}\\ &= 1 - \frac{1}{3^\ell}\frac{(b_3^\ell)_{00} + (b_3^\ell)_{\nu,\nu} + (b_3^\ell)_{\eta+2,\eta+2} + 2\left\{(b_3^\ell)_{0,\nu} + (b_3^\ell)_{0,\eta+2} + (b_3^\ell)_{\nu,\eta+2}\right\}}{\|\boldsymbol{f}\|^2}.\end{aligned}$$

Now, using corollary V.14 and theorem V.15, a straightforward calculation shows that

$$\begin{aligned}(b_3^\ell)_{\nu,\nu} = (b_3^\ell)_{00} = (b_3^\ell)_{\eta+2,\eta+2} = 1 + 2\ell\\ (b_3^\ell)_{0,\nu} = (b_3^\ell)_{\nu,\eta+2} = (b_3^\ell)_{0,\eta+2} = 1 - \ell,\end{aligned}$$

and so,

$$\varepsilon^*_{3,\ell}(f) = 1 - \frac{1}{3^\ell}\frac{3(1+2\ell) + 2\,(3(1-\ell))}{3} = 1 - \frac{1}{3^{\ell-1}}.$$

4.2 Extreme values of normalized epistasis

It follows from theorem V.9 that for any positive fitness function f defined on ternary strings of length ℓ, we have

$$0 \leq \varepsilon^*(f) \leq 1 - \frac{1}{3^{\ell-1}}.$$

We will conclude this chapter by briefly considering the extreme values of normalized epistasis when restricting to generalized unitation functions.

As far as the maximal value is concerned, recall that theorem V.10 characterizes the functions f with maximal normalized epistasis. Among them, we have just shown (in the previous section for the generalized camel functions) that the maximal normalized epistasis may be realized by generalized unitation functions.

Let us now concentrate on the minimal value. First, note that from the expression of normalized epistasis, in terms of $\boldsymbol{h}$ and $\boldsymbol{B}_{3,\ell}$, it immediately follows that

$$\varepsilon^*(f) = 1 - \frac{(2\ell+1)}{3^\ell} + \frac{2}{3^\ell}\frac{H(h_0,\ldots,h_m)}{\sum_{p=0}^{m}\binom{\ell}{k}\binom{k}{r}h_p 2}$$

where

$$H(h_0,\ldots,h_m) = \frac{1}{2}\left(\sum_{p=0}^{m}\left((2\ell+1)\binom{\ell}{k}\binom{k}{r} - (b_{3,\ell})_{pp}\right)h_p 2 - \sum_{\substack{p,q\\p\neq q}}(b_{3,\ell})_{pq}h_p h_q\right).$$

To obtain the extreme values of the normalized epistasis, we will analyze the critical points of the function H, with the restriction that $\|f\| = 1$, i.e., we have to calculate the critical points of

$$F(h_0,\ldots,h_m,\lambda) = H(h_0,\ldots,h_m) + \lambda\left(\sum_{p=0}^{m}\binom{\ell}{k}\binom{k}{r}h_p^2 - 1\right).$$

We will do this by using the method of Lagrange multipliers, which leads to the following homogeneous system of $m+1$ equations:

$$\frac{\partial F}{\partial h_0} = 2\lambda h_0 - \sum_{q=1}^{m}(b_{3,\ell})_{0q}h_q = 0$$

$$\frac{\partial F}{\partial h_1} = \left(\binom{\ell}{1}(2\ell+1+2\lambda) - (b_{3,\ell})_{11}\right)h_1 - \sum_{\substack{q=0\\q\neq 1}}^{m} b_{1q}h_q = 0$$

$$\vdots$$

$$\frac{\partial F}{\partial h_p} = \left(\binom{\ell}{k}\binom{k}{r}(2\ell+1+2\lambda) - (b_{3,\ell})_{pp}\right)h_p - \sum_{\substack{q=0\\q\neq p}}^{m}(b_{3,\ell})_{pq}h_q = 0$$

$$\vdots$$

$$\frac{\partial F}{\partial h_m} = 2\lambda h_m - \sum_{p=0}^{m-1}(b_{3,\ell})_{pm}h_p = 0.$$

Its solutions are the eigenvectors $\boldsymbol{h} \in \mathbb{R}^{m+1}$ associated with the eigenvalues of the matrix $\widehat{\boldsymbol{B}}_{3,\ell}$ whose elements $(\widehat{b}_{3,\ell})_{pq}$ satisfy, for any $0 \le p, q \le m$,

$$(\widehat{b}_{3,\ell})_{pq} = \frac{(b_{3,\ell})_{pq}}{\binom{\ell}{k}\binom{k}{r}}.$$

We are interested in the eigenvectors with $h_i \ge 0$ for all i.

In order to obtain the eigenvalues of $\widehat{\mathbf{B}}_{3,\ell}$, we will use the following two lemmas:

Lemma V.16. *For any integers $0 \le u, v \le \ell$ and writing $z = \frac{u(u+1)}{2} + v$ and $m = [3, \ell] - 1$, we have*

i) $\sum_{z=0}^{m} \binom{\ell}{u}\binom{u}{v} = 3^\ell$,

ii) $\sum_{z=0}^{m} u\binom{\ell}{u}\binom{u}{v} = 2\sum_{z=0}^{m} v\binom{\ell}{u}\binom{u}{v} = 2\ell\, 3^{\ell-1}$,

iii) $\sum_{z=0}^{m} u^2\binom{\ell}{u}\binom{u}{v} = 2\ell(2\ell+1)3^{\ell-2}$, $\qquad \sum_{z=0}^{m} v^2\binom{\ell}{u}\binom{u}{v} = \ell(\ell+2)3^{\ell-2}$,

iv) $\sum_{z=0}^{m} uv\binom{\ell}{u}\binom{u}{v} = \ell(2\ell+1)3^{\ell-2}$,

Proof. To prove i), note that $(x+1)^\ell = \sum_{u=0}^{\ell} \binom{\ell}{u} x^u$ implies, in particular,

$$\begin{aligned}
3^\ell &= \sum_{u=0}^{\ell} \binom{\ell}{u} 2^u = \sum_{u=0}^{\ell} \binom{\ell}{u} \left(\sum_{v=0}^{u} \binom{u}{v}\right) \\
&= \binom{\ell}{0}\binom{0}{0} + \binom{\ell}{1}\left(\binom{1}{0} + \binom{1}{1}\right) + \dots \\
&\quad + \binom{\ell}{\ell}\left(\binom{\ell}{0} + \binom{\ell}{1} + \dots + \binom{\ell}{\ell}\right) \\
&= \sum_{z=0}^{m} \binom{\ell}{u}\binom{u}{v}.
\end{aligned}$$

On the other hand, if we differentiate the function $\varphi(x) = (x+1)^\ell$ at $x = 2$, we immediately obtain

$$\sum_{z=0}^{m} u\binom{\ell}{u}\binom{u}{v} = 2\ell\, 3^{\ell-1}.$$

Moreover,

$$\begin{aligned}
\sum_{z=0}^{m} v\binom{\ell}{u}\binom{u}{v} &= \sum_{u=0}^{\ell} \binom{\ell}{u}\left(\sum_{v=0}^{u} v\binom{u}{v}\right) = \frac{1}{2}\sum_{u=0}^{\ell} u\binom{\ell}{u} 2^u \\
&= \frac{1}{2}\sum_{u=0}^{\ell} u\binom{\ell}{u}\left(\sum_{v=0}^{u}\binom{u}{v}\right) = \frac{1}{2}\sum_{z=0}^{m} u\binom{\ell}{u}\binom{u}{v}.
\end{aligned}$$

This proves ii).

To check the first expression in iii), note that calculating the second derivative of the function φ at $x = 2$ yields

$$\begin{aligned}\ell(\ell-1)3^{\ell-2} &= \frac{1}{4}\sum_{u=0}^{\ell} u(u-1)\binom{\ell}{u}2^u \\ &= \frac{1}{4}\left(\sum_{u=0}^{\ell} u^2\binom{\ell}{u}2^u - \sum_{u=0}^{\ell} u\binom{\ell}{u}2^u\right).\end{aligned}$$

Applying ii) then does the trick. The second expression may be derived similarly, using the fact that $\sum_{v=0}^{u} v^2\binom{u}{v} = u(u+1)2^{u-2}$. Finally, note that iv) is obtained in a straightforward manner from the second expression in iii). □

Lemma V.17. *For any positive integer ℓ, we have $\widehat{\boldsymbol{B}}_{3,\ell}^2 = 3^\ell \widehat{\boldsymbol{B}}_{3,\ell}$.*

Proof. Let us denote by $(\widehat{b}_{3,\ell})^2_{pq}$ the generic element of the matrix $\widehat{\boldsymbol{B}}_{3,\ell}^2$. Then

$$(\widehat{b}_{3,\ell})^2_{pq} = \sum_{z=0}^{m} (\widehat{b}_{3,\ell})_{pz}(\widehat{b}_{3,\ell})_{zq}.$$

We proceed by expanding this expression using theorem V.15 and the previous lemma, which results in a tedious and uninteresting calculation leading to the required equality. □

Note that lemma V.17 implies the eigenvalues of $\widehat{\boldsymbol{B}}_{3,\ell}$ to be 0 and 3^ℓ. In order to describe the eigenspaces of $\widehat{\boldsymbol{B}}_{3,\ell}$, let us calculate its rank.

Lemma V.18. *For any positive integer ℓ, we have $\mathrm{rk}(\widehat{\boldsymbol{B}}_{3,\ell}) = 3$.*

Proof. From the relationship between the elements of the matrices $\boldsymbol{B}_{3,\ell}$ and $\widehat{\boldsymbol{B}}_{3,\ell}$, it follows that $\mathrm{rk}(\widehat{\boldsymbol{B}}_{3,\ell}) = \mathrm{rk}(\boldsymbol{B}_{3,\ell})$. Moreover, for any $0 \le p, q < [3, \ell]$,

$$(b_{3,\ell})_{pq} = \alpha_p(b_{3,\ell})_{0q} + \beta_p(b_{3,\ell})_{1q} + \gamma_q(b_{3,\ell})_{2q},$$

with $\alpha_p, \beta_p, \gamma_q \in \mathbb{R}$. Indeed,

$$\begin{aligned}(b_{3,\ell})_{pq} &= \binom{\ell}{k}\binom{k}{r}\binom{\ell}{s}\binom{s}{t}\left(1+\tfrac{(\ell-2k+r)(\ell-2s+t)+(\ell-k-r)(\ell-s-t)+(k-2r)(s-2t)}{\ell}\right)\\ &= \binom{\ell}{k}\binom{k}{r}(1-k)\binom{\ell}{s}\binom{s}{t}(1+2\ell-3s)\\ &+\binom{\ell}{k}\binom{k}{r}\tfrac{(k-r)}{\ell}\binom{\ell}{1}\binom{\ell}{s}\binom{s}{t}\left(1+\tfrac{(\ell-2)(\ell-2s+t)+(\ell-1)(\ell-s-t)+(s-2t)}{\ell}\right)\\ &+\binom{\ell}{k}\binom{k}{r}\tfrac{r}{\ell}\binom{\ell}{1}\binom{\ell}{s}\binom{s}{t}\left(1+\tfrac{(\ell-1)(\ell-2s+t)+(\ell-2)(\ell-s-t)-(s-2t)}{\ell}\right)\\ &= \binom{\ell}{k}\binom{k}{r}(1-k)(b_{3,\ell})_{0q}+\binom{\ell}{k}\binom{k}{r}\tfrac{(k-r)}{\ell}(b_{3,\ell})_{1q}+\binom{\ell}{k}\binom{k}{r}\tfrac{r}{\ell}(b_{3,\ell})_{2q}.\end{aligned}$$

However,

$$\det\begin{pmatrix}(b_{3,\ell})_{00} & (b_{3,\ell})_{01} & (b_{3,\ell})_{02}\\ (b_{3,\ell})_{10} & (b_{3,\ell})_{11} & (b_{3,\ell})_{12}\\ (b_{3,\ell})_{20} & (b_{3,\ell})_{21} & (b_{3,\ell})_{22}\end{pmatrix} = 27\ell^2,$$

which finishes the proof. □

If we denote by $W_0^{3,\ell}$ and $W_1^{3,\ell}$ the eigenspaces in $\mathbb{R}^{\binom{\ell+2}{\ell}}$ corresponding to the eigenvalues 0 and 3^ℓ of $\widehat{\boldsymbol{B}}_{3,\ell}$, then $\mathbb{R}^{\binom{\ell+2}{\ell}} = W_0^{3,\ell} \oplus W_1^{3,\ell}$. As $W_0^{3,\ell} = \ker(\widehat{\boldsymbol{B}}_{3,\ell})$ and $W_1^{3,\ell} = \operatorname{Im}(\widehat{\boldsymbol{B}}_{3,\ell})$, it is clear that $\dim(W_0^{3,\ell}) = \frac{\ell(\ell+3)}{2} - 2$ and $\dim(W_1^{3,\ell}) = 3$.

Let us prove:

Proposition V.19. *The vectors $\boldsymbol{v}_1$, $\boldsymbol{v}_2$, and $\boldsymbol{v}_3$ given by*

$$\begin{aligned}{}^t\boldsymbol{v}_1 &= (1,0,0,-1,-1,-1,-2,-2,-2,-2,\dots,-\overbrace{(\ell-1),\dots,-(\ell-1)}^{\ell+1})\\ {}^t\boldsymbol{v}_2 &= (0,1,0,2,1,0,3,2,1,\dots,0,\ell,(\ell-1),\dots,2,1,0)\\ {}^t\boldsymbol{v}_3 &= (0,0,1,0,1,2,0,1,2,3,\dots,0,1,2,\dots,(\ell-1),\ell),\end{aligned}$$

form a basis for $W_1^{3,\ell}$.

Proof. As the three vectors are obviously linearly independent, it clearly suffices to show that they belong to $W_1^{3,\ell}$. We will verify this only for the vector $\boldsymbol{v}_3$. A similar reasoning then works for the remaining two vectors.

First note that if $p = \frac{k(k+1)}{2} + r$, $(0 \leq k, r \leq \ell)$, then the p-th component of $\boldsymbol{v}_1$, $\boldsymbol{v}_2$ and $\boldsymbol{v}_3$ is, respectively, $1-k$, $k-r$ and r. So,

$$\begin{aligned}
(\widehat{\boldsymbol{B}}_{3,\ell}\boldsymbol{v}_3)_p &= \sum_{q=0}^{m} \widehat{b}_{pq}(\boldsymbol{v}_3)_q \\
&= \sum \binom{\ell}{s}\binom{s}{t} t \left(1 + \tfrac{(\ell-2k+r)(\ell-2s+t)+(\ell-k-r)(\ell-s-t)+(k-2r)(s-2t)}{\ell}\right) \\
&= \sum t\binom{\ell}{s}\binom{s}{t} + \tfrac{(\ell-2k+r)}{\ell}\sum(\ell t - 2st + t^2)\binom{\ell}{s}\binom{s}{t} \\
&\quad + \tfrac{(\ell-k-r)}{\ell}\sum(\ell t - st - t^2)\binom{\ell}{s}\binom{s}{t} + \tfrac{(k-2r)}{\ell}\sum(st - 2t^2)\binom{\ell}{s}\binom{s}{t}.
\end{aligned}$$

Now, using lemma V.16, it easily follows that

$$(\widehat{\boldsymbol{B}}_{3,\ell}\boldsymbol{v}_3)_p = 3^\ell r = 3^\ell(\boldsymbol{v}_3)_p.$$

□

Note that thanks to theorem V.9, it is easy to prove that the eigenvectors associated with the eigenvalue zero do not contribute positive solutions to the system. Indeed:

Proposition V.20. *If $\boldsymbol{h} = {}^t(h_0, \ldots, h_m) \in W_0^{3,\ell}$ then at least one of the components of $\boldsymbol{h}$ is negative.*

Finally, the following result completely characterizes the generalized unitation functions on ternary alphabets with minimal epistasis:

Proposition V.21. *Let f be a positive-valued generalized unitation function, defined on a multary encoding with an alphabet of three elements. Then f has zero epistasis if and only if the p-th component (with $p = \frac{k(k+1)}{2} + r$) of its associated vector $\boldsymbol{h}$ equals $(1-k)h_0 + (k-r)h_1 + rh_2$.*

Proof. We show that ${}^t\boldsymbol{h}\boldsymbol{B}_{3,\ell}\mathbf{h} = \mathbf{3}^\ell \left\|\boldsymbol{f}\right\|^2$. A direct calculation yields

$$
\begin{aligned}
\left\|\boldsymbol{f}\right\|^2 &= \sum_{p=0}^{m} \binom{\ell}{k}\binom{k}{r} h_p^2 \\
&= \sum_{p=0}^{m} \binom{\ell}{k}\binom{k}{r} \left((1-k)h_0 + (k-r)h_1 + rh_2\right)^2 \\
&= h_0^2 \sum_{p=0}^{m} \binom{\ell}{k}\binom{k}{r}(1-k)^2 + h_1^2 \sum_{p=0}^{m} \binom{\ell}{k}\binom{k}{r}(k-r)^2 \\
&\quad + h_2^2 \sum_{p=0}^{m} \binom{\ell}{k}\binom{k}{r} r^2 + 2h_0h_1 \sum_{p=0}^{m} \binom{\ell}{k}\binom{k}{r}(1-k)(k-r) \\
&\quad + 2h_0h_2 \sum_{p=0}^{m} \binom{\ell}{k}\binom{k}{r}(1-k)r + 2h_1h_2 \sum_{p=0}^{m} \binom{\ell}{k}\binom{k}{r}(k-r)r
\end{aligned}
$$

and

$$
\begin{aligned}
{}^t\boldsymbol{h}\boldsymbol{B}_{3,\ell}\boldsymbol{h} &= \sum_{p=0}^{m} (b_{3,\ell})_{pp} h_p^2 + \sum_{\substack{p,q \\ p\neq q}} (b_{3,\ell})_{pq} h_p h_q \\
&= \sum_{p=0}^{m} (b_{3,\ell})_{pp} \left((1-k)h_0 + (k-r)h_1 + rh_2\right)^2 \\
&\quad + \sum_{\substack{p,q \\ p\neq q}} (b_{3,\ell})_{pq} \left((1-k)h_0 + (k-r)h_1 + rh_2\right)\left((1-s)h_0 + (s-t)h_1 + th_2\right),
\end{aligned}
$$

where the q-th component has been represented, as always, by $q = \frac{s(s+1)}{2} + t$, with $0 \leq s, t \leq \ell$. We prove that they are equal by showing that the coefficients of h_0^2, h_1^2, h_2^2, h_0h_1, h_0h_2 and h_1h_2 are equal in both expressions.

In particular, the coefficient of h_0^2 in ${}^t\boldsymbol{h}\boldsymbol{B}_{3,\ell}\boldsymbol{h}$ is

$$
\begin{aligned}
&\textstyle\sum_{p=0}^{m} (b_{3,\ell})_{pp}(1-k)^2 + \sum_{\substack{p,q \\ p\neq q}} (b_{3,\ell})_{pq}(1-k)(1-s) \\
&= \textstyle\sum_{p=0}^{m} \binom{\ell}{k}^2\binom{k}{r}^2 \left(1 + \frac{(\ell-2k+r)^2 + (\ell-k-r)^2 + (k-2r)^2}{\ell}\right)(1-k)^2 \\
&+ \textstyle\sum_{\substack{p,q \\ p\neq q}} \binom{\ell}{k}\binom{k}{r}\binom{\ell}{s}\binom{s}{t} \left(1 + \frac{(\ell-2k+r)(\ell-2s+t) + (\ell-k-r)(\ell-s-t) + (k-2r)(s-2t)}{\ell}\right)(1-k)(1-s).
\end{aligned}
$$

Fixing the p-th term, applying lemma V.16 and extracting the common factor

$$\tbinom{\ell}{k}\tbinom{k}{r}(1-k),$$

we obtain

$$\binom{\ell}{k}\binom{k}{r}\left(1+\frac{(\ell-2k+r)^2+(\ell-k-r)^2+(k-2r)^2}{\ell}\right)(1-k)$$
$$+\sum_{\substack{q\\q\neq p}}\binom{\ell}{s}\binom{s}{t}\left(1+\frac{(\ell-2k+r)(\ell-2s+t)+(\ell-k-r)(\ell-s-t)+(k-2r)(s-2t)}{\ell}\right)(1-s)$$
$$=3^\ell(1-k).$$

A similar reasoning applies to compare the coefficients of h_1^2, h_2^2, h_0h_1, h_0h_2 and h_1h_2, which finishes the proof. □

Note that one can also obtain the vector $\boldsymbol{h}$ of the previous proposition directly. By theorem V.8, any function f on Ω_3 has epistasis zero if and only if

$$f(s_{\ell-1}\dots s_0)=\sum_{i=0}^{\ell-1}g_i(s_i)=\sum_{i_1\,:\,s_{i_1}=0}g_{i_1}(0)+\sum_{i_2\,:\,s_{i_2}=1}g_{i_2}(1)+\sum_{i_3\,:\,s_{i_3}=2}g_{i_3}(2).$$

The function f is a generalized unitation function if and only if

$$f(s_{\ell-1}\dots s_0)=h_p=f(\overbrace{0\dots0}^{\ell-k}\,1\dots1\,\overbrace{2\dots2}^{r})=\dots=f(\overbrace{2\dots2}^{r}\,1\dots1\,\overbrace{0\dots0}^{\ell-k})$$

for some $0\leq p<\binom{\ell+2}{\ell}$. Combining both expressions, we have

$$g_i(1)-g_i(0)=\alpha$$
$$g_i(2)-g_i(0)=\beta$$

for all $0\leq i<\ell$ when f is a generalized unitation function with zero epistasis. This implies

$$f(s)=h_p=r\beta+(k-r)\alpha+f(0\dots0).$$

In particular,

$$h_0=f(0\dots0),\ h_1=\alpha+f(0\dots0)\text{ and }\ h_2=\beta+f(0\dots0)$$

and from this it follows that $\alpha=h_1-h_0$ and $\beta=h_2-h_0$. So for all $0\leq p<\binom{\ell+2}{\ell}$ we have $h_p=(1-k)h_0+(k-r)h_1+rh_2$, as claimed.

Chapter VI

Generalized Walsh transforms

We saw in chapter IV that the Walsh coefficients associated to schemata over a binary alphabet include, in a very natural way, the basic properties of these schemata. They allow for a very efficient calculation of normalized epistasis for many classes of functions. In this chapter, we show how the same point of view may be applied in the non-binary case by introducing two suitable generalizations of the Walsh transform. We use one of these generalizations to show that epistasis can be computed efficiently in the multary case as well. The same transform is then used to generalize some results of Heckendorn and Whitley [31, 32] about the moments of schema fitness distributions.

1 Generalized Walsh transforms

In this section, we present two possible generalizations of the Walsh transform, making it applicable as a basic tool in the multary case as well. It appears that each of these generalizations yields a nice, efficient way of calculating epistasis, when dealing with fitness functions over a multary alphabet. Moreover, some examples will be included, aiming to give an indication of the strength of this approach.

1.1 First generalization to the multary case

In this section, we will see how the use of *partition coefficients* will be useful in order to generalize the "classical" Walsh transform to the multary case.

So, let us start by defining the partition coefficient $\varepsilon(H)$ of a schema H over a multary alphabet. Exactly as in the binary case, we want these $\varepsilon(H)$ to satisfy the *general partition equation*

$$f(H) = \sum_{H' \supset H} \varepsilon(H'),$$

where, as before, $H' \supset H$ means that H and H' agree on the fixed positions of H. As in the binary case, the partition coefficient of the schema $\# \ldots \#$ will be exactly the average of the fitness function, i.e., $\varepsilon(\# \ldots \#) = f(\# \ldots \#) = \bar{f}$. On the other hand, the fitness value of any schema H of order 1 can be approximated by the schema $\# \ldots \#$, and the difference between this and its correct value will be the partition coefficient of H. So, $\varepsilon(\# \ldots \# i) = f(\# \ldots \#) - f(\# \ldots \# i)$, for example. Iterating, let us consider a schema H of order 2, for example $H = \# \ldots \# ij$ (with $i, j \in \Sigma$). Then we may interpret H as a combination of the schemata H' with order 1 and such that $H' \supset H$. So, as a first approximation, we could calculate the fitness value of H as follows:

$$f(H) \approx f(\# \ldots \#) + (f(\# \ldots \#) - f(\# \ldots i\#)) + (f(\# \ldots \#) - f(\# \ldots \# j)).$$

Obviously, in this approximation we ignore the interaction between the fixed positions in H. In fact, the difference between the actual value of $f(H)$ and the above approximation is exactly the partition coefficient of $H = \# \ldots \# ij$:

$$f(\# \ldots \# ij) = \varepsilon(\# \ldots \#) + \varepsilon(\# \ldots i\#) + \varepsilon(\# \ldots \# j) + \varepsilon(\# \ldots \# ij).$$

Since $\varepsilon(\# \ldots \#) = f(\# \ldots \#)$, rearranging the terms in the general partition equation yields for any schema H,

$$\varepsilon(H) = f(H) - \sum_{H' \supsetneq H} \varepsilon(H').$$

In order to recover the original fitness function from the partition coefficients, we will need the following two results of Mason (see [57] for more details):

Theorem VI.1. *For any* $h_0, \ldots, h_{i-1}, h_{i+1}, \ldots, h_{\ell-1} \in \Sigma \cup \{\#\}$,

$$\sum_{a \in \Sigma} \varepsilon(h_{\ell-1} \ldots h_{i+1} a h_{i-1} \ldots h_0) = 0.$$

Theorem VI.2. *If there exists* $\varepsilon(\cdot)$ *satisfying theorem VI.1, then the general partition equation is satisfied for the fitness values*

$$f(s) = \sum_{s \in H} \varepsilon(H).$$

Theorem VI.1 shows that the partition coefficients are not independent of each other. We will consider a minimal collection of partition coefficients which generates the set of all partition coefficients and, which, by theorem VI.2, also permits to recover the original fitness function. This minimal set will consist of the so-called *general Walsh coefficients*.

We proceed as follows. Denote by Σ' the alphabet Σ augmented with the "don't care" symbol #, and consider the function β which associates with any schema $H = h_{\ell-1} \ldots h_0 \in (\Sigma' \setminus \{0\})^\ell$ the string $\beta(H) = \beta_{\ell-1} \ldots \beta_0 \in \Sigma^\ell$, with

$$\beta_i = \begin{cases} 0 & \text{if } h_i = \#, \\ h_i & \text{otherwise.} \end{cases}$$

For example, if $n = 3$ we have that $\beta(21\#) = 210_3 = 21$.

With each of these n^ℓ schemata H or, equivalently, the associated values $\beta(H)$, we associate a *Walsh coefficient*

$$w_H = w_{\beta(H)} = n^{\frac{\ell}{2}} \varepsilon(H).$$

(Note that $H \neq H'$ if and only if $\beta(H) \neq \beta(H')$, since we restrict to schemata $H, H' \in (\Sigma' \setminus \{0\})^\ell$.) For example, the Walsh coefficient associated to the schema $H = 21\#$ over a ternary alphabet is

$$w_{21} = w_{210_3} = w_{\beta(21\#)} = 3^{\frac{3}{2}} \varepsilon(21\#).$$

We will also need the values $v_H = v_{\beta(H)} = n^{-\frac{\ell}{2}} w_{\beta(H)}$, which, since they only differ from the "correct" Walsh coefficients by a constant factor, will be referred to as Walsh coefficients as well.

We may now define recursively, for any schema $H = h_{\ell-1}\dots h_0 \in \Sigma'^{\ell}$, the value $v(H)$ (or $v_\ell(H)$ if ambiguity arises) by

$$v(H) = \begin{cases} (-1)^{o(H)} v_{\beta(H)} & \text{if } H \in (\Sigma' \setminus \{0\})^{\ell} \\ \displaystyle\sum_{a \in \{1,\dots,n-1\}} -v(h_{\ell-1}\dots h_{i+1} a h_{i-1} \dots h_0) & \text{if } h_i = 0, \end{cases}$$

where $o(H)$ denotes the order of H. The value $\varepsilon(H)$ is then given by

$$\varepsilon(H) = v(H).$$

The function v satisfies the condition of theorem VI.1 and, by theorem VI.2, the general partition equation holds, i.e.,

$$f(s) = \sum_{s \in H} v(H).$$

It thus follows that the original function f may be reconstructed when the Walsh coefficients $v(H)$ are known.

Let us give some examples. For $n = 3$ and $\ell = 1$ the connection between the schemata and the Walsh coefficients is given by

H	$v(H)$
#	v_0
0	$v_1 + v_2$
1	$-v_1$
2	$-v_2$

and for $n = 3$ and $\ell = 2$

H	$v(H)$	H	$v(H)$
##	v_0	1#	$-v_3$
#0	$v_1 + v_2$	10	$-v_4 - v_5$
#1	$-v_1$	11	v_4
#2	$-v_2$	12	v_5
0#	$v_3 + v_6$	2#	$-v_6$
00	$v_4 + v_5 + v_7 + v_8$	20	$-v_7 - v_8$
01	$-v_4 - v_7$	21	v_7
02	$-v_5 - v_8$	22	v_8

It is clear that there exists a unique matrix $\boldsymbol{W}_\ell$ (or $\boldsymbol{W}_{n,\ell}$, if we want to be more precise) such that for any fitness function f, we have $\boldsymbol{f} = \boldsymbol{W}_\ell \boldsymbol{w}$, where $\boldsymbol{f}$ is the vector associated with f, as before, and where the vector $\boldsymbol{w} = {}^t(w_0, \ldots, w_{n^\ell - 1})$ consists of its Walsh coefficients.

If we denote by $\boldsymbol{V}_{n,\ell}$ the matrix given by $\boldsymbol{V}_{n,\ell} = n^{\frac{\ell}{2}} \boldsymbol{W}_{n,\ell}$, the matrices that correspond to the previous examples are

$$\boldsymbol{V}_{3,1} = \begin{pmatrix} 1 & 1 & 1 \\ 1 & -1 & 0 \\ 1 & 0 & -1 \end{pmatrix}$$

and

$$\boldsymbol{V}_{3,2} = \begin{pmatrix} 1 & 1 & 1 & 1 & 1 & 1 & 1 & 1 & 1 \\ 1 & -1 & 0 & 1 & -1 & 0 & 1 & -1 & 0 \\ 1 & 0 & -1 & 1 & 0 & -1 & 1 & 0 & -1 \\ 1 & 1 & 1 & -1 & -1 & -1 & 0 & 0 & 0 \\ 1 & -1 & 0 & -1 & 1 & 0 & 0 & 0 & 0 \\ 1 & 0 & -1 & -1 & 0 & 1 & 0 & 0 & 0 \\ 1 & 1 & 1 & 0 & 0 & 0 & -1 & -1 & -1 \\ 1 & -1 & 0 & 0 & 0 & 0 & -1 & 1 & 0 \\ 1 & 0 & -1 & 0 & 0 & 0 & -1 & 0 & 1 \end{pmatrix},$$

respectively. Note that for an alphabet of size n, we have

$$\boldsymbol{W}_{n,1} = n^{-\frac{1}{2}} \begin{pmatrix} 1\,1 & \ldots & 1 \\ 1 & & \\ \vdots & -\boldsymbol{I}_{n-1} & \\ 1 & & \end{pmatrix},$$

where $\boldsymbol{I}_{n-1}$ is the identity matrix of dimensions $n-1 \times n-1$. In particular, $\boldsymbol{W}_{n,1}$ is a symmetric matrix. Note that $\boldsymbol{W}_{2,1} = n^{-\frac{1}{2}} \left(\begin{smallmatrix} 1 & 1 \\ 1 & -1 \end{smallmatrix}\right) = \boldsymbol{W}_1$, the "ordinary" Walsh matrix defined in section 1.3, chapter IV.

Note also that $\boldsymbol{W}_{3,2} = \boldsymbol{W}_{3,1} \otimes \boldsymbol{W}_{3,1} = \boldsymbol{W}_{3,1}^{\otimes 2}$. This result remains valid in the general case:

Lemma VI.3. *For every positive integer ℓ, we have*

$$\boldsymbol{W}_{n,\ell} = \boldsymbol{W}_{n,1}^{\otimes \ell}.$$

Proof. Firstly, we will prove that $\boldsymbol{W}_{n,2} = \boldsymbol{W}_{n,1}^{\otimes 2}$. In fact, with $a, b \neq 0$, we have

$$\begin{aligned} v(\#\#) &= (-1)^{o(\#\#)} v_{\beta(\#\#)} = v_0, \\ v(\#b) &= (-1)^{o(\#b)} v_{\beta(\#b)} = -v_b, \end{aligned}$$

$$\begin{aligned} v(\#0) &= -\sum_{b=1}^{n-1} v(\#b) = v_1 + \cdots + v_{n-1}, \\ v(a\#) &= (-1)^{o(a\#)} v_{\beta(a\#)} = -v_{an}, \\ v(0\#) &= -\sum_{a=1}^{n-1} v(a\#) = v_n + v_{2n} + \ldots v_{(n-1)n}, \\ v(ab) &= (-1)^{o(ab)} v_{\beta(ab)} = v_{an+b}, \\ v(a0) &= -\sum_{b=1}^{n-1} v(ab) = -\sum_{b=1}^{n-1} v_{an+b}, \\ v(0b) &= -\sum_{a=1}^{n-1} v(ab) = -\sum_{a=1}^{n-1} v_{an+b}, \end{aligned}$$

and

$$v(00) = -\sum_{a=1}^{n-1} v(a0) = \sum_{a,b=1}^{n-1} v(ab) = \sum_{a,b=1}^{n-1} v_{an+b}.$$

So, taking into account these expressions and the fact that

$$f(ab) = \sum_{ab \in H} v(H) = v(\#\#) + v(\#b) + v(a\#) + v(ab),$$

we obtain for all $a, b \neq 0$, that

$$f(ab) = v_0 - v_b - v_{an} + v_{an+b}$$

and

$$\begin{aligned} f(a0) &= \sum_{a0 \in H} v(H) = v(\#\#) + v(\#0) + v(a\#) + v(a0) \\ &= \sum_{i=0}^{n-1} v_i - \sum_{i=0}^{n-1} v_{an+i}. \end{aligned}$$

It follows that

$$\begin{pmatrix} f(a0) \\ \vdots \\ f(a(n-1)) \end{pmatrix} = \begin{pmatrix} v_0 + v_1 + \cdots + v_{n-1} \\ v_0 - v_1 \\ \vdots \\ v_0 - v_{n-1} \end{pmatrix} - \begin{pmatrix} v_{an} + v_{an+1} + \cdots + v_{an+(n-1)} \\ v_{an} - v_{an+1} \\ \vdots \\ v_{an} - v_{an+(n-1)} \end{pmatrix}$$

$$= \boldsymbol{W}_{n,1} \left[\begin{pmatrix} w_0 \\ \vdots \\ w_{n-1} \end{pmatrix} - \begin{pmatrix} w_{an} \\ \vdots \\ w_{an+(n-1)} \end{pmatrix} \right].$$

On the other hand, as for all $b \neq 0$,

$$f(0b) = \sum_{i=0}^{n-1} v_{in} - \sum_{i=0}^{n-1} v_{in+b}$$

and

$$f(00) = \sum_{i,j=0}^{n-1} v_{in+j},$$

we obtain that

$$\begin{pmatrix} f(00) \\ \vdots \\ f(0(n-1)) \end{pmatrix} = \begin{pmatrix} v_0 + v_1 + \cdots + v_{n^2-1} \\ (v_0 - v_1) + (v_n - v_{n+1}) + \cdots + (v_{n(n-1)} - v_{n(n-1)+1}) \\ \vdots \\ (v_0 - v_{n-1}) + (v_n - v_{n+(n-1)}) + \cdots + (v_{n(n-1)} - v_{n(n-1)+(n-1)}) \end{pmatrix}$$

$$= \boldsymbol{W}_{n,1} \left[\begin{pmatrix} w_0 \\ \vdots \\ w_{n-1} \end{pmatrix} + \begin{pmatrix} w_n \\ \vdots \\ w_{2n-1} \end{pmatrix} + \cdots + \begin{pmatrix} w_{(n-1)n} \\ \vdots \\ w_{n^2-1} \end{pmatrix} \right].$$

This proves that $\boldsymbol{W}_{n,2} = \boldsymbol{W}_{n,1}^{\otimes 2}$.

In the general case, the first Walsh coefficient w_0 is essentially the average of the fitness function. The coefficients $w_1, \ldots, w_{n^{\ell-1}-1}$ represent the contributions of the schemata $\#h_{\ell-2} \ldots h_0$ and the coefficients $w_{an^{\ell-1}}$ to $w_{(a+1)n^{\ell-1}-1}$ represent the contributions of the schemata $ah_{\ell-2}\dots h_0$, for $a \in \{1,\dots,n-1\}$. Then, as

$$\begin{aligned} f(as_{\ell-2}s_{\ell-3}\dots s_0) &= \sum_{as_{\ell-2}\dots s_0 \in H} v(H) \\ &= \sum_{s_{\ell-2}s_{\ell-3}\dots s_0 \in H'} v(\#H') + \sum_{s_{\ell-2}s_{\ell-3}\dots s_0 \in H'} v(aH') \end{aligned}$$

for all $a \neq 0$, and

$$\begin{aligned} f(0s_{\ell-2}s_{\ell-3}\dots s_0) &= \sum_{0s_{\ell-2}\dots s_0 \in H} v(H) \\ &= \sum_{s_{\ell-2}s_{\ell-3}\dots s_0 \in H'} \left\{ v(\#H') - \sum_{s_{\ell-2}s_{\ell-3}\dots s_0 \in H'} v(aH') \right\}, \end{aligned}$$

where $v(\#H')$ and $v(aH')$, $a \neq 0$, are given recursively by

$$\begin{aligned} v_\ell(\#H') &= \begin{cases} (-1)^{o(H')} \cdot v_{\beta(H')} & \text{if } H' \in (\Sigma' \setminus \{0\})^{\ell-1} \\ \displaystyle\sum_{b \in \{1,\dots,n-1\}} -v(\#h_{\ell-2}\dots h_{i+1}bh_{i-1}\dots h_0) & \text{if } h_i = 0, \end{cases} \\ &= v_{\ell-1}(H') \end{aligned}$$

and

$$\begin{aligned} v_\ell(aH') &= \begin{cases} (-1)^{o(aH')} \cdot v_{an^{\ell-1}+\beta(H')} & \text{if } H' \in (\Sigma' \setminus \{0\})^{\ell-1} \\ \displaystyle\sum_{b \in \{1,\dots,n-1\}} -v(ah_{\ell-2}\dots h_{i+1}bh_{i-1}\dots h_0) & \text{if } h_i = 0 \end{cases} \\ &= -\begin{cases} (-1)^{o(H')} \cdot v_{an^{\ell-1}+\beta(H')} & \text{if } H' \in (\Sigma' \setminus \{0\})^{\ell-1} \\ - \displaystyle\sum_{b \in \{1,\dots,n-1\}} (-1)^{o(H')} \cdot v_{an^{\ell-1}+\beta(h_{\ell-2}\dots h_{i+1}bh_{i-1}\dots h_0)} & \text{if } h_i = 0, \end{cases} \end{aligned}$$

respectively, we thus see that

$$\begin{pmatrix} f_{0\dots 0} \\ \vdots \\ f_{0(n-1)\dots(n-1)} \end{pmatrix} = \boldsymbol{W}_{n,\ell-1} \cdot \left[\begin{pmatrix} w_0 \\ \vdots \\ w_{n^{\ell-1}-1} \end{pmatrix} + \begin{pmatrix} w_{n^{\ell-1}} \\ \vdots \\ w_{2n^{\ell-1}-1} \end{pmatrix} + \dots + \begin{pmatrix} w_{(n-1)n^{\ell-1}} \\ \vdots \\ w_{n^\ell-1} \end{pmatrix} \right]$$

and

$$\begin{pmatrix} f_{a0\dots 0} \\ \vdots \\ f_{a(n-1)\dots(n-1)} \end{pmatrix} = \boldsymbol{W}_{n,\ell-1} \cdot \left[\begin{pmatrix} w_0 \\ \vdots \\ w_{n^{\ell-1}-1} \end{pmatrix} - \begin{pmatrix} w_{an^{\ell-1}} \\ \vdots \\ w_{(a+1)n^{\ell-1}-1} \end{pmatrix} \right],$$

for all $a \in \{1, \ldots, n-1\}$ and so, it follows that $\boldsymbol{W}_{n,\ell} = \boldsymbol{W}_{n,\ell-1} \otimes \boldsymbol{W}_{n,1} = \boldsymbol{W}_{n,1}^{\otimes \ell}$. □

To calculate the epistasis of a function f in terms of its Walsh coefficients, we will use the recursion relation

$$\boldsymbol{G}_{n,\ell} = \boldsymbol{U}_{n,1} \otimes \boldsymbol{G}_{n,\ell-1} + (\boldsymbol{G}_{n,1} - \boldsymbol{U}_{n,1}) \otimes \boldsymbol{U}_{n,\ell-1}.$$

We find that

$${}^t\boldsymbol{W}_{n,\ell} \boldsymbol{G}_{n,\ell} \boldsymbol{W}_{n,\ell} = \boldsymbol{A} + \boldsymbol{B}$$

with

$$\begin{aligned}
\boldsymbol{A} &= {}^t\boldsymbol{W}_{n,\ell} (\boldsymbol{U}_{n,1} \otimes \boldsymbol{G}_{n,\ell-1}) \boldsymbol{W}_{n,\ell} \\
\boldsymbol{B} &= {}^t\boldsymbol{W}_{n,\ell} \left((\boldsymbol{G}_{n,1} - \boldsymbol{U}_{n,1}) \otimes \boldsymbol{U}_{n,\ell-1} \right) \boldsymbol{W}_{n,\ell}.
\end{aligned}$$

In order to calculate $\boldsymbol{A}$, note that

$$\begin{aligned}
{}^t\boldsymbol{W}_{n,\ell} \boldsymbol{W}_{n,\ell} &= ({}^t\boldsymbol{W}_{n,1} \otimes {}^t\boldsymbol{W}_{n,\ell-1})(\boldsymbol{W}_{n,1} \otimes \boldsymbol{W}_{n,\ell-1}) \\
&= ({}^t\boldsymbol{W}_{n,1} \boldsymbol{W}_{n,1}) \otimes ({}^t\boldsymbol{W}_{n,\ell-1} \boldsymbol{W}_{n,\ell-1}) \\
&= (\boldsymbol{W}_{n,1} \boldsymbol{W}_{n,1})^{\otimes \ell}
\end{aligned}$$

and

$$\begin{aligned}
\boldsymbol{W}_{n,1} \boldsymbol{W}_{n,1} &= n^{-\frac{1}{2}} \begin{pmatrix} 1\,1 & \ldots & 1 \\ 1 & & \\ \vdots & -\boldsymbol{I}_{n-1} & \\ 1 & & \end{pmatrix} n^{-\frac{1}{2}} \begin{pmatrix} 1\,1 & \ldots & 1 \\ 1 & & \\ \vdots & -\boldsymbol{I}_{n-1} & \\ 1 & & \end{pmatrix} \\
&= n^{-1} \begin{pmatrix} n & 0 & \ldots & \ldots & 0 \\ 0 & 2 & 1 & \ldots & 1 \\ \vdots & 1 & 2 & \ddots & \vdots \\ \vdots & \vdots & \ddots & \ddots & 1 \\ 0 & 1 & \ldots & 1 & 2 \end{pmatrix}.
\end{aligned}$$

This permits us to calculate

$$
\begin{aligned}
{}^t\boldsymbol{W}_{n,1}\boldsymbol{G}_{n,1}\boldsymbol{W}_{n,1} &= n\,{}^t\boldsymbol{W}_{n,1}\boldsymbol{I}_{n,1}\boldsymbol{W}_{n,1} \\
&= \begin{pmatrix} n & 0 & \ldots & \ldots & \ldots & 0 \\ 0 & 2 & 1 & \ldots & & 1 \\ \vdots & 1 & \ddots & & \ddots & \vdots \\ \vdots & \vdots & & \ddots & \ddots & 1 \\ 0 & 1 & \ldots & & 1 & 2 \end{pmatrix}.
\end{aligned}
$$

On the other hand, we have

$$
\begin{aligned}
{}^t\boldsymbol{W}_{n,\ell}\boldsymbol{U}_{n,\ell}\boldsymbol{W}_{n,\ell} &= ({}^t\boldsymbol{W}_{n,1} \otimes {}^t\boldsymbol{W}_{n,\ell-1})(\boldsymbol{U}_{n,1} \otimes \boldsymbol{U}_{n,\ell-1})(\boldsymbol{W}_1 \otimes \boldsymbol{W}_{n,\ell-1}) \\
&= ({}^t\boldsymbol{W}_{n,1}\boldsymbol{U}_{n,1}\boldsymbol{W}_{n,1}) \otimes ({}^t\boldsymbol{W}_{n,\ell-1}\boldsymbol{U}_{n,\ell-1}\boldsymbol{W}_{n,\ell-1}) \\
&= (\boldsymbol{W}_{n,1}\boldsymbol{U}_{n,1}\boldsymbol{W}_{n,1})^{\otimes \ell} \\
&= n^{\ell} \begin{pmatrix} 1 & 0 & \ldots & 0 \\ 0 & 0 & \ldots & 0 \\ \vdots & \vdots & \ddots & \vdots \\ 0 & 0 & \ldots & 0 \end{pmatrix},
\end{aligned}
$$

since

$$
\boldsymbol{W}_{n,1}\boldsymbol{U}_{n,1}\boldsymbol{W}_{n,1} = \boldsymbol{W}_{n,1} \cdot n^{-\frac{1}{2}} \begin{pmatrix} n & 0 & \ldots & 0 \\ \vdots & \vdots & \ddots & \vdots \\ n & 0 & \ldots & 0 \end{pmatrix} = \begin{pmatrix} n & 0 & \ldots & 0 \\ 0 & 0 & \ldots & 0 \\ \vdots & \vdots & \ddots & \vdots \\ 0 & 0 & \ldots & 0 \end{pmatrix}.
$$

These preliminary calculations allow us to compute the matrix $\boldsymbol{A}$:

$$
\begin{aligned}
\boldsymbol{A} &= {}^t\boldsymbol{W}_{n,\ell}(\boldsymbol{U}_{n,1} \otimes \boldsymbol{G}_{n,\ell-1})\boldsymbol{W}_{n,\ell} \\
&= ({}^t\boldsymbol{W}_{n,1} \otimes {}^t\boldsymbol{W}_{n,\ell-1})(\boldsymbol{U}_{n,1} \otimes \boldsymbol{G}_{n,\ell-1})(\boldsymbol{W}_{n,1} \otimes \boldsymbol{W}_{n,\ell-1}) \\
&= ({}^t\boldsymbol{W}_{n,1}\boldsymbol{U}_{n,1}\boldsymbol{W}_{n,1}) \otimes ({}^t\boldsymbol{W}_{n,\ell-1}\boldsymbol{G}_{n,\ell-1}\boldsymbol{W}_{n,\ell-1}) \\
&= \begin{pmatrix} n & 0 & \ldots & 0 \\ 0 & 0 & \ldots & 0 \\ \vdots & \vdots & \ddots & \vdots \\ 0 & 0 & \ldots & 0 \end{pmatrix} \otimes ({}^t\boldsymbol{W}_{n,\ell-1}\boldsymbol{G}_{n,\ell-1}\boldsymbol{W}_{n,\ell-1}).
\end{aligned}
$$

The matrix $\boldsymbol{B}$ may be calculated as follows:

$$\begin{aligned}
\boldsymbol{B} &= {}^t\boldsymbol{W}_{n,\ell}\left((\boldsymbol{G}_{n,1} - \boldsymbol{U}_{n,1}) \otimes \boldsymbol{U}_{n,\ell-1}\right)\boldsymbol{W}_{n,\ell} \\
&= {}^t\boldsymbol{W}_{n,1}((\boldsymbol{G}_{n,1} - \boldsymbol{U}_{n,1})\boldsymbol{W}_{n,1}) \otimes ({}^t\boldsymbol{W}_{n,\ell-1}\boldsymbol{U}_{n,\ell-1}\boldsymbol{W}_{n,\ell-1}) \\
&= (\boldsymbol{W}_{n,1}\boldsymbol{G}_{n,1}\boldsymbol{W}_{n,1} - \boldsymbol{W}_{n,1}\boldsymbol{U}_{n,1}\boldsymbol{W}_{n,1}) \otimes {}^t\boldsymbol{W}_{n,\ell-1}\boldsymbol{U}_{n,\ell-1}\boldsymbol{W}_{n,\ell-1} \\
&= \left[\begin{pmatrix} n & 0 & \dots & \dots & 0 \\ 0 & 2 & 1 & \dots & 1 \\ \vdots & 1 & \ddots & \ddots & \vdots \\ \vdots & \vdots & \ddots & \ddots & 1 \\ 0 & 1 & \dots & 1 & 2 \end{pmatrix} - \begin{pmatrix} n & 0 & \dots & 0 \\ 0 & 0 & \dots & 0 \\ \vdots & \vdots & \ddots & \vdots \\ 0 & 0 & \dots & 0 \end{pmatrix}\right] \otimes \begin{pmatrix} n^{\ell-1} & 0 & \dots & 0 \\ 0 & 0 & \dots & 0 \\ \vdots & \vdots & \ddots & \vdots \\ 0 & 0 & \dots & 0 \end{pmatrix} \\
&= \begin{pmatrix} 0 & 0 & \dots & \dots & 0 \\ 0 & 2 & 1 & \dots & 1 \\ \vdots & 1 & \ddots & \ddots & \vdots \\ \vdots & \vdots & \ddots & \ddots & 1 \\ 0 & 1 & \dots & 1 & 2 \end{pmatrix} \otimes \begin{pmatrix} n^{\ell-1} & 0 & \dots & 0 \\ 0 & 0 & \dots & 0 \\ \vdots & \vdots & \ddots & \vdots \\ 0 & 0 & \dots & 0 \end{pmatrix}.
\end{aligned}$$

Combining the previous results yields an expression for ${}^t\boldsymbol{W}_{n,\ell}\boldsymbol{G}_{n,\ell}\boldsymbol{W}_{n,\ell}$. On the other hand, as $\boldsymbol{f} = \boldsymbol{W}_{n,\ell}\boldsymbol{w}$, we find

$$\begin{aligned}
{}^t\boldsymbol{f}\boldsymbol{G}_{n,\ell}\boldsymbol{f} &= {}^t(\boldsymbol{W}_{n,\ell}\boldsymbol{w})\boldsymbol{G}_{n,\ell}(\boldsymbol{W}_{n,\ell}\boldsymbol{w}) \\
&= {}^t\boldsymbol{w}({}^t\boldsymbol{W}_{n,\ell}\boldsymbol{G}_{n,\ell}\boldsymbol{W}_{n,\ell})\boldsymbol{w},
\end{aligned}$$

and we obtain:

Lemma VI.4. *Let f be a fitness function defined on Ω_n and denote by $w_0, \dots, w_{n^\ell-1}$ its Walsh coefficients. Then*

$${}^t\boldsymbol{f}\boldsymbol{G}_{n,\ell}\boldsymbol{f} = n^\ell w_0^2 + 2n^{\ell-1}\sum_{i=0}^{\ell-1}\left(\sum_{\substack{p,q=1\\ p\leq q}}^{n-1} w_{pn^i}w_{qn^i}\right).$$

Proof. For $\ell = 1$, we find that

$$
\begin{aligned}
{}^t\boldsymbol{w}({}^t\boldsymbol{W}_{n,1}\boldsymbol{G}_{n,1}\boldsymbol{W}_{n,1})\boldsymbol{w} &= n\,{}^t\boldsymbol{w}(\boldsymbol{W}_1\boldsymbol{W}_1)\boldsymbol{w} \\
&= (w_0, \ldots, w_{n-1}) \begin{pmatrix} n\,0 \ldots \ldots 0 \\ 0\,2\;\;1\;\ldots\;1 \\ \vdots\;1\;\ddots\;\;\ddots\;\vdots \\ \vdots\;\vdots\;\ddots\;\;\ddots\;1 \\ 0\,1 \ldots\;\;1\;\;2 \end{pmatrix} \begin{pmatrix} w_0 \\ \vdots \\ w_{n-1} \end{pmatrix} \\
&= nw_0^2 + 2 \sum_{\substack{p,q=1 \\ p \leq q}}^{n-1} w_p w_q.
\end{aligned}
$$

We prove the general case by induction on ℓ. First, note that

$$
\begin{aligned}
&(w_0, \ldots, w_{n^\ell-1}) \left[\begin{pmatrix} n\;0 \ldots 0 \\ 0\;0 \ldots 0 \\ \vdots\;\vdots\;\ddots\;\vdots \\ 0\;0 \ldots 0 \end{pmatrix} \otimes \left({}^t\boldsymbol{W}_{n,\ell-1}\boldsymbol{G}_{n,\ell-1}\boldsymbol{W}_{n,\ell-1}\right) \right] \begin{pmatrix} w_0 \\ \vdots \\ w_{n^\ell-1} \end{pmatrix} \\
&= n(w_0, \ldots, w_{n^{\ell-1}-1}) \left({}^t\boldsymbol{W}_{n,\ell-1}\boldsymbol{G}_{n,\ell-1}\boldsymbol{W}_{n,\ell-1}\right) \begin{pmatrix} w_0 \\ \vdots \\ w_{n^{\ell-1}-1} \end{pmatrix} \\
&= n \left[n^{\ell-1} w_0^2 + 2n^{(\ell-1)-1} \sum_{i=0}^{(\ell-1)-1} \left(\sum_{\substack{p,q=1 \\ p \leq q}}^{n-1} w_{pn^i} w_{qn^i} \right) \right] \\
&= n^\ell w_0^2 + 2n^{\ell-1} \sum_{i=0}^{\ell-2} \left(\sum_{\substack{p,q=1 \\ p \leq q}}^{n-1} w_{pn^i} w_{qn^i} \right).
\end{aligned}
$$

On the other hand, we also have

$$
\begin{aligned}
&(w_0,\dots,w_{n^\ell-1})\left[\begin{pmatrix} 0 & 0 & \dots & \dots & 0\\ 0 & 2 & 1 & \dots & 1\\ \vdots & 1 & \ddots & \ddots & \vdots\\ \vdots & \vdots & \ddots & \ddots & 1\\ 0 & 1 & \dots & 1 & 2\end{pmatrix} \otimes \begin{pmatrix} n^{\ell-1} & 0 & \dots & 0\\ 0 & 0 & \dots & 0\\ \vdots & \vdots & \ddots & \vdots\\ 0 & 0 & \dots & 0\end{pmatrix}\right]\begin{pmatrix} w_0\\ \vdots\\ w_{n^\ell-1}\end{pmatrix}\\
&= (w_0, w_{n^{\ell-1}}\dots, w_{(n-1)n^{\ell-1}})\begin{pmatrix} 0 & 0 & \dots & \dots & 0\\ 0 & 2n^{\ell-1} & n^{\ell-1} & \dots & n^{\ell-1}\\ \vdots & n^{\ell-1} & \ddots & \ddots & \vdots\\ \vdots & \vdots & \ddots & \ddots & n^{\ell-1}\\ 0 & n^{\ell-1} & \dots & n^{\ell-1} & 2n^{\ell-1}\end{pmatrix}\begin{pmatrix} w_0\\ w_{n^{\ell-1}}\\ \vdots\\ w_{(n-1)n^{\ell-1}}\end{pmatrix}\\
&= n^{\ell-1}(w_{n^{\ell-1}},\dots,w_{(n-1)n^{\ell-1}})\begin{pmatrix} 2 & 1 & \dots & 1\\ 1 & \ddots & \ddots & \vdots\\ \vdots & \ddots & \ddots & 1\\ 1 & \dots & 1 & 2\end{pmatrix}\begin{pmatrix} w_n^{\ell-1}\\ \vdots\\ w_{(n-1)n^{\ell-1}}\end{pmatrix}\\
&= 2n^{\ell-1}\sum_{\substack{p,q=1\\ p\le q}}^{n-1} w_{pn^{\ell-1}}w_{qn^{\ell-1}}.
\end{aligned}
$$

Combining these results, i.e., applying the remarks preceding the statement of the present result, we now indeed obtain:

$$
\begin{aligned}
\boldsymbol{w}({}^t\boldsymbol{W}_{n,\ell}\boldsymbol{G}_{n,\ell}\boldsymbol{W}_{n,\ell})\boldsymbol{w} &= n^\ell w_0^2 + 2n^{\ell-1}\sum_{i=0}^{\ell-2}\left(\sum_{\substack{p,q=1\\ p\le q}}^{n-1} w_{pn^i}w_{qn^i}\right)\\
&\quad + 2n^{\ell-1}\sum_{\substack{p,q=1\\ p\le q}}^{n-1} w_{pn^{\ell-1}}w_{qn^{\ell-1}}\\
&= n^\ell w_0^2 + 2n^{\ell-1}\sum_{i=0}^{\ell-1}\left(\sum_{\substack{p,q=1\\ p\le q}}^{n-1} w_{pn^i}w_{qn^i}\right).
\end{aligned}
$$

□

It remains to calculate the norm of f (or of its associated vector $\boldsymbol{f}$) in terms of its Walsh coefficients. This may be done as follows:

$$\begin{aligned}
{}^t\boldsymbol{f}\boldsymbol{f} &= {}^t(\boldsymbol{W}_{n,\ell}\boldsymbol{w})(\boldsymbol{W}_{n,\ell}\boldsymbol{w}) \\
&= {}^t\boldsymbol{w}({}^t\boldsymbol{W}_{n,\ell}\boldsymbol{W}_{n,\ell})\boldsymbol{w} \\
&= {}^t\boldsymbol{w}({}^t\boldsymbol{W}_{n,1}^{\otimes\ell}\boldsymbol{W}_{n,1}^{\otimes\ell})\boldsymbol{w} \\
&= {}^t\boldsymbol{w}(\boldsymbol{W}_{n,1}\boldsymbol{W}_{n,1})^{\otimes\ell}\boldsymbol{w} \\
&= n^{-\ell}\,{}^t\boldsymbol{w}\begin{pmatrix} n & 0 & \dots & \dots & 0 \\ 0 & 2 & 1 & \dots & 1 \\ \vdots & 1 & \ddots & \ddots & \vdots \\ \vdots & \vdots & \ddots & \ddots & 1 \\ 0 & 1 & \dots & 1 & 2 \end{pmatrix}^{\otimes\ell}\boldsymbol{w}.
\end{aligned}$$

The explicit expression of the epistasis of some fitness function f in terms of the associated Walsh coefficients now trivially follows by combining the previous calculation with lemma VI.4.

1.2 Second generalization to the multary case

As we saw in chapter IV, the Walsh coefficients of a fitness function f may be introduced in terms of partition coefficients, or as a kind of discrete Fourier transform. In the previous section, we have shown that the first approach functions in the multary case as well, although it does not lead to an elegant expression for $\varepsilon^*(f)$. In the present section we introduce a second generalization of the Walsh transform, inspired by the Fourier-like approach in the binary setting. We will show that this one does allow for an elegant description of normalized epistasis.

Just as in the binary case, the whole set-up essentially reduces to recursively defining a suitable transformation matrix $\boldsymbol{W}$. In view of the fact that we will have to use the n-th roots of unity $e^{\frac{2\pi}{n}i}$ (and not just the square roots of unity $+1$ and -1 as in the binary case) it should come as no surprise that the transformation matrices used in the multary case are actually complex.

Let us therefore assume r to be a primitive root of unity, i.e., $r = e^{\frac{2\pi}{n}i} = \cos(\frac{2\pi}{n}) + i\sin(\frac{2\pi}{n})$. We define the set of complex vectors $\{\boldsymbol{v}_0, \boldsymbol{v}_1, \ldots, \boldsymbol{v}_{n-1}\}$ by putting

$$\boldsymbol{v}_k = {}^t\left(1, r^k, r^{2k}, \ldots, r^{(n-1)k}\right) \in \mathbb{C}^n$$

for all $0 \leq k < n$. Let us denote by $\boldsymbol{V}_{n,1}$ the symmetric n-dimensional complex matrix given by

$$\boldsymbol{V}_{n,1} = (\boldsymbol{v}_0\boldsymbol{v}_1 \ldots \boldsymbol{v}_{n-1}) .$$

For small values of n, we have $\boldsymbol{V}_{1,1} = (1)$ and

$$\boldsymbol{V}_{2,1} = \begin{pmatrix} 1 & 1 \\ 1 & -1 \end{pmatrix} \in M_2(\mathbb{C}), \qquad \boldsymbol{V}_{3,1} = \begin{pmatrix} 1 & 1 & 1 \\ 1 & r & r^2 \\ 1 & r^2 & r \end{pmatrix} \in M_3(\mathbb{C}),$$

where $r = -\frac{1}{2} + i\frac{\sqrt{3}}{2}$.

Let us put $\boldsymbol{V}_{n,\ell} = \boldsymbol{V}_{n,1}^{\otimes \ell}$, for any positive integer ℓ.

Lemma VI.5. *For any positive integer ℓ, we have*

$$\overline{\boldsymbol{V}}_{n,\ell}\boldsymbol{V}_{n,\ell} = n^\ell \boldsymbol{I}_{n,\ell},$$

where $\boldsymbol{I}_{n,\ell}$ is the identity matrix of dimension n^ℓ and where $\overline{\boldsymbol{V}}_{n,\ell}$ denotes the conjugate complex matrix of $\boldsymbol{V}_{n,\ell}$.

Proof. The assertion is true for $\ell = 1$ because

$$(\overline{\boldsymbol{V}}_{n,1}\boldsymbol{V}_{n,1})_{ij} = \sum_{k=0}^{n-1} r^{n-ik} r^{jk} = \sum_{k=0}^{n-1} r^{k(j-i)} = \begin{cases} n & \text{if } i = j \\ 0 & \text{otherwise.} \end{cases}$$

The general case follows from a straightforward induction argument on ℓ, using the previous result:

$$\begin{aligned}
\overline{\boldsymbol{V}}_{n,\ell}\boldsymbol{V}_{n,\ell} = \overline{\boldsymbol{V}_{n,1}^{\otimes \ell}}\boldsymbol{V}_{n,1}^{\otimes \ell} &= \overline{\left(\boldsymbol{V}_{n,1}^{\otimes(\ell-1)} \otimes \boldsymbol{V}_{n,1}\right)} \left(\boldsymbol{V}_{n,1}^{\otimes(\ell-1)} \otimes \boldsymbol{V}_{n,1}\right) \\
&= \left(\overline{\boldsymbol{V}}_{n,\ell-1} \otimes \overline{\boldsymbol{V}}_{n,1}\right) \left(\boldsymbol{V}_{n,\ell-1} \otimes \boldsymbol{V}_{n,1}\right) \\
&= \left(\overline{\boldsymbol{V}}_{n,\ell-1}\boldsymbol{V}_{n,\ell-1}\right) \otimes \left(\overline{\boldsymbol{V}}_{n,1}\boldsymbol{V}_{n,1}\right) \\
&= n^{\ell-1}\boldsymbol{I}_{n,\ell-1} \otimes n\boldsymbol{I}_{n,1} = n^\ell \boldsymbol{I}_{n,\ell}.
\end{aligned}$$

□

We now define the *complex Walsh functions* as $\psi_t(s) = r^{s \cdot t}$, where $s \cdot t$ denotes the pointwise product of s and t.

In order to prepare for the calculation of normalized epistasis in terms of generalized Walsh coefficients, let us first prove:

Lemma VI.6. *For any positive integer ℓ, we have*

$$\overline{\boldsymbol{V}}_{n,\ell} \boldsymbol{U}_{n,\ell} \boldsymbol{V}_{n,\ell} = n^{2\ell} \begin{pmatrix} 1 0 \dots 0 \\ 0 0 \dots 0 \\ \vdots \vdots \ddots \vdots \\ 0 0 \dots 0 \end{pmatrix}.$$

Proof. Let us again argue by induction on ℓ. The statement holds true for $\ell = 1$, because

$$\begin{aligned} \overline{\boldsymbol{V}}_{n,1} \boldsymbol{U}_{n,1} \boldsymbol{V}_{n,1} &= \begin{pmatrix} 1 & 1 & \dots & 1 \\ 1 & r^{n-1} & \dots & r \\ \vdots & \vdots & \ddots & \vdots \\ 1 & r & \dots & r^{n-1} \end{pmatrix} \begin{pmatrix} 1 1 \dots 1 \\ 1 1 \dots 1 \\ \vdots \vdots \ddots \vdots \\ 1 1 \dots 1 \end{pmatrix} \begin{pmatrix} 1 & 1 & \dots & 1 \\ 1 & r & \dots & r^{n-1} \\ \vdots & \vdots & \ddots & \vdots \\ 1 & r^{n-1} & \dots & r^{(n-1)^2} \end{pmatrix} \\ &= \begin{pmatrix} n^2 & 0 \dots 0 \\ 0 & 0 \dots 0 \\ \vdots & \vdots \ddots \vdots \\ 0 & 0 \dots 0 \end{pmatrix}. \end{aligned}$$

Now, assume the assertion to hold true up to $\ell - 1$ and let us prove it for ℓ. Then

$$\begin{aligned}
\overline{\boldsymbol{V}}_{n,\ell}\boldsymbol{U}_{n,\ell}\boldsymbol{V}_{n,\ell} &= (\overline{\boldsymbol{V}_{n,1}^{\otimes \ell}})\boldsymbol{U}_{n,\ell}(\boldsymbol{V}_{n,1}^{\otimes \ell}) \\
&= \overline{\left(\boldsymbol{V}_{n,1}^{\otimes(\ell-1)} \otimes \boldsymbol{V}_{n,1}\right)} (\boldsymbol{U}_{n,\ell-1} \otimes \boldsymbol{U}_{n,1}) \left(\boldsymbol{V}_{n,1}^{\otimes(\ell-1)} \otimes \boldsymbol{V}_{n,1}\right) \\
&= \left(\overline{\boldsymbol{V}}_{n,\ell-1}\boldsymbol{U}_{n,\ell-1}\boldsymbol{V}_{n,\ell-1}\right) \otimes \left(\overline{\boldsymbol{V}}_{n,1}\boldsymbol{U}_{n,1}\boldsymbol{V}_{n,1}\right) \\
&= n^{2(\ell-1)} \begin{pmatrix} 1\,0 \dots 0 \\ 0\,0 \dots 0 \\ \vdots\,\vdots \ddots \vdots \\ 0\,0 \dots 0 \end{pmatrix} \otimes n^2 \begin{pmatrix} 1\,0 \dots 0 \\ 0\,0 \dots 0 \\ \vdots\,\vdots \ddots \vdots \\ 0\,0 \dots 0 \end{pmatrix} \\
&= n^{2\ell} \begin{pmatrix} 1\,0 \dots 0 \\ 0\,0 \dots 0 \\ \vdots\,\vdots \ddots \vdots \\ 0\,0 \dots 0 \end{pmatrix}.
\end{aligned}$$

□

We may now prove:

Lemma VI.7. *With notations as before, we have:*

$$\overline{\boldsymbol{V}}_{n,\ell}\boldsymbol{G}_{n,\ell}\boldsymbol{V}_{n,\ell} = \boldsymbol{D}_{n,\ell},$$

where $\boldsymbol{D}_{n,\ell}$ is the diagonal matrix whose only non-zero diagonal entries d_{ii} have value $n^{2\ell}$ and are situated at $i = kn^j$, for values $0 \leq k < n$ and $0 \leq j < \ell$.

Proof. Using the recursion relation

$$\boldsymbol{G}_{n,\ell} = \boldsymbol{U}_{n,1} \otimes \boldsymbol{G}_{n,\ell-1} + (\boldsymbol{G}_{n,1} - \boldsymbol{U}_{n,1}) \otimes \boldsymbol{U}_{n,\ell-1},$$

we can write

$$\overline{\boldsymbol{V}}_{n,\ell}\boldsymbol{G}_{n,\ell}\boldsymbol{V}_{n,\ell} = \boldsymbol{A} + \boldsymbol{B},$$

where

$$\boldsymbol{A} = \overline{\boldsymbol{V}}_{n,\ell}\left(\boldsymbol{U}_{n,1} \otimes \boldsymbol{G}_{n,\ell-1}\right)\boldsymbol{V}_{n,\ell}$$

and

$$B = \overline{V}_{n,\ell}\left(\left(G_{n,1} - U_{n,1}\right) \otimes U_{n,\ell-1}\right) V_{n,\ell}.$$

Let us now calculate these matrices A and B. First, using the fact that

$$V_{n,\ell} = V_{n,1} \otimes V_{n,\ell-1},$$

we note that

$$\begin{aligned}
A &= \left(\overline{V}_{n,1} \otimes \overline{V}_{n,\ell-1}\right)\left(U_{n,1} \otimes G_{n,\ell-1}\right)\left(V_{n,1} \otimes V_{n,\ell-1}\right) \\
&= \left(\overline{V}_{n,1} U_{n,1} V_{n,1}\right) \otimes \left(\overline{V}_{n,\ell-1} G_{n,\ell-1} V_{n,\ell-1}\right) \\
&= n^2 \begin{pmatrix} 1\,0 \ldots 0 \\ 0\,0 \ldots 0 \\ \vdots\,\vdots \ddots \vdots \\ 0\,0 \ldots 0 \end{pmatrix} \otimes \left(\overline{V}_{n,\ell-1} G_{n,\ell-1} V_{n,\ell-1}\right) \\
&= n^2 \begin{pmatrix} \overline{V}_{n,\ell-1} G_{n,\ell-1} V_{n,\ell-1} & 0 \ldots 0 \\ 0 & 0 \ldots 0 \\ \vdots & \vdots \ddots \vdots \\ 0 & 0 \ldots 0 \end{pmatrix}.
\end{aligned}$$

On the other hand,

$$\begin{aligned}
B &= \left(\overline{V}_{n,1} \otimes \overline{V}_{n,\ell-1}\right)\left(\left(G_{n,1} - U_{n,1}\right) \otimes U_{n,\ell-1}\right)\left(V_{n,1} \otimes V_{n,\ell-1}\right) \\
&= \left(\overline{V}_{n,1}(G_{n,1} - U_{n,1}) V_{n,1}\right) \otimes \left(\overline{V}_{n,\ell-1} U_{n,\ell-1} V_{n,\ell-1}\right) \\
&= \left(n \overline{V}_{n,1} V_{n,1} - \overline{V}_{n,1} U_{n,1} V_{n,1}\right) \otimes \left(\overline{V}_{n,\ell-1} U_{n,\ell-1} V_{n,\ell-1}\right) \\
&= \left[n^2 I_{n,1} - \begin{pmatrix} n^2\, 0 \ldots 0 \\ 0\;0 \ldots 0 \\ \vdots\;\vdots \ddots \vdots \\ 0\;0 \ldots 0 \end{pmatrix} \right] \otimes n^{2(\ell-1)} \begin{pmatrix} 1\,0 \ldots 0 \\ 0\,0 \ldots 0 \\ \vdots\,\vdots \ddots \vdots \\ 0\,0 \ldots 0 \end{pmatrix} \\
&= n^{2\ell} \begin{pmatrix} 0\,0 \ldots 0 \\ 0\,1 \ldots 0 \\ \vdots\,\vdots \ddots \vdots \\ 0\,0 \ldots 1 \end{pmatrix} \otimes \begin{pmatrix} 1\,0 \ldots 0 \\ 0\,0 \ldots 0 \\ \vdots\,\vdots \ddots \vdots \\ 0\,0 \ldots 0 \end{pmatrix}.
\end{aligned}$$

Using this, another straightforward induction argument finishes the proof. □

By analogy with the binary case, we define the (generalized) Walsh transform w of f by $\boldsymbol{w} = \boldsymbol{W}_{n,\ell}\boldsymbol{f}$, with $\boldsymbol{W}_{n,\ell} = n^{-\frac{\ell}{2}}\boldsymbol{V}_{n,\ell}$. The (complex!) components $w_i = w_i(f)$ of $\boldsymbol{w}$ will be called *Walsh coefficients* of f. These coefficients, of course, easily permit to recover f, since it follows from $\overline{\boldsymbol{W}}_{n,\ell}\boldsymbol{W}_{n,\ell} = \boldsymbol{I}_{n,\ell}$ that

$$\boldsymbol{f} = \overline{\boldsymbol{W}}_{n,\ell}(\boldsymbol{W}_{n,\ell}\boldsymbol{f}) = \overline{\boldsymbol{W}}_{n,\ell}\boldsymbol{w}.$$

In particular, we have for any $s \in \Omega_n$ that

$$f(s) = n^{-\frac{\ell}{2}} \sum_{t\in\Omega_n} \overline{\psi_t(s)} w_t = n^{-\frac{\ell}{2}} \sum_{t\in\Omega_n} \psi_t(s)\overline{w_t}.$$

Of course, $\overline{\psi_t(s)} = r^{-s\cdot t} = \psi_{-t}(s)$.

We may now prove:

Proposition VI.8. *If $w_0, \ldots, w_{n^\ell-1}$ are the Walsh coefficients of the fitness function f, then the normalized epistasis $\varepsilon^*(f)$ of f is given by*

$$\varepsilon^*(f) = 1 - \frac{|w_0|^2 + \sum_{i=0}^{\ell-1}\sum_{k=1}^{n-1} |w_{kn^i}|^2}{\sum_{i=0}^{n^\ell-1} |w_i|^2}.$$

Proof. Since $\boldsymbol{f} = \overline{\boldsymbol{W}}_{n,\ell}\boldsymbol{w} = \boldsymbol{W}_{n,\ell}\overline{\boldsymbol{w}}$, we obtain that

$${}^t\boldsymbol{f}\boldsymbol{f} = {}^t(\overline{\boldsymbol{W}}_{n,\ell}\boldsymbol{w})\boldsymbol{W}_{n,\ell}\overline{\boldsymbol{w}} = {}^t\boldsymbol{w}\overline{\boldsymbol{W}}_{n,\ell}\boldsymbol{W}_{n,\ell}\overline{\boldsymbol{w}} = {}^t\boldsymbol{w}\overline{\boldsymbol{w}},$$

as $\boldsymbol{W}_{n,\ell}$ is symmetric and $\overline{\boldsymbol{W}}_{n,\ell}\boldsymbol{W}_{n,\ell} = \boldsymbol{I}_{n,\ell}$.

On the other hand,

$${}^t\boldsymbol{f}\boldsymbol{E}_{n,\ell}\boldsymbol{f} = \frac{1}{n^\ell}\,{}^t(\overline{\boldsymbol{W}}_{n,\ell}\boldsymbol{w})\boldsymbol{G}_{n,\ell}\boldsymbol{W}_{n,\ell}\overline{\boldsymbol{w}} = \frac{1}{n^{2\ell}}\,{}^t\boldsymbol{w}\boldsymbol{D}_{n,\ell}\overline{\boldsymbol{w}}.$$

It thus follows that

$$\varepsilon^*(f) = 1 - \frac{{}^t\boldsymbol{f}\boldsymbol{E}_{n,\ell}\boldsymbol{f}}{{}^t\boldsymbol{f}\boldsymbol{f}} = 1 - \frac{{}^t\boldsymbol{w}\boldsymbol{D}_{n,\ell}\overline{\boldsymbol{w}}}{n^{2\ell}\,{}^t\boldsymbol{w}\overline{\boldsymbol{w}}}$$

and this equals

$$1 - \frac{|w_0|^2 + \sum_{i=0}^{\ell-1}\sum_{k=1}^{n-1} |w_{kn^i}|^2}{\sum_{i=0}^{n^\ell-1} |w_i|^2}.$$

□

It appears that calculating epistasis by using the complex Walsh transform is far more easy than through the use of the "natural" Walsh transform. Of course, when the partition coefficients of a given fitness function are known, this is no longer true. Moreover, in many practical applications one is really interested in knowing exactly or, at least, being able to calculate these partition coefficients.

In view of these remarks, it makes sense to try and find a method of directly translating one transform into the other.

Consider a fitness function f with associated "natural" and complex Walsh coefficients $\boldsymbol{w}_{nat}$ and $\boldsymbol{w}_c$, respectively. Denoting by $\boldsymbol{W}_{n,\ell}^{nat}$ and $\boldsymbol{W}_{n,\ell}^{c}$ the natural complex Walsh matrix, respectively, we obtain

$$
\begin{aligned}
\boldsymbol{w}_c &= \boldsymbol{W}_{n,\ell}^{c} \boldsymbol{f} \\
&= \boldsymbol{W}_{n,\ell}^{c} (\boldsymbol{W}_{n,\ell}^{nat} \boldsymbol{w}_{nat}) \\
&= (\boldsymbol{W}_{n,\ell}^{c} \boldsymbol{W}_{n,\ell}^{nat}) \boldsymbol{w}_{nat} \\
&= (\boldsymbol{W}_{n,1}^{c} \boldsymbol{W}_{n,1}^{nat})^{\otimes \ell} \boldsymbol{w}_{nat},
\end{aligned}
$$

where

$$
\boldsymbol{W}_{n,1}^{c} \boldsymbol{W}_{n,1}^{nat} = n^{-1} \begin{pmatrix} n & 0 & \dots & 0 \\ 0 & 1-r & \dots & 1-r^{n-1} \\ \vdots & \vdots & \ddots & \vdots \\ 0 & 1-r^{(n-1)} & \dots & 1-r^{(n-1)^2} \end{pmatrix}
$$

i.e.,

$$
(\boldsymbol{W}_{n,1}^{c} \boldsymbol{W}_{n,1}^{nat})_{ij} = \begin{cases} 1 & \text{if } i = j = 0 \\ \dfrac{1-r^{ij}}{n} & \text{otherwise.} \end{cases}
$$

2 Examples

In this section we describe some examples which illustrate how the previous results may be applied in order to effectively calculate the epistasis of some given fitness function.

2.1 Minimal epistasis

As we pointed out in chapters II and V, an arbitrary fitness function f has $\varepsilon^*(f) = 0$ if and only if f is of the form

$$f(s_{\ell-1} \dots s_0) = \sum_{i=0}^{\ell-1} g_i(s_i),$$

for some real-valued functions g_i which only depend on one position. Let us show how our approach allows for an easy proof of this result.

From the expression of normalized epistasis in terms of generalized (complex) Walsh coefficients, it follows that $\varepsilon^*(f) = 0$ is equivalent to $w_j = 0$ for all $j \neq 0, kn^i$ with $1 \leq k < n$ and $0 \leq i < \ell$. So,

$$\begin{aligned} w_{kn^i} &= n^{-\frac{\ell}{2}} \sum_{t \in \Omega_n} r^{kn^i \cdot t} f(t) = n^{-\frac{\ell}{2}} \sum_{t \in \Omega_n} r^{kt_i} f(t) \\ &= n^{-\frac{\ell}{2}} \left(\sum_{t \in \Omega_n(i,0)} f(t) + \sum_{t \in \Omega_n(i,1)} r^k f(t) + \dots + \sum_{t \in \Omega_n(i,n-1)} r^{k(n-1)} f(t) \right) \\ &= n^{\frac{\ell}{2}-1} \sum_{j=0}^{n-1} r^{kj} f_{(i,j)}, \end{aligned} \tag{VI.1}$$

where $\Omega_n(i,j)$ consists of all $s \in \Omega_n$ with $s_i = j$ and $f_{(i,j)} = \frac{1}{n^{\ell-1}} \sum_{t \in \Omega_n(i,j)} f(t)$ for all $0 \leq i < \ell$ and $0 \leq j < n$.

It thus follows that $\varepsilon^*(f) = 0$ is equivalent to

$$\begin{aligned} f(s) &= n^{-\frac{\ell}{2}} \sum_{t\in\Omega_n} \overline{\psi_t(s)} w_t \\ &= n^{-\frac{\ell}{2}} \sum_{t\in\Omega_n} r^{-s\cdot t} w_t \\ &= n^{-\frac{\ell}{2}} \left(w_0 + \sum_{i=0}^{\ell-1} \sum_{k=1}^{n-1} r^{-s\cdot kn^i} w_{kn^i} \right) \\ &= n^{-\frac{\ell}{2}} \left(w_0 + \sum_{i=0}^{\ell-1} \sum_{k=1}^{n-1} r^{-ks_i} w_{kn^i} \right) \\ &= n^{-\frac{\ell}{2}} \sum_{i=0}^{\ell-1} \left(\frac{w_0}{\ell} + \sum_{k=1}^{n-1} r^{-ks_i} w_{kn^i} \right) \\ &= \sum_{i=0}^{\ell-1} h_i(s_i), \end{aligned}$$

where

$$h_i(s_i) = n^{-\frac{\ell}{2}} \left(\frac{w_0}{\ell} + \sum_{k=1}^{n-1} r^{-ks_i} w_{kn^i} \right) \in \mathbb{C},$$

for all i. Actually,

$$f(s) = \sum_{i=0}^{\ell-1} h_i(s_i) = \sum_{i=0}^{\ell-1} \overline{h_i(s_i)} \in \mathbb{R},$$

so, $f(s) = \sum_{i=0}^{\ell-1} g_i(s_i)$, with

$$\begin{aligned} g_i(s_i) &= \frac{1}{2}\left(h_i(s_i) + \overline{h_i(s_i)} \right) \\ &= n^{-\frac{\ell}{2}} \left[\frac{w_0}{\ell} + \frac{1}{2} \sum_{k=1}^{n-1} \left(r^{-ks_i} w_{kn^i} + r^{ks_i} \overline{w_{kn^i}} \right) \right] \\ &= n^{-\frac{\ell}{2}} \frac{w_0}{\ell} + \frac{1}{2n} \sum_{k=1}^{n-1} \left(\sum_{j=0}^{n-1} \left(r^{k(j-s_i)} + r^{k(s_i-j)} \right) f_{(i,j)} \right) \\ &= n^{-\frac{\ell}{2}} \frac{w_0}{\ell} + \frac{1}{n} \sum_{k=1}^{n-1} \sum_{j=0}^{n-1} \left(\cos \frac{2k(j-s_i)\pi}{n} \right) f_{(i,j)}, \end{aligned}$$

which clearly belongs to $\mathbb{R}$.

On the other hand, as

$$\begin{aligned} f_{(i,a)} &= \frac{1}{n^{\ell-1}} \sum_{t\in\Omega_n(i,a)} f(t) = \frac{1}{n^{\ell-1}} \sum_{t\in\Omega_n(i,a)} \sum_{j=0}^{\ell-1} g_j(t_j) \\ &= \frac{1}{n^{\ell-1}} \left(n^{\ell-1} g_i(a) + \sum_{\substack{j=0\\ j\neq i}}^{\ell-1} \sum_{b\in\Sigma} n^{\ell-2} g_j(b) \right) \\ &= g_i(a) + \frac{1}{n} \sum_{\substack{j=0\\ j\neq i}}^{\ell-1} \sum_{b\in\Sigma} g_j(b), \end{aligned}$$

we can rewrite the first order Walsh coefficients w_{kn^i}, for all $0 \leq k < n$ and $0 \leq i < \ell$, as

$$\begin{aligned} w_{kn^i} &= n^{\frac{\ell}{2}-1} \sum_{a\in\Sigma} r^{ka} f_{(i,a)} = n^{\frac{\ell}{2}-1} \sum_{a\in\Sigma} r^{ka} \left(g_i(a) + \frac{1}{n} \sum_{\substack{j=0\\ j\neq i}}^{\ell-1} \sum_{b\in\Sigma} g_j(b) \right) \\ &= n^{\frac{\ell}{2}-1} \sum_{a\in\Sigma} r^{ka} g_i(a), \end{aligned}$$

because

$$\sum_{a\in\Sigma} r^{ka} \left(\sum_{\substack{j=0\\ j\neq i}}^{\ell-1} \sum_{b\in\Sigma} g_j(b) \right) = \left(\sum_{\substack{j=0\\ j\neq i}}^{\ell-1} \sum_{b\in\Sigma} g_j(b) \right) \cdot \sum_{a\in\Sigma} r^{ka} = 0.$$

Moreover, with this notation, the average of a first order function is given by

$$\begin{aligned} \tilde{w}_0 &= \frac{1}{n^\ell} \sum_{t\in\Omega_n} f(t) = \frac{1}{n^\ell} \sum_{t\in\Omega_n} \left(\sum_{i=0}^{\ell-1} g_i(t_i) \right) \\ &= \frac{1}{n^\ell} \sum_{i=0}^{\ell-1} n^{\ell-1} \sum_{a\in\Sigma} g_i(a) \\ &= \frac{1}{n} \sum_{\substack{0\leq j<\ell\\ a\in\Sigma}} g_i(a), \end{aligned}$$

and all other Walsh coefficients are zero, as we just pointed out.

2.2 Generalized camel functions

Let us start by considering the canonical basis $\left\{ \boldsymbol{e}_0, \boldsymbol{e}_{\frac{n^\ell}{n-1}}, \ldots, \boldsymbol{e}_{k\frac{n^\ell-1}{n-1}}, \ldots, \boldsymbol{e}_{n-1} \right\}$ of $\mathbb{R}^{n^\ell}$, i.e.,

$$\boldsymbol{e}_{k\frac{n^\ell-1}{n-1}} = {}^t(0, \ldots, 0, 1, 0, \ldots, 0)$$

where 1 appears as $k\frac{n^\ell-1}{n-1}$-th coordinate, for $0 \leq k < n$. If we inductively define the set of complex vectors $\{\boldsymbol{v}_{0,\ell}, \ldots, \boldsymbol{v}_{n^\ell-1,\ell}\}$ in $\mathbb{C}^{n^\ell}$ by putting $\boldsymbol{v}_{k,0} = 1$ for all k and, inductively,

$$\boldsymbol{v}_{k,\ell} = \begin{pmatrix} \boldsymbol{v}_{k,\ell-1} \\ r^k \boldsymbol{v}_{k,\ell-1} \\ r^{2k} \boldsymbol{v}_{k,\ell-1} \\ \vdots \\ r^{(n-1)k} \boldsymbol{v}_{k,\ell-1} \end{pmatrix},$$

then $\boldsymbol{V}_{n,\ell} \boldsymbol{e}_{k\frac{n^\ell-1}{n-1}} = \boldsymbol{v}_{k\frac{n^\ell-1}{n-1},\ell}$, for all $0 \leq k < n$.

Let us now consider generalized camel functions, i.e., functions f with the property that they are zero everywhere except for $f(c_0) = f(c_1) = \cdots = f(c_{n-1}) = \frac{\sqrt{n}}{n}$, where $c_0, c_1, \ldots, c_{n-1}$ form a set of strings in Ω_n which are at pairwise Hamming distances of ℓ from each other. In order to calculate their normalized epistasis, we may assume $c_0 = 0$ and so $c_k = k\frac{n^\ell-1}{n-1}$, with $1 \leq k < n$. Then, with the notations of section 3.2, chapter V, the vector associated to f is

$$\boldsymbol{q}_{0,\ldots,k\frac{n^\ell-1}{n-1},\ldots,(n^\ell-1)} = n^{-\frac{1}{2}} \sum_{k=0}^{n-1} \boldsymbol{e}_{k\frac{n^\ell-1}{n-1}},$$

and the complex Walsh coefficients of f are given by

$$\begin{aligned} \boldsymbol{w} = \boldsymbol{W}_{n,\ell} \boldsymbol{f} = \boldsymbol{W}_{n,\ell} \boldsymbol{q}_{0,\ldots,k\frac{n^\ell-1}{n-1},\ldots,(n^\ell-1)} &= n^{-\frac{1}{2}} \boldsymbol{W}_{n,\ell} \sum_{k=0}^{n-1} \boldsymbol{e}_{k\frac{n^\ell-1}{n-1}} \\ = n^{-\frac{(\ell+1)}{2}} \sum_{k=0}^{n-1} \boldsymbol{V}_{n,\ell} \boldsymbol{e}_{k\frac{n^\ell-1}{n-1}} &= n^{-\frac{(\ell+1)}{2}} \sum_{k=0}^{n-1} \boldsymbol{v}_{k\frac{n^\ell-1}{n-1},\ell}. \end{aligned}$$

One may thus verify, for example through an induction argument, that $w_{kn^i} = 0$ for all $1 \leq k < n$ and $0 \leq i < \ell$.

As $w_0 = n^{\frac{-\ell+1}{2}}$ and ${}^t\boldsymbol{w}\overline{\boldsymbol{w}} = {}^t\boldsymbol{f}\overline{\boldsymbol{f}} = \|f\|^2 = 1$, we finally have that

$$\varepsilon^*(f) = 1 - \frac{1}{n^{\ell-1}},$$

as claimed in chapter V.

2.3 Generalized unitation functions

As a final example, we consider generalized unitation functions. Recalling their definition given in section 4 of chapter V, these are fitness functions f with the property that $f(s) = \sum_{i=0}^{n-1} g_i(u_i(s))$ for some functions g_i on $\{0, \ldots, \ell-1\}$ and where $u_i(s)$ denotes, for any $s \in \Omega_n$, the number of i's in the n-ary representation of s. Since for any permutation $\sigma(s)$ of s we obviously have $f(s) = f(\sigma(s))$, it is clear that $f_{i,j}$ is independent of i. From the definition of w_{kn^i}, it follows that

$$w_k = w_{kn} = w_{kn^2} = \cdots = w_{kn^{\ell-1}}.$$

In order to give a general expression for $f_{i,j}$, let us consider the case $n = 3$ and then deduce the formula for the general case. Let us start with

$$f_{i,0} = \frac{1}{3^{\ell-1}} \sum_{s \in \Omega_3(0,0)} f(s) = \frac{1}{3^{\ell-1}} \sum_{s \in \Omega_3(0,0)} \left(g_0(u_0) + g_1(u_1) + g_2(u_2)\right),$$

with $u_i = u_i(s)$, for $i = 0, 1, 2$.

If we take any $0 \le a \le \ell$ and $0 \le b \le \ell - a$, then it is clear that there are $\binom{\ell-1}{a-1}\binom{\ell-a}{b}$ strings in $\Omega_3(0,0)$ with $u_0 = a$, $u_1 = b$ and $u_2 = \ell - a - b$. This yields that

$$\begin{aligned} f_{i,0} &= \frac{1}{3^{\ell-1}} \sum_{s \in \Omega_3(0,0)} f(s) \\ &= \frac{1}{3^{\ell-1}} \sum_{\substack{u_0, u_1, u_2 \\ u_0+u_1+u_2=\ell}} \binom{\ell-1}{u_0-1}\binom{\ell-u_0}{u_1} \left(g_0(u_0) + g_1(u_1) + g_2(u_2)\right) \end{aligned}$$

Of course, the same argument shows that

$$\begin{aligned} f_{i,1} &= \frac{1}{3^{\ell-1}} \sum_{s \in \Omega_3(0,1)} f(s) \\ &= \frac{1}{3^{\ell-1}} \sum_{\substack{u_0, u_1, u_2 \\ u_0+u_1+u_2=\ell}} \binom{\ell-1}{u_1-1}\binom{\ell-u_1}{u_2} \left(g_0(u_0) + g_1(u_1) + g_2(u_2)\right) \end{aligned}$$

and

$$f_{i,2} = \frac{1}{3^{\ell-1}} \sum_{s\in\Omega_3(0,2)} f(s)$$
$$= \frac{1}{3^{\ell-1}} \sum_{\substack{u_0,u_1,u_2 \\ u_0+u_1+u_2=\ell}} \binom{\ell-1}{u_2-1}\binom{\ell-u_2}{u_0} (g_0(u_0)+g_1(u_1)+g_2(u_2)).$$

So, if $k=1,2$, we have that

$$w_{k3^i} = 3^{-\frac{\ell}{2}} \sum_{\substack{u_0,u_1,u_2 \\ u_0+u_1+u_2=\ell}} \left\{\binom{\ell-1}{u_0-1}\binom{\ell-u_0}{u_1} + r^k\binom{\ell-1}{u_1-1}\binom{\ell-u_1}{u_2}\right.$$
$$\left. + r^{2k}\binom{\ell-1}{u_2-1}\binom{\ell-u_2}{u_0}\right\} (g_0(u_0)+g_1(u_1)+g_2(u_2)).$$

In the general case, we find

$$w_{kn^i} = n^{-\frac{\ell}{2}} \sum_{\substack{u_0,\dots,u_{n-1} \\ u_0+\dots+u_{n-1}=\ell}} \alpha^k_{u_0\dots u_{n-1}} \sum_{i=0}^{n-1} g_i(u_i)$$

with

$$\alpha^k_{u_0\dots u_{n-1}} = \sum_{i=0}^{n-1} r^{ki}\binom{\ell-1}{u_i-1}\binom{\ell-u_i}{u_{i+1}}\cdots\binom{\ell-u_i-u_{i+1}-\dots-u_{i+n-3}}{u_{i+n-2}},$$

where we work with coefficients modulo n.

On the other hand, note that

$$w_0 = n^{-\ell}\sum_{s\in\Omega_{n,\ell}} f(s) = n^{-\ell} \sum_{\substack{u_0,\dots,u_{n-1} \\ u_0+\dots+u_{n-1}=\ell}} \sum_{\substack{s\in\Omega_{n,\ell} \\ u_i(s)=u_i}} f(s).$$

Since f is a unitation function,

$$w_0 = n^{-\ell} \sum_{\substack{u_0,\dots,u_{n-1} \\ u_0+\dots+u_{n-1}=\ell}} \binom{\ell}{u_0}\binom{\ell-u_0}{u_1}\cdots\binom{\ell-u_0-\dots-u_{n-3}}{u_{n-2}} \sum_{i=0}^{n-1} g_i(u_i).$$

The normalized epistasis of f is thus finally given by

$$\varepsilon^*(f) = 1 - \frac{|w_0|^2 + \ell\sum_{k=1}^{n-1}|w_k|^2}{\sum_{i=0}^{n^\ell-1}|w_i|^2},$$

with coefficients as above.

2.4 Second order functions

Early in this book, equation I.4 defined the class of second order functions as those f with the functional form

$$f : \{0,1\}^\ell \to \mathbb{R} : s \mapsto \sum_{0\le i<\ell} g_i(s_i) + \sum_{0\le i<j<\ell} g_{ij}(s_i, s_j),$$

where the g_i and g_{ij} are real valued functions that depend on only one and only two variables, respectively. In the multary setting, the class contains the binary constraint satisfaction problem (defined in chapter I, section 6) as a large subclass. Thanks to its simple functional form, it is not too difficult to compute the Walsh coefficients and epistasis of a generic second order function. We will do so in this section, following the results of [95], and use them in section 3.5 as an example of the computation of the moments of schema fitness distributions.

To avoid having to include the normalization factor $n^{\frac{\ell}{2}}$ in most of what follows, we work with modified Walsh coefficients $\widetilde{w}_i = n^{-\frac{\ell}{2}} w_i$. So, the average of a second order function f is

$$\begin{aligned}
\tilde{w}_0 &= \frac{1}{n^\ell} \sum_{s\in\Omega_n} f(s) \\
&= \frac{1}{n^\ell} \sum_{s\in\Omega_n} \sum_{0\le i<\ell} g_i(s_i) + \frac{1}{n^\ell} \sum_{s\in\Omega_n} \sum_{0\le i<j<\ell} g_{ij}(s_i, s_j) \\
&= \frac{1}{n} \sum_{a\in\Sigma} \sum_{0\le i<\ell} g_i(a) + \frac{1}{n^\ell} \sum_{a,b\in\Sigma} \sum_{0\le i<j<\ell} n^{\ell-2} g_{ij}(a, b) \\
&= \frac{1}{n} \sum_{a\in\Sigma} \sum_{0\le i<\ell} g_i(a) + \frac{1}{n^2} \sum_{a,b\in\Sigma} \sum_{0\le i<j<\ell} g_{ij}(a, b).
\end{aligned}$$

Note that because $\boldsymbol{W}(\boldsymbol{f}+\boldsymbol{g}) = \boldsymbol{W}\boldsymbol{f} + \boldsymbol{W}\boldsymbol{g}$, the formal separation between the first and the second order part of f is respected by the Walsh transform. This allows us to calculate each of these components separately. In particular, we will denote by $f^{(1)}$ and $f^{(2)}$ the first and second order part of f, respectively, i.e., we write $f = f^{(1)} + f^{(2)}$.

In order to calculate the Walsh coefficients w_{kn^i} (also called the *first order Walsh coefficients*), we need the average of the fitness value over the strings with a fixed bit a at position i for $f^{(2)}$:

$$\begin{aligned}
f^{(2)}_{(i,a)} &= \frac{1}{n^{\ell-1}} \sum_{t\in\Omega_n(i,a)} f^{(2)}(t) = \frac{1}{n^{\ell-1}} \sum_{t\in\Omega_n(i,a)} \sum_{0\le j<k<\ell} g_{jk}(t_j,t_k) \\
&= \frac{1}{n^{\ell-1}} \Bigg[\sum_{0\le p<i<\ell} \sum_{c\in\Sigma} n^{\ell-2} g_{pi}(c,a) + \sum_{i<q<\ell} \sum_{d\in\Sigma} n^{\ell-2} g_{iq}(a,d) \\
&\qquad + \sum_{\substack{0\le p<q<\ell \\ p,q\neq i}} \sum_{c,d\in\Sigma} n^{\ell-3} g_{pq}(c,d) \Bigg] \\
&= \frac{1}{n} \sum_{\substack{0\le p<i<\ell \\ c\in\Sigma}} g_{pi}(c,a) + \frac{1}{n} \sum_{\substack{i\le q<\ell \\ d\in\Sigma}} g_{iq}(a,d) + \frac{1}{n^2} \sum_{\substack{0\le p<q<\ell \\ p,q\neq i}} \sum_{c,d\in\Sigma} g_{pq}(c,d).
\end{aligned}$$

Since section 2.1 already gave the coefficients for $f^{(1)}$, it remains to use the previous equality and (VI.1) to calculate the coefficients of $f^{(2)}$:

$$\begin{aligned}
\tilde{w}_{kn^i}(f^{(2)}) &= n^{-\frac{\ell}{2}} w_{kn^i}(f^{(2)}) = \frac{1}{n} \sum_{a\in\Sigma} r^{ka} f^{(2)}_{(i,a)} \\
&= \frac{1}{n^2} \sum_{a\in\Sigma} r^{ka} \left[\sum_{\substack{0\le p<i<\ell \\ c\in\Sigma}} g_{pi}(c,a) + \sum_{\substack{i<q<\ell \\ d\in\Sigma}} g_{iq}(a,d) \right] \\
&\quad + \frac{1}{n^3} \sum_{a\in\Sigma} r^{ka} \left[\sum_{\substack{0\le p<q<\ell \\ p,q\neq i}} \sum_{c,d\in\Sigma} g_{pq}(c,d) \right] \\
&= \frac{1}{n^2} \sum_{a\in\Sigma} r^{ka} \left[\sum_{\substack{0\le p<i<\ell \\ c\in\Sigma}} g_{pi}(c,a) + \sum_{\substack{i<q<\ell \\ d\in\Sigma}} g_{iq}(a,d) \right],
\end{aligned}$$

because $\sum_{a\in\Sigma} r^{ka} = 0$.

Let us now calculate the *second order Walsh coefficients* $w_{an^i+bn^j}$. If $t = an^i + bn^j$, then

$$\begin{aligned}\tilde{w}_t &= n^{-\frac{\ell}{2}} w_t = n^{-\ell} \sum_{s\in\Omega_n} r^{ts} f^{(2)}(s) = n^{-\ell} \sum_{s\in\Omega_n} r^{as_i+bs_j} f^{(2)}(s) \\ &= n^{-\ell} \sum_{c,d\in\Sigma} \sum_{s\in\Omega_n(ij,cd)} r^{ac+bd} f^{(2)}(s) \\ &= n^{-\ell} \sum_{c,d\in\Sigma} r^{ac+bd} \sum_{s\in\Omega_n(ij,cd)} f^{(2)}(s) = \frac{1}{n^2} \sum_{c,d\in\Sigma} r^{ac+bd} f^{(2)}_{(ij,cd)},\end{aligned}$$

where $\Omega_n(ij, cd)$ denotes the subset of Ω_n which contains all strings with alleles c and d in loci i and j, respectively, and

$$\begin{aligned}f^{(2)}_{(ij,ab)} &= \frac{1}{n^{\ell-2}} \sum_{s\in\Omega_n(ij,ab)} f^{(2)}(s) \\ &= g_{ij}(a,b) + \frac{1}{n} \sum_{0\le p<i} \sum_{c\in\Sigma} g_{pi}(c,a) + \frac{1}{n} \sum_{\substack{i<q<\ell \\ q\neq j}} \sum_{d\in\Sigma} g_{iq}(a,d) \\ &+ \frac{1}{n} \sum_{\substack{0\le p<j \\ p\neq i}} \sum_{c\in\Sigma} g_{pj}(c,b) + \frac{1}{n} \sum_{j<q<\ell} \sum_{d\in\Sigma} g_{jq}(b,d) \\ &+ \frac{1}{n^2} \sum_{\substack{0\le p<q<\ell \\ p,q\neq i,j}} \sum_{c,d\in\Sigma} g_{pq}(c,d).\end{aligned}$$

Finally, as

$$\frac{1}{n^2} \sum_{c,d\in\Sigma} r^{ac+bd} \left(\frac{1}{n} \sum_{\substack{i<q<\ell \\ q\neq j}} \sum_{e\in\Sigma} g_{iq}(a,e) \right) = \frac{1}{n^3} \sum_{c\in\Sigma} \left(r^{ac} \sum_{\substack{i<q<\ell \\ q\neq j}} \sum_{e\in\Sigma} g_{iq}(a,e) \right) \sum_{d\in\Sigma} r^{bd} = 0,$$

(because $\sum_{d\in\Sigma} r^{bd} = 0$) and, as all of the other double sums also vanish, we have

$$\begin{aligned}\tilde{w}_{an^i+bn^j}(f^{(2)}) &= \frac{1}{n^2} \sum_{c,d\in\Sigma} r^{ac+bd} f^{(2)}_{(ij,cd)} \\ &= \frac{1}{n^2} \sum_{c,d\in\Sigma} r^{ac+bd} g_{ij}(c,d).\end{aligned}$$

Moreover, $\tilde{w}_{an^i+bn^j}(f^{(2)}) = \tilde{w}_{an^i+bn^j}(f)$ because, as we pointed out in section 2.1, the second order coefficients of first order functions are zero. In fact, for $t = an^i + bn^j$,

$$\begin{aligned}
\tilde{w}_t(f^{(1)}) &= n^{-\frac{\ell}{2}} w_t = n^{-\ell} \sum_{s\in\Omega_n} r^{ts} f^{(1)}(s) \\
&= n^{-\ell} \sum_{s\in\Omega_n} r^{as_i+bs_j} f^{(1)}(s) \\
&= n^{-\ell} \sum_{c,d\in\Sigma} \sum_{s\in\Omega_n(ij,cd)} r^{ac+bd} f^{(1)}(s) \\
&= n^{-\ell} \sum_{c,d\in\Sigma} r^{ac+bd} \sum_{s\in\Omega_n(ij,cd)} \left(\sum_{k=0}^{\ell-1} g_k(s) \right) \\
&= n^{-\ell} \sum_{c,d\in\Sigma} r^{ac+bd} \left[n^{\ell-1} g_i(c) + n^{\ell-1} g_j(d) + \sum_{\substack{p=0\\ p\neq i,j}}^{\ell-1} n^{\ell-2} \sum_{e\in\Sigma} g_p(e) \right] \\
&= \frac{1}{n} \sum_{c,d\in\Sigma} r^{ac+bd} \left(g_i(c) + g_j(d)\right) + \frac{1}{n^2} \sum_{c,d\in\Sigma} r^{ac+bd} \sum_{\substack{p=0\\ p\neq i,j}}^{\ell-1} \sum_{e\in\Sigma} g_p(e) = 0,
\end{aligned}$$

because

$$\begin{aligned}
&\sum_{c,d\in\Sigma} r^{ac+bd} \left(g_i(c) + g_j(d)\right) \\
&= \sum_{c,d\in\Sigma} r^{ac+bd} g_i(c) + \sum_{c,d\in\Sigma} r^{ac+bd} g_j(d) \\
&= \sum_{c\in\Sigma} r^{ac} g_i(c) \left(\sum_{d\in\Sigma} r^{bd} \right) + \sum_{d\in\Sigma} r^{bd} g_j(d) \left(\sum_{c\in\Sigma} r^{ac} \right) = 0.
\end{aligned}$$

The same argument yields that

$$\frac{1}{n^2} \sum_{c,d\in\Sigma} r^{ac+bd} \sum_{\substack{p=0\\ p\neq i,j}}^{\ell-1} \sum_{e\in\Sigma} g_p(e) = 0.$$

As an example, let us now include the binary case ($n = 2$). With our notation,

$$\begin{aligned}
\tilde{w}_0(f) &= \tilde{w}_0(f^{(1)}) + \tilde{w}_0(f^{(2)}) \\
&= \frac{1}{2} \sum_{0\leq i<\ell} \left(g_i(0) + g_i(1)\right) + \frac{1}{4} \sum_{0\leq i<j<\ell} \left(g_{ij}(0,0) + g_{ij}(0,1) + g_{ij}(1,0) + g_{ij}(1,1)\right),
\end{aligned}$$

and

$$\begin{aligned}\tilde{w}_{2^i}(f^{(1)}+f^{(2)}) =& \frac{1}{2}g_i(0)-\frac{1}{2}g_i(1)\\ &+\frac{1}{4}\sum_{0\le p<i<\ell}(g_{pi}(0,0)+g_{pi}(1,0))+\frac{1}{4}\sum_{i<q<\ell}(g_{iq}(0,0)+g_{iq}(0,1))\\ &-\frac{1}{4}\sum_{0\le p<i<\ell}(g_{pi}(0,1)+g_{pi}(1,1))-\frac{1}{4}\sum_{i<q<\ell}(g_{iq}(1,0)+g_{iq}(1,1))\\ =& \frac{1}{2}g_i(0)-\frac{1}{2}g_i(1)\\ &+\frac{1}{4}\sum_{0\le p<i<\ell}(g_{pi}(0,0)+g_{pi}(1,0)-g_{pi}(0,1)-g_{pi}(1,1))\\ &+\frac{1}{4}\sum_{i<q<\ell}(g_{iq}(0,0)+g_{iq}(0,1)-g_{iq}(1,0)-g_{iq}(1,1)),\end{aligned}$$

for all $i\in\{0,\ldots,\ell-1\}$. The expression for the second order Walsh coefficients is

$$\tilde{w}_t=\frac{1}{4}\sum_{c,d\in\{0,1\}}r^{c+d}g_{ij}(c,d)=\frac{1}{4}\left(g_{ij}(0,0)+g_{ij}(1,1)-g_{ij}(0,1)-g_{ij}(1,0)\right),$$

where, in this case, $t=n^i+n^j$ for some $0\le i,j<\ell$.

Finally let us prove that the complex Walsh coefficients w_t are zero for $t\neq 0, kn^i, an^i+bn^j$, with $0\le i,j<\ell$ and $1\le k,a,b<n$.

The multary representation of t has a zero at all of its loci, except for $i_0,\ldots,i_q$, which respectively contain the alleles $a_0,\ldots,a_q\in\{1,\ldots,n-1\}$. So,

$$\begin{aligned}w_t &= n^{-\frac{\ell}{2}}\sum_{s\in\Omega_n}r^{t\cdot s}f^{(2)}(s)=n^{-\frac{\ell}{2}}\sum_{s\in\Omega_n}r^{\sum_j a_j s_{i_j}}f^{(2)}(s)\\ &= n^{-\frac{\ell}{2}}\sum_{\forall j:\, b_j\in\Sigma}\ \sum_{\substack{s\in\Omega_n\\ \forall j:\, s_{i_j}=b_j}}r^{\sum_j a_j b_j}f^{(2)}(s)\\ &= n^{-\frac{\ell}{2}}\sum_{\forall j:\, b_j\in\Sigma}r^{\sum_j a_j b_j}\sum_{\substack{s\in\Omega_n\\ \forall j:\, s_{i_j}=b_j}}\left(\sum_{0\le p,q<\ell}g_{pq}(s)\right)\end{aligned}$$

$$
\begin{aligned}
&= n^{-\frac{\ell}{2}} \sum_{\forall j:\, b_j \in \Sigma} r^{\sum_j a_j b_j} \Bigg[n^{\ell-q} \sum_{0 \le u,v \le q} g_{i_u i_v}(a_u, b_v) \\
&\quad + \sum_{0 \le u \le q} \left(\sum_{\substack{0 \le p < i_u \\ \forall j:\, p \neq i_j}} \sum_{c \in \Sigma} n^{\ell-q-1} g_{p i_u}(c, a_u) + \sum_{\substack{i_u < p < \ell \\ \forall j:\, p \neq i_j}} \sum_{c \in \Sigma} n^{\ell-q-1} g_{i_u p}(a_u, c) \right) \\
&\quad + \sum_{\substack{0 \le m,y < \ell \\ \forall j:\, m,y \neq i_j}} \sum_{c,d \in \Sigma} n^{\ell-q-1} g_{my}(c, d) \Bigg] \\
&= 0,
\end{aligned}
$$

using the same arguments as in the first order case. Once the Walsh coefficients are known, we can immediately compute the normalized epistasis as

$$
\begin{aligned}
\varepsilon^*(f) &= 1 - \frac{|w_0|^2 + \sum_{0 \le i < \ell} \sum_{1 \le k < n} |w_{kn^i}|^2}{\sum_{0 \le i < n^\ell} |w_i|^2} \\
&= 1 - \frac{|w_0|^2 + \sum_{0 \le i < \ell} \sum_{1 \le k < n} |w_{kn^i}|^2}{|w_0|^2 + \sum_{0 \le i < \ell} \sum_{1 \le k < n} |w_{kn^i}|^2 + \sum_{0 \le i,j < \ell} \sum_{1 \le a,b < n} |w_{an^i + bn^j}|^2} \\
&= \frac{\sum_{0 \le i,j < \ell} \sum_{1 \le a,b < n} |w_{an^i + bn^j}|^2}{|w_0|^2 + \sum_{0 \le i < \ell} \sum_{1 \le k < n} |w_{kn^i}|^2 + \sum_{0 \le i,j < \ell} \sum_{1 \le a,b < n} |w_{an^i + bn^j}|^2} \\
&= \frac{\sum_{0 \le i,j < \ell} \sum_{1 \le a,b < n} |w_{an^i + bn^j}|^2}{\sum_{0 \le i < n^\ell} |w_i|^2}.
\end{aligned}
$$

3 Odds and ends

This section generalizes a number of results recently published by Heckendorn and co-authors about the moments of schema fitness distributions expressed in terms of Walsh coefficients. We rewrite their results about these summary statistics, which are restricted to functions on binary strings, to functions on multary strings. We also show an application of the results in the context of randomly generated binary constraint satisfaction problems.

To prove the results about summary statistics, we need generalized versions of the balanced sum theorems that appeared in section 2 of chapter IV. The proofs that we present here are significantly shorter than the ones for the binary case. This is due to a better notation, which we also exploit to elegantly link the partition coefficients with the complex Walsh coefficients.

3.1 Notations and terminology

The fitness distribution of a schema $H = h_{\ell-1} \dots h_0 \in (\Sigma \cup \{\#\})^\ell = \Sigma'^\ell$ is defined as the distribution of fitness values of the strings belonging to the schema. The number of strings generated by schema H is, as usual, denoted by $|H|$.

The definitions of the functions β and J, first used in chapter IV, section 2, are trivially extended to multary alphabets:

$$\beta(H)_i = \begin{cases} 0 & \text{if } h_i = \# \text{ or } h_i = 0 \\ h_i & \text{otherwise.} \end{cases}$$

$$J(H) = \{t \in \Omega_n;\ \forall 0 \leq i < \ell :\ h_i = \# \Rightarrow t_i = 0\}.$$

The function α gets a new definition:

$$\alpha(H) = \{t \in \Omega_n;\ \forall 0 \leq i < \ell :\ h_i = \# \Leftrightarrow t_i = 0\}.$$

3.2 Balanced sum theorems

Corollary IV.18 is called the *balanced sum theorem* for binary alphabets. Here we prove it for multary alphabets.

Theorem VI.9 (Balanced sum theorem for multary alphabets). *With notations as before, we have*

$$\sum_{x=0}^{n^\ell - 1} \psi_j(x) = \begin{cases} 0 & \text{if } j \neq 0 \\ n^\ell & \text{otherwise.} \end{cases}$$

Proof. This follows immediately from lemma VI.5 where the multiplication of the first row of $\overline{\boldsymbol{V}}_{n,\ell}$ with $\mathbf{V}_{n,\ell}$ yields the first row of $n^\ell \boldsymbol{I}_{n,\ell}$. □

Before proceeding with the balanced sum theorem for hyperplanes, we prove the following lemma:

Lemma VI.10. *If $j \in J(H)$ and $s \in H$, then $\psi_j(s) = \psi_j(\beta(H))$.*

Proof. Knowing that $\psi_j(s) = \prod_i r^{s_i j_i}$, we consider two cases for h_i:

1. $h_i = \#$. This implies that $j_i = 0$, and hence $1 = r^{s_i j_i} = r^{\beta(h_i) j_i}$.

2. $h_i \neq \#$. It follows that $\beta(h_i) = h_i = s_i$, and therefore $r^{s_i j_i} = r^{\beta(h_i) j_i}$.

As a result, $\psi_j(s) = \prod_i r^{s_i j_i} = \prod_i r^{\beta(h_i) j_i} = \psi_j(\beta(H))$. □

Theorem VI.11 (Balanced sum theorem for hyperplanes and multary alphabets). *Let $H \in \Sigma'^\ell$ be a schema with at least one position containing a $\#$. We then have*

$$\sum_{s \in H} \psi_j(s) = \begin{cases} 0 & \textit{if } j \notin J(H) \\ |H| \psi_j(\beta(H)) & \textit{otherwise.} \end{cases}$$

Proof. Let us first assume that $j \in J(H)$. Then

$$\sum_{s \in H} \psi_j(s) = \sum_{s \in H} \psi_j(\beta(H)) = |H| \psi_j(\beta(H))$$

by the previous lemma.

To prove that $\sum_{s \in H} \psi_j(s) = 0$ when $j \notin J(H)$, we proceed as follows. There must exist at least one position where j takes a value different from 0, and H contains a $\#$. For notation's sake, let us assume $\ell - 1$ to be such a position.

Observing that

$$\begin{aligned} \sum_{s \in H} \psi_j(s) &= \sum_{\substack{s \in H \\ s_{\ell-1}=0}} \sum_{\substack{s \in H \\ s_{\ell-1}=1}} \cdots \sum_{\substack{s \in H \\ s_{\ell-1}=n-1}} \psi_j(s), \\ &= \sum_{\substack{s \in H \\ s_{\ell-1}=0}} \sum_{\substack{s \in H \\ s_{\ell-1}=1}} \cdots \sum_{\substack{s \in H \\ s_{\ell-1}=n-1}} r^{s_{\ell-1} j_{\ell-1}} r^{\sum_{i=0}^{\ell-2} s_i j_i}, \end{aligned}$$

we only need to rearrange the terms, using the notation $H = \#\hat{H}$, to obtain

$$= \sum_{\hat{s} \in \hat{H}} \psi_{\hat{j}}(\hat{s}) \left(1 + r^{j_{\ell-1}} + r^{2j_{\ell-1}} + \cdots + r^{(n-1)j_{\ell-1}}\right)$$
$$= 0.$$

□

Finally, we generalize in a straightforward way the hyperplane averaging theorem for binary alphabets (corollary IV.10) to multary alphabets:

Theorem VI.12 (Hyperplane averaging theorem for multary alphabets). *Let $H \in \Sigma'^{\ell}$ be a schema with at least one position containing $\#$. Then*

$$f(H) = \frac{1}{|H|} \sum_{x \in H} f(x) = n^{-\frac{\ell}{2}} \sum_{j \in J(H)} \overline{\psi_j(\beta(H))} w_j.$$

Proof.

$$\begin{aligned}
\frac{1}{|H|} \sum_{x \in H} f(x) &= \frac{1}{|H|} \sum_{x \in H} n^{-\frac{\ell}{2}} \sum_{j=0}^{n^\ell - 1} \overline{\psi_j(x)} w_j \\
&= \frac{n^{-\frac{\ell}{2}}}{|H|} \sum_{j=0}^{n^\ell - 1} \left(\sum_{x \in H} \overline{\psi_j(x)} \right) w_j \\
&= \frac{n^{-\frac{\ell}{2}}}{|H|} \sum_{j \in J(H)} \left(\sum_{x \in H} \overline{\psi_j(x)} \right) w_j \\
&= \frac{n^{-\frac{\ell}{2}}}{|H|} \sum_{j \in J(H)} |H| \overline{\psi_j(\beta(H))} w_j \\
&= n^{-\frac{\ell}{2}} \sum_{j \in J(H)} \overline{\psi_j(\beta(H))} w_j.
\end{aligned}$$

□

3.3 Partition coefficients revisited

To write the partition coefficients in terms of complex Walsh coefficients, we start with three lemmas:

Lemma VI.13. *For all schemata $H, H' \in \Sigma'^{\ell}$, we have*

1. $H' \supset H \Rightarrow \alpha(H') \subset J(H)$,
2. *for every $t \in J(H)$, there exists exactly one $H' \supset H$ with $t \in \alpha(H')$.*

Proof. To prove the first statement, we observe that $t \in \alpha(H')$ if and only if $h'_i = \# \Leftrightarrow t_i = 0$ for all $0 \leq i < \ell$. Now $h_i = \#$ implies $h'_i = \#$, hence $t_i = 0$ for all i, and hence $t \in J(H)$.

To prove the existence of an H' for the second statement, we define H' by

$$h'_i = \begin{cases} \# & \text{if } t_i = 0 \\ h_i & \text{if } t_i \neq 0, \end{cases}$$

for all $0 \leq i < \ell$. Clearly $H' \supset H$. On the other hand, to show that $t \in \alpha(H')$ we consider both cases for each t_i. If $t_i = 0$, then $h'_i = \#$, by definition. If $t_i \neq 0$, then $h_i \neq \#$, and by definition $h'_i = h_i \neq \#$, as required.

The unicity of H' is shown as follows. If $t_i = 0$, then $t \in \alpha(H')$ implies $h'_i = \#$. If $t_i \neq 0$, then $h_i \neq \#$, as $t \in J(H)$. Now suppose that $h'_i = \#$. Then $t \in \alpha(H')$ implies $t_i = 0$, a contradiction. Hence $h'_i \neq \#$, and because $H' \supset H$, we must have $h'_i = h_i$. □

Lemma VI.14. *For all schemata $H \in \Sigma'^{\ell}$, we have*

$$J(H) = \bigcup_{H' \supset H} \alpha(H'),$$

where the union is one of disjoint sets.

Proof. First observe that if $H' \supset H$, then $\alpha(H') \subset J(H)$. Next, if $t \in J(H)$ then there exists exactly one H' for which $t \in \alpha(H')$. Finally, we claim that $H' \neq H''$ if and only if $\alpha(H') \cup \alpha(H'') = \emptyset$. Let $t \in \alpha(H') \cap \alpha(H'')$, then $t \in J(H)$. By the unicity statement of the previous lemma, we have necessarily $H' = H''$. The other implication is trivial. □

Lemma VI.15. *For all schemata $H, H' \in \Sigma'^{\ell}$ and strings $t \in \Omega_n$, we have*

$$H' \supset H \text{ and } t \in \alpha(H') \Rightarrow \psi_t(\beta(H)) = \psi_t(\beta(H')).$$

Proof. Let us check that $t \cdot \beta(H) = t \cdot \beta(H')$. It suffices to verify that for all i, $t_i \beta(H)_i = t_i \beta(H')_i$. We consider two cases.

If $h_i = \#$, then $\beta(H)_i = 0$. But because $H' \supset H$, we also have $h'_i = \#$, and $\beta(H')_i = 0$.

If $h_i \neq \#$, then either $h'_i = h_i$ or $h'_i = \#$. In the first case, $\beta(H)_i = \beta(H')_i$. In the second case, $t_i = 0$ because $t \in \alpha(H')$. □

We can now finally present the link between partition coefficients and complex Walsh coefficients:

Theorem VI.16. *For all schemata $H \in \Sigma'^{\ell}$, we have*

$$\varepsilon(H) = n^{-\frac{\ell}{2}} \sum_{t \in \alpha(H)} \overline{\psi_t(\beta(H))} w_t.$$

Note that in the binary case ($n = 2$), $|\alpha(H)| = 1$, and (as all ψ_t are real-valued) by lemma IV.14,

$$\varepsilon(H) = 2^{-\frac{\ell}{2}} \psi_{\alpha(H)}(\beta(H)) w_{\alpha(H)} = 2^{-\frac{\ell}{2}} (-1)^{u(\beta(H))} w_{\alpha(H)}.$$

Proof. Starting from the hyperplane averaging theorem, and subsequently applying lemma VI.14 and lemma VI.15, we obtain

$$\begin{aligned} f(H) &= n^{-\frac{\ell}{2}} \sum_{t \in J(H)} \overline{\psi_t(\beta(H))} w_t \\ &= n^{-\frac{\ell}{2}} \sum_{H' \supset H} \sum_{t \in \alpha(H')} \overline{\psi_t(\beta(H))} w_t \\ &= n^{-\frac{\ell}{2}} \sum_{H' \supset H} \sum_{t \in \alpha(H')} \overline{\psi_t(\beta(H'))} w_t. \end{aligned}$$

The theorem clearly holds for $H = \Omega_n$, as $\varepsilon(\Omega_n) = f(\Omega_n)$, $\alpha(\Omega_n) = \{0\}$, $\beta(\Omega_n)=0$, $\psi_0(0) = 1$ and $w_0 = n^{\frac{\ell}{2}} f(\Omega_n)$. We can therefore proceed recursively, and assume that

$$\varepsilon(H') = n^{-\frac{\ell}{2}} \sum_{t \in \alpha(H')} \overline{\psi_t(\beta(H'))} w_t$$

for all $H' \supsetneqq H$. Then

$$\begin{aligned}\varepsilon(H) &= f(H) - \sum_{H' \supsetneqq H} \varepsilon(H') \\ &= n^{-\frac{\ell}{2}} \sum_{H' \supset H} \sum_{t \in \alpha(H')} \overline{\psi_t(\beta(H'))} w_t - \sum_{H' \supsetneqq H} \left(n^{-\frac{\ell}{2}} \sum_{t \in \alpha(H')} \overline{\psi_t(\beta(H'))} w_t \right) \\ &= n^{-\frac{\ell}{2}} \sum_{t \in \alpha(H)} \overline{\psi_t(\beta(H))} w_t.\end{aligned}$$

□

3.4 Application: moments of schemata and fitness function

Given a mean μ and a discrete random variable X, the r-th moment of X around μ is given by

$$\begin{aligned}\mu_r &= E[(X - \mu)^r] \\ &= \sum_{x \in X} (x - \mu)^r p(x),\end{aligned}$$

where $p(x)$ denotes the probability of the event $x \in X$. In [32], Heckendorn, Rana and Whitley compute the moments of the fitness function around the function mean for functions on binary strings. We follow their line of proof to generalize the following result. (As in section 2.4, we write $\tilde{w}_s = n^{-\frac{\ell}{2}} w_s$ to avoid including normalization factors. Note again that $\mu = \tilde{w}_0$ is the average of f.)

Theorem VI.17. *The r-th moment about the function mean in terms of the Walsh coefficients of the function is given by*

$$\mu_r = \sum_{\substack{0 \neq a_i \in \Omega_n \\ a_1 \oplus \cdots \oplus a_r = 0}} \tilde{w}_{a_1} \ldots \tilde{w}_{a_r} = n^{-\frac{\ell r}{2}} \sum_{\substack{0 \neq a_i \in \Omega_n \\ a_1 \oplus \cdots \oplus a_r = 0}} w_{a_1} \ldots w_{a_r},$$

where $\oplus$ denotes the addition in $(\mathbb{Z}/n\mathbb{Z})^\ell$.

Proof. In our situation, where all strings are considered equally likely, we have

$$\begin{aligned}\mu_r &= \frac{1}{n^\ell}\sum_{x\in\Omega_n}(f(x)-\mu)^r \\ &= \frac{1}{n^\ell}\sum_{x=0}^{n^\ell-1}\left(\sum_{i=1}^{n^\ell-1}\widetilde{w}_i\overline{\psi_i(x)}\right)^r \\ &= \frac{1}{n^\ell}\sum_{a_1=1}^{n^\ell-1}\cdots\sum_{a_r=1}^{n^\ell-1}\widetilde{w}_{a_1}\ldots\widetilde{w}_{a_r}\sum_{x=0}^{n^\ell-1}\overline{\psi_{a_1}(x)}\ldots\overline{\psi_{a_r}(x)}.\end{aligned}$$

Using the fact that for arbitrary p and q,

$$\psi_p(x)\psi_q(x) = \psi_{p\oplus q}(x),$$

we obtain

$$\mu_r = \frac{1}{n^\ell}\sum_{a_1=1}^{n^\ell-1}\cdots\sum_{a_r=1}^{n^\ell-1}\widetilde{w}_{a_1}\ldots\widetilde{w}_{a_r}\sum_{x=0}^{n^\ell-1}\overline{\psi_{a_1\oplus\cdots\oplus a_r}(x)}.$$

According to the balanced sum theorem, the inner sum is non-zero only when $a_1 \oplus a_2 \oplus \cdots \oplus a_r = 0$. Therefore,

$$\begin{aligned}\mu_r &= \frac{1}{n^\ell}\sum_{\substack{0\neq a_i\in\Omega_n\\ a_1\oplus\cdots\oplus a_r=0}}\widetilde{w}_{a_1}\ldots\widetilde{w}_{a_r}n^\ell \\ &= \sum_{\substack{0\neq a_i\in\Omega_n\\ a_1\oplus\cdots\oplus a_r=0}}\widetilde{w}_{a_1}\ldots\widetilde{w}_{a_r}.\end{aligned}$$

□

We also generalize two other theorems of [31]:

Theorem VI.18. *The r-th moment of a schema H about the function mean in terms of the Walsh coefficients of the function is given by*

$$\mu_r = \sum_{\substack{0\neq a_i\in\Omega_n\\ a_1\oplus\cdots\oplus a_r\in J(H)}}\widetilde{w}_{a_1}\ldots\widetilde{w}_{a_r}\overline{\psi_{a_1\oplus\cdots\oplus a_r}(\beta(H))}.$$

Proof. As in [31] and in the spirit of the previous proof, we write

$$\begin{aligned}\mu_r(H) &= \frac{1}{|H|}\sum_{x\in H}(f(x)-\mu)^r \\ &= \frac{1}{|H|}\sum_{a_1=1}^{n^\ell-1}\cdots\sum_{a_r=1}^{n^\ell-1}\tilde{w}_{a_1}\ldots\tilde{w}_{a_r}\sum_{x=0}^{n^\ell-1}\overline{\psi_{a_1\oplus\cdots\oplus a_r}(x)}.\end{aligned}$$

Applying the balanced sum theorem, we obtain

$$\mu_r = \sum_{\substack{0\neq a_i\in\Omega_n \\ a_1\oplus\cdots\oplus a_r\in J(H)}}\tilde{w}_{a_1}\ldots\tilde{w}_{a_r}\overline{\psi_{a_1\oplus\cdots\oplus a_r}(\beta(H))}.$$

□

Theorem VI.19. *The r-th moment of a schema H about the mean of the schema in terms of the Walsh coefficients of the function is given by*

$$\mu_r = \sum_{\substack{a_i\in\Omega_n\setminus J(H) \\ a_1\oplus\cdots\oplus a_r\in J(H)}}\tilde{w}_{a_1}\ldots\tilde{w}_{a_r}\overline{\psi_{a_1\oplus\cdots\oplus a_r}(\beta(H))}.$$

Proof. Given that the hyperplane averaging theorem has been generalized to multary alphabets, the same line of proof as in [31] can be followed. □

3.5 Application: summary statistics for binary CSPs

As an application of the above results, we summarize in this section some results of Schoofs and Naudts [68, 84] related to the use of summary statistics to predict the problem difficulty that randomly generated *binary constraint satisfaction problems* (*binary CSPs,* e.g. [96]) induce on a GA.

We start by defining a binary CSP as

- a set of variables x_i with $0 \leq i < \ell$,
- a domain or alphabet $\Sigma = \{0, \ldots, n-1\}$,
- a subset $C_{ij} \subset \Sigma \times \Sigma$ for each pair of variables (i, j), with $0 \leq i < j < \ell$, which represents a *constraint* when it differs from the Cartesian product $\Sigma \times \Sigma$.

The goal of a CSP is to *assign* to the variables x_i a value from their domain Σ in such a way that all constraints are satisfied. Formally, we say that a constraint C_{ij} is satisfied if and only if $(x_i, x_j) \in C_{ij}$. The couple (x_i, x_j) is then called a *valid assignment.* When $(x_i, x_j) \notin C_{ij}$ we say that the assignment (x_i, x_j) *violates* the constraint C_{ij}. Each element of $\Sigma \times \Sigma$ that is not in a constraint set is called a *conflict* for that constraint.

A typical example of a CSP that can be found in any artificial intelligence textbook, is the *N-queens problem.* The objective of this problem is to position N queens on a chess board of size N in such a way that they cannot attack each other. A representation for this problem that fits the above definition is the following: let each row of the board be represented by one variable, and let each variable take the value of the position of the queen in that row. This representation implicitly assumes that each queen has to be in a different row, but that is fine because otherwise two queens would be able to attack each other. The two other classes of constraints have to be implemented by filling in the constraint tables: (1) no two queens should be in the same column, and (2) no two queens should be on the same diagonal.

Binary CSP instances can be randomly generated, as we saw in the first chapter; one possible way is known as *model E* [54]: select uniformly, independently and with repetition, $pn^2\ell(\ell-1)/2$ conflicts out of the $n^2\ell(\ell-1)/2$ possible. The parameter p clearly controls the density of conflicts in the problem instance, which has a serious effect on the problem difficulty it imposes on a GA: when there are too few conflicts, the problem is trivially solvable; when there are too many, it becomes rapidly clear that no solution exists. Interestingly, a *phase transition* from solvable to unsolvable can be shown to exist in the limit of an infinite number of variables [109]. In practice, a *mushy region* is observed, for some small interval for p, where solvable and unsolvable instances co-exist, and where the expected number of solutions is close to 1. Problem instances from this region have been studied in the above mentioned work, the question being whether the first two moments of the fitness distributions of low order schemata could be used to predict the number of generations to a solution of a GA on problem instances with at least one solution.

Because a CSP does not have an explicit fitness function, one is usually constructed

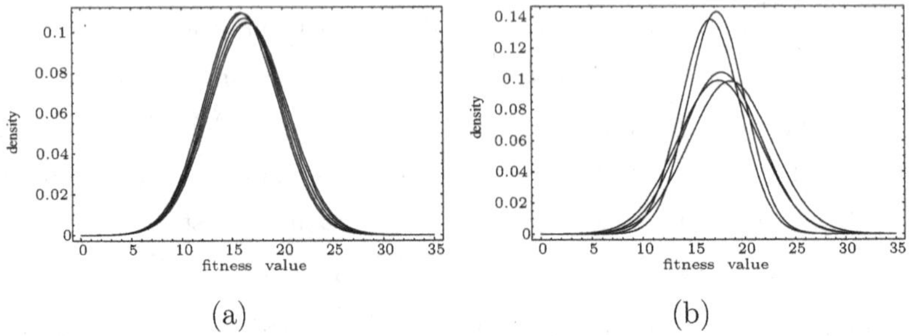

Figure VI.1: Normal approximation of the fitness distributions of the schemata $a\#\ldots\#$, $0 \leq a < 5$, for a for a randomly generated instance with 15 variables and an alphabet of size 5. (a) shows the distributions computed over an initial population, (b) records the distributions near the end of a typical GA run.

by counting the number of violated constraints. By doing so, we reduce a CSP to a second order function. Let $g_{ij}(s_i, s_j)$ denote the interaction between the positions i and j, with the respective values s_i and s_j. In a binary CSP, it equals 1 if there is a conflict between the values on the positions i and j, and 0 otherwise.

Using the material of section 2.4, we are able to compute the first few moments of the fitness distribution of an arbitrary schema efficiently. Given that for binary CSPs, at most $O(\ell^2)$ Walsh coefficients are nonzero, we can compute the mean and variance of the fitness distribution, and the mean of arbitrary schema fitness distributions in $O(\ell^2)$ time. The variance of arbitrary schema fitness distributions is computed in $O(\ell^4)$ time.

Figure VI.1(a) shows, for a randomly generated instance with 15 variables and an alphabet of size 5, that the mean and standard deviation of the distributions of the schemata $a\#\ldots\#$, $0 \leq a < 5$, are very close to each other. Near the end of a GA run (figure VI.1(b)), the overlap is less prominent but still visible in two subsets of schemata.

The distributions presented in the plot are based on empirical values for mean and standard deviation, obtained from the population of one typical run. In the case of the initial population, these summary statistics correspond closely to the ones predicted by theorem VI.19, since they are dominated by average fitness strings. The situation at the end of a run cannot be predicted by the theory of this chapter.

The fact that the distributions overlap shows that not all GA dynamics follow Goldberg's decomposition (section 4, chapter I). It is not clear what building blocks are in the context of randomly generated CSPs. Still the GA proceeds, albeit slower than in situations where building blocks can be clearly distinguished, and reaches a situation where finding the optimum depends on many other factors than the lack of building blocks only.

Appendix A

The schema theorem (variations on a theme)

An open question for a long time in the theoretical study of GAs was whether there exist subsets other than "Holland schemata" which, under appropriate genetic operators, behave according to the schema theorem. In [5], Battle and Vose answered the question positively by proposing schemata based on transformation matrices connected to isomorphisms of GAs. Their generalization of traditional schemata preserves the algebraic structure and may be thought of as an explicit example of the more general definition of a schema considered by Vose [104], who generalized the notion of schema to arbitrary predicates.

Recall that Holland originally defined a schema as certain linear varieties of strings (a subset of strings with "similarities" at certain string positions). For example, considering strings of length four over the binary alphabet, the "similarity" of the strings 0011, 0001, 1001 and 1011 may be expressed by a schema $H \in \{0, 1, \#\}^4$ ($H = \#0\#1$), where # is a *don't care* character. It is thus clear that schemata may be used to structurally describe "strong" or "weak" populations with respect to some given fitness function.

The point of view of Vose in [104] is that a schema is a predicate. Clearly, the schema $H = \#0\#1$ may be considered as the function

$$H : \{0,1\}^4 \to \{true, false\}$$

which behaves according to the rule

$$H(x) = \text{true} \Leftrightarrow x \text{ matches } H = \#0\#1 \text{ at every position not containing } \#.$$

Vose proved the schema theorem in this set-up, and applied it to introduce concepts like locality, monotonicity and stability, which allows GAs to be viewed as constrained random walks. His work also yielded new insights concerning deceptive problems (we refer the reader to [104] for details).

Although the use of general predicates or subsets of $\{0,1\}^\ell$ as schemata has proven to be a very adequate tool, practical applications require our set-up to be a little more general.

Indeed, the fitness function on $\{0,1\}^\ell$ often associates to strings an experimentally obtained, measured value, with a certain inherent amount of vagueness and measuring errors. This implies that extreme values for f may never be *correctly* determined, but only approximated. If we just wish describe a near optimal set H of length ℓ binary strings as a solution for some optimization problem, this is of course, not of great importance. It becomes important, however, if we want to use schemata to indicate the structural reasons why this set H is optimal.

In fact, due to the experimental vagueness of the fitness value of individual strings, at best, we will only be able to guess or approximate the likelihood of a string belonging to H, i.e., one should view H as a *fuzzy subset* of $\{0,1\}^\ell$.

In the next section we introduce *fuzzy schemata*, a notion encompassing both traditional schemata and Vose's arbitrary predicates. The fuzzy schemata allow to describe structural information of imprecise data. Moreover, they satisfy a suitable version of the schema theorem, thus answering the question raised in [34] concerning the existence of (generalized) schemata possessing this property.

1 A Fuzzy Schema Theorem

Fuzzy subsets generalize ordinary subsets (see for example [15] for more details). Whereas an ordinary subset H of $\{0,1\}^\ell$ partitions $\{0,1\}^\ell$ in a "black" and a "white" zone (the black consisting of the points belonging to H), a fuzzy subset allows for shades of "grey", where the shade of a point measures its degree of belonging to H. Black points are members of H, white points are not, while the other points "more or less" belong to H. Fuzzy schemata work in essentially the same way: a black point certainly possesses the "correct" structure, a white one does not, while a grey point possesses it up to a certain degree.

Let us consider a (finite or infinite) universe Ω — in practice Ω will usually be the set $\{0,1\}^\ell$ of binary strings of length ℓ. A *fuzzy schema* is a fuzzy subset H of Ω, i.e., a map $H : \Omega \to [0,1]$. If H only takes the values 0 and 1, we speak of a *crisp schema* in Ω. Of course, this just corresponds to the ordinary subset of Ω consisting of all $p \in \Omega$ with $H(p) = 1$. A population is defined to be a multiset P of Ω (hence repetitions are allowed). We put

$$|H|_P = \sum_{p \in P} H(p).$$

It is clear that $|H|_P$ "counts" the elements of P belonging to H. Indeed, if H is a crisp schema in Ω, then $|H|_P$ is just the cardinality $|P \cap H|$ of the intersection of P and H, (counting multiplicities, as always).

Consider a fitness function $f : \Omega \to \mathbb{R}$. For any fuzzy schema H of Ω, we write

$$f_P(H) = \frac{1}{|H|_P} \sum_{p \in P} H(p) f(p)$$

(putting $f_P(H) = 0$, if $|H|_P = 0$). If H is a crisp subset of Ω, then

$$f_P(H) = \frac{1}{|P \cap H|} \sum_{p \in P} f(p),$$

i.e., $f_P(H)$ is the average fitness of $P \cap H$. In particular, if $H = \Omega$ (the constant function on Ω with value 1), then $f_P(H) = f_P(\Omega)$ is the average fitness of the population P.

Let us consider an evolving population $A(t)$, where the positive integer t may be viewed as a discrete time parameter. We write $m(H,t) = A(t)$, i.e., $m(H,t)$ is the "number" of elements of the population "belonging" to H at time t. Indeed, in the crisp case, $m(H,t) = |A(H,t)|$, is the cardinality of the set $A(H,t)$ of all $p \in A(t)$, which belong to H.

Let us also put

$$f(H,t) = \frac{1}{m(H,t)} \sum_{p \in A(t)} H(p) f(p) = f_{A(t)}(H).$$

In particular,

$$f(\Omega,t) = \frac{1}{m(\Omega,t)} \sum_{p \in A(t)} \Omega(p) f(p) = \frac{1}{m(\Omega,t)} \sum_{p \in A(t)} f(p),$$

where $m(\Omega,t) = |\Omega|_{A(t)}$.

Note that in the crisp case, $f(H,t)$ is the average fitness of $A(H,t)$.

Let us now fix $\Omega = \{0,1\}^\ell$ and assume that the population $A(t) \subseteq \Omega$ evolves through the application of genetic operators. Although other operators (like mutation or inversion) may be taken into account as well, we will only consider selection and crossover.

The selection operator picks strings in the population $A(t)$ with a probability of being selected proportional to their fitness. Let us first consider selection separately. Since $m(H,t+1) = \sum_{p \in A(t+1)} H(p)$, it follows for any fuzzy schema H that

$$\mathbb{E}(m(H,t+1)) = \sum_{p \in A(t)} H(p) \frac{f(p)}{f(\Omega,t)} = \frac{f(H,t)}{f(\Omega,t)} m(H,t),$$

where $\mathbb{E}$ denotes the expectation operator.

In order to include crossover as well, we have to modify this as follows. Recall that the crossover operator starts from a string $p = p_0 \ldots p_{\ell-1}$, and selects a second string $q = q_0 \ldots q_{\ell-1}$ and a random crossover site $0 < z < \ell$. It then produces two new strings, $p_0 \ldots p_{z-1} q_z \ldots q_{\ell-1}$ and $q_0 \ldots q_{z-1} p_z \ldots p_{\ell-1}$, and replaces p by one of them. We denote the latter by $p \otimes q$.

Moreover, denote by $\tau_H(p,q)$ the probability that $p \otimes q$ belongs to the schema H, when crossover is applied to p and q. Then,

$$\tau_H(p,q) = \sum_{r \in X(p,q)} \frac{H(r)}{|X(p,q)|},$$

where $X(p,q)$ is the full (multi)set of potential offspring that may be produced by applying crossover to p and q.

Since the probability $\pi_{p,q}$ of selecting a certain pair (p,q) of strings in $A(t)$ is

$$\pi_{p,q} = \frac{f(p)}{m(\Omega,t)f(\Omega,t)} \frac{f(q)}{m(\Omega,t)f(\Omega,t)},$$

we obtain

$$\mathbb{E}(m(H,t+1)) = m(\Omega,t)\frac{\sum_{p,q} f(p)f(q)}{(m(\Omega,t)f(\Omega,t))^2}\left((1-p_c)\frac{H(p)+H(q)}{2} + p_c\tau_H(p,q)\right),$$

where p_c denotes the probability that crossover occurs.

It thus follows:

Theorem A.1 (The Fuzzy Schema Theorem). *For any fuzzy schema H in Ω,*

$$\mathbb{E}(m(H,t+1)) \geq m(\Omega,t)\frac{f(H,t)}{f(\Omega,t)}\left(1 - p_c\alpha(H,t)\right),$$

where

$$\alpha(H,t) = \sum_p \frac{H(p)f(p)}{m(H,t)f(H,t)} \sum_q \frac{f(q)}{m(\Omega,t)f(\Omega,t)}(1-\tau_H(p,q)).$$

Let us take a closer look at $\alpha(H,t)$. Consider a stochastic operator F on a finite set M which, for any (random value) $m \in M$, produces a string $F(m) \in \Omega$. If H is a crisp schema, $\mathbb{E}(F \in H)$ is the probability of $F(m)$ belonging to H, i.e.,

$$\mathbb{E}(F \in H) = \frac{|\{m \in M;\ F(m) \in H\}|}{|M|}.$$

For a fuzzy schema H, it thus makes sense to use

$$\mathbb{E}(H \circ F) = \frac{\sum_{m \in M} H(F(m))}{|M|}.$$

For example, view $p \otimes q$ as a stochastic operator (depending on p and q) which arbitrarily selects a string in $X(p,q)$. Then $\tau_H(p,q) = \mathbb{E}(H \circ (p \otimes q))$ is the expected membership value of $p \otimes q$ in H.

Since

$$\alpha(H,t) = \sum_p \frac{H(p)f(p)}{m(H,t)f(H,t)}\left(1 - \sum_q \frac{f(q)}{m(\Omega,t)f(\Omega,t)}\tau_H(p,q)\right)$$

and $\frac{f(q)}{m(\Omega,t)f(\Omega,t)}$ is the probability of selecting q in $A(t)$, it follows that

$$1 - \sum_q \frac{f(q)}{m(\Omega,t)f(\Omega,t)} \tau_H(p,q) = \mathbb{E}((1-H) \circ (p \otimes -))$$

is the expected non-membership value of $p \otimes q$ in H for fixed p and random $q \in A(t)$. On the other hand, let us denote by η the selection operator in $A(t)$. Then

$$\mathbb{E}(\eta = p) = \frac{f(p)}{\sum_p f(p)}$$

since the probability of p being selected in $A(t)$ is proportional to its fitness. If H is a crisp schema in Ω, it follows for the restriction $\eta|H$ of η to H that

$$\mathbb{E}(\eta|H = p) = \begin{cases} 0 & \text{if } p \notin H \\ \frac{f(p)}{\sum_{p \in A(H,t)} f(p)} & \text{if } p \in H, \end{cases}$$

the probability of selecting p within H.

It thus makes sense to put

$$\mathbb{E}(\eta|H = p) = \frac{H(p)f(p)}{\sum_{p \in A(H,t)} H(p)f(p)},$$

for any fuzzy schema H in Ω. With this definition, it follows that

$$\alpha(H,t) = \sum_p \mathbb{E}(\eta|H = p)\mathbb{E}((1-H) \circ (p \otimes -)),$$

i.e., $\alpha(H,t)$ may be viewed as the "probability" that the child of any $p \in A(t)$ does not belong to H, provided that p does. This is in accordance with the conclusions of [104].

Note that if mutation is also applied with probability p_m, in each bit, then theorem A.1 changes to

$$\mathbb{E}(m(H,t+1)) \geq m(\Omega,t)\frac{f(H,t)}{f(\Omega,t)}\left(1 - p_c\alpha(H,t)\right)\left(1 - p_m\beta(H,t)\right),$$

where $\beta(H,t)$ takes into account the effect of mutation.

2 The schema theorem on measure spaces

In this section, we describe a "global" approach to the previous set-up, in the non-fuzzy case. We start from a search space Ω, which is just assumed to be a measure space (like $\mathbb{R}^n$, for example), and we shift attention from strings of symbols to points in this space. The fitness function will be an integrable function on Ω, whereas the crossover operator will be replaced by a suitable, rather general, Ω-valued stochastic operator (in two variables) on Ω.

In particular, this degree of generality will permit us to take into account the distance between points in Ω (if it is endowed with a metric), for example, if we want to produce new points out of them.

Let Ω be a measure space, e.g., $\Omega = \mathbb{R}^n$, and consider a measurable subset P of Ω. We will call P a *population* in Ω. Let us also consider a bounded, integrable map $f : \Omega \to \mathbb{R}$; we call f a fitness function on Ω, if it is *positively* valued. Let us fix a measurable subset $H \subseteq \Omega$. We will call H a *schema* and identify it with its characteristic function $H : \Omega \to \{0, 1\}$, given by

$$H(p) = \begin{cases} 1 & \text{if } x \in H \\ 0 & \text{if } x \notin H. \end{cases}$$

The map H is integrable.

Fixing the population P for a moment, we put

$$|H|_P = \int_P H(\omega) d\omega = \mu(H \cap P),$$

where μ is the measure map on Ω. Let us also define

$$f_P(H) = \begin{cases} \frac{1}{|H|_P} \int_P H(\omega) f(\omega) d\omega & \text{if } |H|_P \neq 0 \\ 0 & \text{if } |H|_P = 0. \end{cases}$$

In other words,

$$\mu(H \cap P) f_P(H) = \int_{H \cap P} f(\omega) d\omega.$$

If $P = A(t) \subseteq \Omega$ is a population, which depends upon some (discrete) parameter t, then we write

$$m(H,t) = |H|_{A(t)} = \int_{A(t)} H(\omega)d\omega = \mu(H \cap A(t))$$

and

$$f(H,t) = f_{A(t)}(H),$$

respectively. If $m(H,t) \neq 0$, we thus obtain

$$f(H,t) = \frac{\int_{A(t)} H(\omega)f(\omega)d\omega}{\int_{A(t)} H(\omega)d\omega}.$$

Note that in the previous set-up, we did not allow for multiple occurrences of elements in the population P. If we want to make P into a multiset, we slightly have to change our point of view, for example by defining P to be an integrable map

$$P : \Omega \to \mathbb{R},$$

all of whose values are positive integers. The value $P(x)$ then represents the number of occurrences of $x \in \Omega$ in the population P. Since our results may easily be adapted to take this feature into account, we will not go into this deeper, leaving details to the reader instead.

Let us now consider the selection operator η in Ω which picks elements with a probability proportional to their fitness. Our assumptions imply that η transforms measurable subsets into measurable subsets and that, up to subsets of measure zero, if $x \in \Omega$ is selected, so are points sufficiently close to x.

As

$$m(H,t+1) = \int_{A(t+1)} H(\omega)d\omega,$$

it follows that

$$\mathbb{E}(m(H,t+1)) = \frac{1}{f(\Omega,t)} \int_{A(t)} H(\omega)f(\omega)d\omega = \frac{f(H,t)}{f(\Omega,t)} m(H,t)$$

where $\mathbb{E}$ denotes, as in the fuzzy case, the expectation operator.

On the other hand, crossover should be an operator which, for every pair of elements $p, q \in \Omega$, returns a new element $p \otimes q \in \Omega$. It will thus be a stochastic operator

$$\chi : \Omega \times \Omega \to \Omega$$

which maps measurable subsets onto measurable subsets.

The classical crossover operator on strings (choosing crossover sites randomly) is of this type. Another example is the operator

$$\chi : \mathbb{R}^n \times \mathbb{R}^n \to \mathbb{R}^n : (p, q) \mapsto \alpha p + (1 - \alpha) q,$$

where the real random variable α is normally distributed around $\alpha_0 = 0.5$.

We already know that the probability of selecting a particular couple $(p, q) \in A(t) \times A(t)$ is given by

$$\frac{f(p)}{\int_{A(t)} f(\omega) d\omega} \frac{f(q)}{\int_{A(t)} f(\omega) d\omega} = \frac{f(p) f(q)}{(f(\Omega, t) m(\Omega, t))^2}.$$

Denote by $\tau_H(p, q)$ the probability that the output of the crossover operator applied to the selected pair p and q belongs to H, and assume

$$\tau_H : \Omega \times \Omega \to [0, 1] \subseteq \mathbb{R}$$

to be integrable. The probability of obtaining an element in H, starting from any couple $(p, q) \in A(t) \times A(t)$, after selection and crossover is thus given by

$$\gamma(p, q) = p_c \tau_H(p, q) + \frac{1 - p_c}{2} \left(H(p) + H(q) \right),$$

where p_c is the probability that crossover is applied. It thus easily follows that $\mathbb{E}(m(H, t + 1)) = A + B$, with

$$A = p_c' \frac{1}{f(\Omega, t)} \int\int_{A(t) \times A(t)} \frac{(H(p) + H(q)) f(p) f(q)}{m(\Omega, t) f(\Omega, t)} dp\, dq,$$

where $p_c' = \frac{1 - p_c}{2}$, and with

$$B = \frac{p_c}{f(\Omega, t)} \int\int_{A(t) \times A(t)} \tau_H(p, q) \frac{f(p) f(q)}{m(\Omega, t) f(\Omega, t)} dp\, dq.$$

In order to calculate the term A, let us first assume that the following condition is satisfied:

$(*)$ the set of discontinuities of the map

$$A(t) \times A(t) \to R : (p,q) \mapsto (Hf)(p)f(q)$$

has measure zero.

Here Hf is defined on $A(t)$ by

$$(Hf)(p) = \begin{cases} f(p) & \text{if } p \in H \cap A(t) \\ 0 & \text{if } p \notin H. \end{cases}$$

If the set of discontinuities of f has measure zero, in particular, if f is continuous, then Hf is continuous, except, possibly, on the border of $H \cap A(t)$ and in the discontinuity points of f, so $(*)$ holds.

Clearly,

$$A = p'_c \frac{2}{f(\Omega,t)} \int\int_{A(t)\times A(t)} \frac{H(p)f(p)f(q)}{m(\Omega,t)f(\Omega,t)} dp\, dq,$$

and this may be rewritten as

$$\frac{1-p_c}{f(\Omega,t)} \int_{A(t)} H(p)f(p) \left(\int_{A(t)} \frac{f(q)}{m(\Omega,t)f(\Omega,t)} dq \right) dp.$$

As

$$\int_{A(t)} \frac{f(q)}{m(\Omega,t)f(\Omega,t)} dq = 1$$

and

$$\int_{A(t)} H(p)f(p)dp = f(H,t)m(H,t),$$

respectively, we find that

$$A = (1-p_c)\frac{f(H,t)}{f(\Omega,t)} m(H,t).$$

The term B may be calculated if we assume that the following condition is satisfied:

$(**)$ the map

$$A(t) \times A(t) \to \mathbb{R} : (p,q) \mapsto f(p)f(q)\tau_H(p,q)$$

has at most a measure zero set of discontinuities.

With

$$\gamma_{A(t)}(p,f) = \int_{A(t)} \frac{f(q)}{m(\Omega,t)f(\Omega,t)} \tau_H(p,q)dq$$

we obtain that

$$\begin{aligned} B &= \frac{p_c}{f(\Omega,t)} \int_{A(t)} f(p)\gamma_{A(t)}(p,f)dp \\ &\geq \frac{p_c}{f(\Omega,t)} \int_{A(t)} H(p)f(p)\gamma_{A(t)}(p,f)dp \\ &= p_c \frac{m(H,t)f(H,t)}{f(\Omega,t)} \int_{A(t)} \frac{H(p)f(p)\gamma_{A(t)}(p,f)}{m(H,t)f(H,t)}dp. \end{aligned}$$

Putting $\theta_H(p,q) = 1 - \tau_H(p,q)$, we define

$$\begin{aligned} \gamma'_{A(t)}(p,f) &= \int_{A(t)} \frac{f(q)}{m(\Omega,t)f(\Omega,t)} \theta_H(p,q)dq \\ &= 1 - \gamma_{A(t)}(p,f). \end{aligned}$$

So,

$$B \geq p_c \frac{m(H,t)f(H,t)}{f(\Omega,t)} \times \left(1 - \int_{A(t)} \frac{H(p)f(p)}{m(H,t)f(H,t)} \gamma'_{A(t)}(p,f)dp\right).$$

Let us now assume that both conditions (*) and (**) are satisfied. Combining the previous calculations, we then finally obtain:

Theorem A.2 (The "global" schema theorem). *With the above notations, we have*

$$\mathbb{E}(m(H,t+1)) \geq m(H,t)\frac{f(H,t)}{f(\Omega,t)}\left(1 - p_c\alpha(H,t)\right),$$

with

$$\alpha(H,t) = \int_{A(t)} \frac{H(p)f(p)}{m(H,t)f(H,t)} \gamma'_{A(t)}(p,f)dp.$$

The latter term may be viewed (like in the fuzzy case) as the "probability" that no offspring of elements belonging to H belongs to H, through the action of crossover, i.e., the probability that no child produced by crossing an element of $H \cap A(t)$ and of $A(t)$ belongs to H.

It has been proved that, for the crossover operator mentioned before, $\alpha(H,t)$ takes small values.

Let us conclude by noting that if we allow mutation, say with probability p_m, then a similar argument as the previous one yields a somewhat more general schema theorem, which states that

$$\mathbb{E}(m(H,t+1)) \geq m(H,t)\frac{f(H,t)}{f(\Omega,t)}\left(1 - p_c\alpha(H,t) - p_m\beta(H,t)\right),$$

for a suitable function $\beta(H,t)$, which depends upon the population $A(t)$ and structural information contained in H. We leave further details to the reader.

Appendix B

Algebraic background

The main purpose of this appendix is to briefly recollect some of the algebraic background that has been used throughout the text. As most of the material included below is amply documented in the literature, no proofs have been included. Moreover, to most readers, what follows will essentially be "standard" mathematics and may thus be viewed as a quick refresher.

1 Matrices

1.1 Generalities

A matrix is any rectangular array

$$\boldsymbol{A} = \begin{pmatrix} a_{11} & \cdots & a_{1n} \\ \vdots & \ddots & \vdots \\ a_{m1} & \cdots & a_{mn} \end{pmatrix} = (a_{ij})$$

of elements called *coefficients*. These may belong to any set A whatsoever, but are usually real or complex numbers or matrices themselves. We endow the coefficients of $\boldsymbol{A}$ with double subscripts, the first one denoting the (horizontal) row and the second one the (vertical) column in which the coefficient is located.

Note that for convenience's sake, we will sometimes start numbering the rows from 0 to $m - 1$ and the columns from 0 to $n - 1$.

If all coefficients are chosen within a fixed set A, which we then call the set of *scalars*, we denote by $M_{m\times n}(A)$ the set of $m \times n$ matrices with coefficients in A, i.e., with m rows and n columns. If $m = n$, then we speak of square matrices of dimension n and the set of these is denoted by $M_n(A)$.

A matrix

$$\boldsymbol{a} = \begin{pmatrix} a_{11} \\ \vdots \\ a_{m1} \end{pmatrix} = \begin{pmatrix} a_1 \\ \vdots \\ a_m \end{pmatrix}$$

is sometimes referred to as a *vector*. In particular, the columns

$$\boldsymbol{a}_j = \begin{pmatrix} a_{1j} \\ \vdots \\ a_{mj} \end{pmatrix}$$

$(j = 1, \ldots, n)$ of any $m \times n$ matrix $\boldsymbol{A}$ are vectors and we will frequently write

$$\boldsymbol{A} = (\boldsymbol{a}_1 \ldots \boldsymbol{a}_n).$$

Similarly, if

$$\boldsymbol{b}_i = \begin{pmatrix} a_{i1} \\ \vdots \\ a_{in} \end{pmatrix}$$

$(i = 1, \ldots, m)$, then we may write

$$\boldsymbol{A} = \begin{pmatrix} {}^t\boldsymbol{b}_1 \\ \vdots \\ {}^t\boldsymbol{b}_m \end{pmatrix}.$$

Here ${}^t(-)$ is the so-called *transposition*, which is defined by associating to any $\boldsymbol{A} = (a_{ij}) \in M_{m\times n}(A)$ the matrix ${}^t\boldsymbol{A} = (a_{ji}) \in M_{n\times m}(A)$.

For example,

$${}^t\begin{pmatrix} 1\,2\,3 \\ 4\,5\,6 \end{pmatrix} = \begin{pmatrix} 1 & 4 \\ 2 & 5 \\ 3 & 6 \end{pmatrix}.$$

Note that we, obviously, always have ${}^t({}^t\boldsymbol{A}) = \boldsymbol{A}$.

A (necessarily square) matrix $\boldsymbol{A}$ with ${}^t\boldsymbol{A} = \boldsymbol{A}$ is called *symmetric.*

From now on, we will always assume coefficients to be chosen in $\mathbb{R}$, the field of real numbers, or $\mathbb{C}$, the field of complex numbers.

We may then define the sum of two $m \times n$ matrices $\boldsymbol{A} = (a_{ij})$ and $\boldsymbol{B} = (b_{ij})$ by

$$\boldsymbol{A} + \boldsymbol{B} = (a_{ij} + b_{ij}).$$

Since this addition of matrices has been defined in terms of the addition of the coefficients, one easily verifies:

Proposition B.1. *Within the set of $m \times n$ matrices, we have:*

1. $\boldsymbol{A} + \boldsymbol{B} = \boldsymbol{B} + \boldsymbol{A}$,
2. $\boldsymbol{A} + (\boldsymbol{B} + \boldsymbol{C}) = (\boldsymbol{A} + \boldsymbol{B}) + \boldsymbol{C}$,
3. *the matrix $\boldsymbol{O}$ with all entries equal to zero has the property that for any matrix $\boldsymbol{A}$ we have*
$$\boldsymbol{A} + \boldsymbol{O} = \boldsymbol{A},$$
4. *for each matrix $\boldsymbol{A}$, there exists another matrix $-\boldsymbol{A} = (-a_{ij})$ such that*
$$\boldsymbol{A} + (-\boldsymbol{A}) = \boldsymbol{O}.$$

In a similar way, one may define the product $r\boldsymbol{A}$ of a matrix $\boldsymbol{A}$ and a scalar r by putting $r\boldsymbol{A} = (ra_{ij})$.

The following result lists the basic properties of this operation:

Proposition B.2. *Let $\boldsymbol{A}$ and $\boldsymbol{B}$ be $m \times n$ matrices and let r, s be scalars. Then:*

1. $r(\boldsymbol{A} + \boldsymbol{B}) = r\boldsymbol{A} + r\boldsymbol{B}$,
2. $(r + s)\boldsymbol{A} = r\boldsymbol{A} + s\boldsymbol{A}$,
3. $(rs)\boldsymbol{A} = r(s\boldsymbol{A})$,
4. $1\boldsymbol{A} = \boldsymbol{A}$,

5. ${}^t(r\boldsymbol{A}) = r({}^t\boldsymbol{A})$.

Next, let us define the product of an $m \times n$ matrix $\boldsymbol{A}$ and an $n \times p$ matrix $\boldsymbol{B}$ to be the $m \times p$ matrix $\boldsymbol{C} = (c_{ik})$, with

$$c_{ik} = a_{i1}b_{1k} + \cdots + a_{in}b_{nk} = \sum_{j=1}^{n} a_{ij}b_{jk},$$

for any $1 \le i \le m$ and $1 \le k \le p$. Note that this product is non-commutative, as the following example shows.

Example B.3. Let $\boldsymbol{A} = \begin{pmatrix} 0 & 1 \\ -2 & 1 \end{pmatrix}$ and $\boldsymbol{B} = \begin{pmatrix} 1 & 4 \\ -2 & 1 \end{pmatrix}$, then

$$\boldsymbol{AB} = \begin{pmatrix} -2 & 1 \\ -4 & -7 \end{pmatrix} \text{ while } \boldsymbol{BA} = \begin{pmatrix} -8 & 5 \\ -2 & -1 \end{pmatrix}.$$

On the other hand, let us point out that the product of matrices satisfies the following basic properties:

Proposition B.4. *Let $\boldsymbol{A}$, $\boldsymbol{B}$ and $\boldsymbol{C}$ be matrices (with suitable dimensions) and denote by $\boldsymbol{I}_n$ the identity matrix of dimension n, i.e.,*

$$(\boldsymbol{I}_n)_{ij} = \begin{cases} 0 & \text{if } i \neq j \\ 1 & \text{if } i = j. \end{cases}$$

Then:

1. $\boldsymbol{A}(\boldsymbol{BC}) = (\boldsymbol{AB})\boldsymbol{C}$,
2. $\boldsymbol{A}(\boldsymbol{B}+\boldsymbol{C}) = \boldsymbol{AB} + \boldsymbol{AC}$,
3. $(\boldsymbol{B}+\boldsymbol{C})\boldsymbol{A} = \boldsymbol{BA} + \boldsymbol{CA}$,
4. $r(\boldsymbol{AB}) = (r\boldsymbol{A})\boldsymbol{B} = \boldsymbol{A}(r\boldsymbol{B})$,
5. *if $\boldsymbol{A}$ has dimension $m \times n$ and $\boldsymbol{B}$ has dimension $n \times p$, then $\boldsymbol{AI}_n = \boldsymbol{A}$ and $\boldsymbol{I}_n\boldsymbol{B} = \boldsymbol{B}$.*

Finally the *Kronecker product* or *tensor product* of an $m \times n$ matrix $\boldsymbol{A} = (a_{ij})$ and a $p \times q$ matrix $\boldsymbol{B} = (b_{ij})$ is defined to be the $np \times mq$ matrix

$$\boldsymbol{A} \otimes \boldsymbol{B} = (a_{ij}\boldsymbol{B})_{ij} = \begin{pmatrix} a_{11}\boldsymbol{B} \cdots a_{1m}\boldsymbol{B} \\ \vdots \quad \ddots \quad \vdots \\ a_{n1}\boldsymbol{B} \cdots a_{nm}\boldsymbol{B} \end{pmatrix}.$$

Example B.5. If $\boldsymbol{A} = \begin{pmatrix} a_{11} & a_{12} \\ a_{21} & a_{22} \end{pmatrix}$ and $\boldsymbol{B} = \begin{pmatrix} b_{11} & b_{12} & b_{13} \\ b_{21} & b_{22} & b_{23} \end{pmatrix}$, then

$$\boldsymbol{A} \otimes \boldsymbol{B} = \begin{pmatrix} a_{11}\boldsymbol{B} & a_{12}\boldsymbol{B} \\ a_{21}\boldsymbol{B} & a_{22}\boldsymbol{B} \end{pmatrix} = \begin{pmatrix} a_{11}b_{11} & a_{11}b_{12} & a_{11}b_{13} & a_{12}b_{11} & a_{12}b_{12} & a_{12}b_{13} \\ a_{11}b_{21} & a_{11}b_{22} & a_{11}b_{23} & a_{12}b_{21} & a_{12}b_{22} & a_{12}b_{23} \\ a_{21}b_{11} & a_{21}b_{12} & a_{21}b_{13} & a_{22}b_{11} & a_{22}b_{12} & a_{22}b_{13} \\ a_{21}b_{21} & a_{21}b_{22} & a_{21}b_{23} & a_{22}b_{21} & a_{22}b_{22} & a_{22}b_{23} \end{pmatrix}.$$

For any $m \times n$ matrices $\boldsymbol{A}$, $\boldsymbol{C}$ and any $p \times q$ matrices $\boldsymbol{B}$, $\boldsymbol{D}$, it is clear that

$${}^t(\boldsymbol{A} \otimes \boldsymbol{B}) = {}^t\boldsymbol{A} \otimes {}^t\boldsymbol{B}$$

and

$$(\boldsymbol{A} \otimes \boldsymbol{B})(\boldsymbol{C} \otimes \boldsymbol{D}) = (\boldsymbol{AC}) \otimes (\boldsymbol{BD}).$$

1.2 Invertible matrices

Any square matrix $\boldsymbol{A}$ of dimension n is said to be *invertible* (or *nonsingular*) if there exists a matrix $\boldsymbol{B}$ (of the same dimension) such that $\boldsymbol{AB} = \boldsymbol{BA} = \boldsymbol{I}_n$. The matrix $\boldsymbol{B}$ is then necessarily unique. It is called the *inverse* of $\boldsymbol{A}$ and is denoted by $\boldsymbol{A}^{-1}$. Note that if $\boldsymbol{AB} = \boldsymbol{I}_n$ and $\boldsymbol{CA} = \boldsymbol{I}_n$, then $\boldsymbol{A}$ is invertible and $\boldsymbol{B} = \boldsymbol{C} = \boldsymbol{A}^{-1}$. The main properties of the inverse of matrices are given by:

Proposition B.6. *For any pair of invertible matrices $\boldsymbol{A}$ and $\boldsymbol{B}$ and any non-zero scalar r, we have*

1. $(\boldsymbol{A}^{-1})^{-1} = \boldsymbol{A}$,
2. $(\boldsymbol{AB})^{-1} = \boldsymbol{B}^{-1}\boldsymbol{A}^{-1}$,

3. $(r\boldsymbol{A})^{-1} = \frac{1}{r}\boldsymbol{A}^{-1}$,

4. ${}^{t}(\boldsymbol{A}^{-1}) = ({}^{t}\boldsymbol{A})^{-1}$.

Let us briefly recall the definition of the determinant of any square matrix $\boldsymbol{A}$. If

$$\boldsymbol{A} = \begin{pmatrix} a_{11} \, a_{12} \\ a_{21} \, a_{22} \end{pmatrix}$$

is a square matrix of dimension 2, we define its determinant by

$$\det(\boldsymbol{A}) = a_{11}a_{22} - a_{12}a_{21}.$$

For higher dimensions, we work by induction. If $\boldsymbol{A}$ is any square matrix of dimension n, the *minor* of the coefficient a_{ij} is the determinant of the submatrix obtained by deleting row i and column j from $\mathbf{A}$. The *cofactor* A_{ij} of a_{ij} is the minor of a_{ij} multiplied by $(-1)^{i+j}$. With these definitions, the determinant of $\mathbf{A}$ is now defined by

$$\det(\boldsymbol{A}) = a_{i1}A_{i1} + \ldots + a_{in}A_{in} = \sum_{j=1}^{n} a_{ij}A_{ij}.$$

One may show that the value of this expression does not depend on the choice of i; it is sometimes referred to as the *Laplace expansion* of the determinant of $\boldsymbol{A}$ by the i-th row. It is also easy to see that the same value is obtained if, for any $1 \leq j \leq n$, we expand by the j-th column, i.e.,

$$\det(\boldsymbol{A}) = a_{1j}A_{1j} + \ldots + a_{nj}A_{nj} = \sum_{i=1}^{n} a_{ij}A_{ij}.$$

The main properties of determinants are given by the following result:

Proposition B.7. *Let* $\boldsymbol{A}$ *and* $\boldsymbol{B}$ *be square matrices. Then:*

1. $\det(\boldsymbol{A}) = \det({}^{t}\boldsymbol{A})$,

2. If $\boldsymbol{A} = (a_{ij})$ *is any* $n \times n$ *upper (or lower) triangular matrix, i.e., if* $a_{ij} = 0$ *if* $i > j$ *(or* $a_{ij} = 0$ *when* $i < j$*), then*

$$\det(\boldsymbol{A}) = a_{11}a_{22}\ldots a_{nn} = \prod_{i=1}^{n} a_{ii},$$

3. $\det(\boldsymbol{I}_n) = 1$,

4. $\det(\boldsymbol{AB}) = \det(\boldsymbol{A})\det(\boldsymbol{B})$.

Define the *adjoint matrix* $\mathrm{adj}(\boldsymbol{A})$ of any square matrix $\boldsymbol{A} = (a_{ij})$ by

$$\mathrm{adj}(\boldsymbol{A}) = \begin{pmatrix} A_{11} \dots A_{n1} \\ \vdots \quad \ddots \quad \vdots \\ A_{1n} \dots A_{nn} \end{pmatrix},$$

where, as before, A_{ij} is the cofactor of a_{ij}.

Proposition B.8. *For any square matrix* $\boldsymbol{A}$ *we have:*

1. $\boldsymbol{A}$ *is invertible if, and only if,* $\det(\boldsymbol{A}) \neq 0$,

2. in this case,

$$\boldsymbol{A}^{-1} = \frac{1}{\det(\boldsymbol{A})}\mathrm{adj}(\boldsymbol{A}).$$

1.3 Generalized inverses

It is clear that not every matrix $\boldsymbol{A}$ is invertible, in particular if it is an $m \times n$ matrix with $m \neq n$. However, one may still try and introduce an "approximate" inverse of $\boldsymbol{A}$. Such a *generalized inverse* or *Moore-Penrose inverse* of $\boldsymbol{A}$ is defined to be an $n \times m$ matrix $\boldsymbol{X}$, with the properties that

1. $\boldsymbol{AXA} = \boldsymbol{A}$,

2. $\boldsymbol{XAX} = \boldsymbol{X}$,

3. both $\boldsymbol{AX}$ and $\boldsymbol{XA}$ are symmetric.

One may show that any $m \times n$ matrix $\boldsymbol{A}$ has a generalized inverse and that this generalized inverse is then necessarily unique. We will denote it by $\boldsymbol{A}^{\dagger}$. Its main properties are given by:

Proposition B.9. *Let* $\boldsymbol{A}$ *be an arbitray* $m \times n$ *matrix and* $\boldsymbol{b}$ *a vector of dimension* m. *Then:*

1. *the linear system* $\boldsymbol{Ax} = \boldsymbol{b}$ *has* $\boldsymbol{A}^{\dagger}\boldsymbol{b}$ *as a solution, whenever solutions exist,*

2. *if* $\boldsymbol{A}$ *is invertible, then* $\boldsymbol{A}^{\dagger} = \boldsymbol{A}^{-1}$.

2 Vector spaces

2.1 Generalities

A non-empty set V is said to be a (real) *vector space* if it is endowed with two operations: a *sum*, which to any $v, w \in V$ associates some $v + w \in V$ and a *scalar product*, which to any real number α and any $v \in V$ associates some $\alpha v \in V$, these operations satisfying:

1. for any u,v and w in V, we have $u + v = v + u$ and $(u + v) + w = u + (v + w)$,

2. there is a some $o \in V$ such that $o + v = v$ for any $v \in V$, moreover, for any $v \in V$, there exists $-v \in V$ such that $v + (-v) = o$,

3. for any pair of real numbers α and β and any $v, w \in V$, we have $\alpha(v + w) = \alpha v + \alpha w$ resp. $(\alpha + \beta)v = \alpha v + \beta v$ and $(\alpha\beta)v = \alpha(\beta v)$,

4. if $v \in V$, then $1v = v$.

The elements of V are called *vectors*, in particular, o is called the zero vector, the elements of $\mathbb{R}$, the field of real numbers, are usually referred to as *scalars*.

Example B.10. It is easy to see that the set $\mathbb{R}^n = \{(x_1, \ldots, x_n), x_i \in \mathbb{R}\}$ is a vector space, when endowed with the operations

$$(x_1, \ldots, x_n) + (y_1, \ldots, y_n) = (x_1 + y_1, \ldots, x_n + y_n)$$

and

$$\alpha(x_1, \ldots, x_n) = (\alpha x_1, \ldots, \alpha x_n).$$

Note that we will frequently view the elements of $\mathbb{R}^n$ as (column) vectors

$$\boldsymbol{x} = \begin{pmatrix} x_1 \\ \vdots \\ x_n \end{pmatrix}.$$

In a similar way, when endowed with the obvious operations, the set of $m \times n$ matrices $M_{m\times n}(\mathbb{R})$ is a vector space as well.

One easily verifies that for any vector v and any scalar α, one has $0v = o$ and $\alpha o = o$. Actually, one may show that $\alpha v = o$ exactly when $v = o$ or $\alpha = 0$.

A non-empty subset U of V is said to be a *subspace* of V if it is a vector space, when endowed with the operations of V. This is clearly equivalent to asserting that for any pair of vectors v, w in U and any pair of scalars α, β, the vector $\alpha v + \beta w$ also belongs to U.

Using this, one easily checks that for any $m \times n$ matrix $\boldsymbol{A}$, the set of solutions of the linear system defined by $\boldsymbol{A}$, i.e., the set

$$\{\boldsymbol{x} \in \mathbb{R}^n, \boldsymbol{A}\boldsymbol{x} = \boldsymbol{0}\}$$

is a a subspace of $\mathbb{R}^n$. Actually, one may show that every subspace of $\mathbb{R}^n$ is of this form.

For any pair of subspaces U, $U' \subset V$, define their *sum* $U + U'$ by

$$U + U' = \{u + u'; u \in U, u' \in U'\}.$$

We then have:

Proposition B.11. *Let U and U' be subspaces of the vector space V. Then $U \cap U'$ and $U + U'$ are also subspaces of V.*

When $U \cap U' = \{o\}$ and $U + U' = V$, we say that V is the *direct sum* of U and U' and we write this as $V = U \oplus U'$.

2.2 Linear independence, generators and bases

Let us fix a vector space V. Consider vectors $v_1, \ldots, v_n \in V$ and scalars $\alpha_1, \ldots, \alpha_n$. The vector

$$v = \alpha_1 v_1 + \ldots + \alpha_n v_n = \sum_{i=1}^{n} \alpha_i v_i$$

is then said to be a *linear combination* of $v_1, \ldots, v_n$.

A subset of V is said to be *linearly independent* if, for any vectors $v_1, \ldots, v_n$ belonging to S, it follows from

$$\alpha_1 v_1 + \ldots + \alpha_n v_n = o$$

that $\alpha_1 = \cdots = \alpha_n = 0$. It is fairly easy to see that S is linearly independent if no vector of S can be expressed as a linear combination of the remaining vectors of S. If S is not linearly independent, then we call it *linearly dependent*. This means that we may find scalars $\alpha_1, \ldots, \alpha_n$ not all zero and vectors $v_1, \ldots, v_n$ in S, such that

$$\alpha_1 v_1 + \ldots + \alpha_n v_n = o;$$

equivalently, there exists a vector in S which may be written as a linear combination of the other vectors in S.

The *rank* of S is denoted by $\mathrm{rk}(S)$ and defined to be the maximal number of linearly independent vectors in S. Some of the main properties of this notion are given in the following result:

Proposition B.12. *Consider vectors $v_1, \ldots, v_n$ in the vector space V and a scalar α. Then:*

1. $\mathrm{rk}(\{v_1, \ldots, v_i, \ldots, v_j, \ldots, v_n\}) = \mathrm{rk}(\{v_1, \ldots, v_j, \ldots, v_i, \ldots, v_n\})$,
2. $\mathrm{rk}(\{v_1, \ldots, v_i, \ldots, v_n\}) = \mathrm{rk}(\{v_1, \ldots, \alpha v_i, \ldots, v_n\})$ *if* $\alpha \neq 0$,
3. $\mathrm{rk}(\{v_1, \ldots, v_i, \ldots, v_j, \ldots, v_n\}) = \mathrm{rk}(\{v_1, \ldots, v_i + \alpha v_j, \ldots, v_j, \ldots, v_n\})$.

Note also:

Lemma B.13. *For any $m \times n$ matrix*

$$\boldsymbol{A} = (\boldsymbol{a}_1 \ldots \boldsymbol{a}_n) = \begin{pmatrix} {}^t\boldsymbol{b}_1 \\ \vdots \\ {}^t\boldsymbol{b}_m \end{pmatrix}$$

we have:

$$\mathrm{rk}(\{\boldsymbol{a}_1, \ldots, \boldsymbol{a}_n\}) = \mathrm{rk}(\{\boldsymbol{b}_1, \ldots, \boldsymbol{b}_m\}).$$

This common value is called the *rank* of the matrix $\boldsymbol{A}$ and is denoted by $\mathrm{rk}(\boldsymbol{A})$. The rank of $\boldsymbol{A}$ is not modified when elementary operations are applied to $\boldsymbol{A}$, like interchanging two rows (or columns), multiplying a row (or column) by a non-zero scalar, or adding to any row (or column) a multiple of another one.
Note also:

Proposition B.14. *A square matrix $\boldsymbol{A}$ of dimension n is invertible if, and only if, $\mathrm{rk}(\boldsymbol{A}) = n$.*

For any subset S of V, we denote by $\langle S \rangle$ the set of all linear combinations of the vectors of S, i.e., $\langle S \rangle$ consists of all vectors of the form $\alpha_1 v_1 + \ldots + \alpha_n v_n$, for some vectors $v_1, \ldots, v_n$ belonging to S and some scalars $\alpha_1, \ldots, \alpha_n$. It is easy to see that $\langle S \rangle$ is a subspace of V and that it is actually the smallest subspace of V containing S. If $V = \langle S \rangle$, then we say that S is a *set of generators* of V; if a finite set of generators exists, then we call V *finitely generated.*
The columns $\boldsymbol{a}_1, \ldots, \boldsymbol{a}_n$ of any $m \times n$ matrix $\boldsymbol{A}$ are vectors in $\mathbb{R}^m$. The subspace generated by these is called the *range* of $\boldsymbol{A}$ and will be denoted by $\mathrm{Im}(\boldsymbol{A})$. It is easy to see that $\mathrm{Im}(\boldsymbol{A})$ does not change when elementary operations are applied to $\boldsymbol{A}$.
Note also:

Proposition B.15. *If $\{v_1, \ldots, v_n\}$ is a set of generators of V and if v_i is a linear combination of the remaining vectors, then the set*

$$\{v_1, \ldots, v_{i-1}, v_{i+1}, \ldots, v_n\}$$

is also a set of generators of V.

The next result links generators to linear independency:

Proposition B.16. *If the set $\{v_1, \ldots, v_m\}$ is linearly independent and if $\{u_1, \ldots, u_n\}$ is a set of generators, then $m \leq n$.*

A linearly independent set of generators of V is said to be a *basis.*

Proposition B.17. *A subset $B = \{e_1, \ldots, e_n\}$ of the vector space V is a basis if, and only if, each vector v of V can be expressed in a unique way as a linear combination of the vectors of B.*

In this case, the uniquely determined scalars $v_1, \ldots, v_n$ with

$$v = v_1 e_1 + \ldots + v_n e_n$$

are called the *coordinates* of v with respect to B. We will also write

$$\boldsymbol{v} = \begin{pmatrix} v_1 \\ \vdots \\ v_n \end{pmatrix}$$

and call this the *coordinate vector* of v (with respect to B).

It is easy to see that if V is finitely generated, say $V = \langle S \rangle$ for some finite set S, then V possesses a basis $B \subset S$.

Note also:

Theorem B.18. *If V has a finite basis B, then any other basis of V has the same cardinality as B.*

In view of the previous result, it makes sense to define the *dimension* of a finitely generated vector space V, denoted by $\dim(V)$, to be the cardinality of any basis of V.

Example B.19.

1. If $V = \langle S \rangle$ with S a finite set, then, obviously, $\dim(V) = \mathrm{rk}(S)$,

2. it is clear that the vector space $\mathbb{R}^n$ has dimension n, since the vectors

$$\{(1, 0, \ldots, 0), (0, 1, \ldots, 0), \ldots, (0, 0, \ldots, 1)\}$$

form a basis for $\mathbb{R}^n$ (we will usually call this the *canonical basis* of $\mathbb{R}^n$),

3. consider in $M_{m \times n}(\mathbb{R})$, the matrices $\boldsymbol{E}_{ij}$, which have 1 at the intersection of row i and column j and 0 in the other positions; then

$$\{\boldsymbol{E}_{ij}\ ;\ i = 1, \ldots, m,\ j = 1, \ldots, n\}$$

is a basis for $M_{m \times n}(\mathbb{R})$, hence $\dim(M_{m \times n}(\mathbb{R})) = mn$.

The following result is sometimes referred to as the "dimension formula":

Proposition B.20. *For any pair of subspaces U and U' of the vector space V we have:*

$$\dim(U) + \dim(U') = \dim(U + U') + \dim(U \cap U').$$

Let us now suppose that both $B = \{e_1 \ldots, e_n\}$ and $B' = \{e'_1, \ldots, e'_n\}$ are bases of V. Then we may find scalars s_{ij}, such that for each $1 \leq j \leq n$ we have

$$e'_j = s_{1j}e_1 + \ldots + s_{nj}e_n = \sum_{i=1}^{n} s_{ij}e_i.$$

Since any vector v may be written uniquely as

$$v = \sum_{i=1}^{n} v_i e_i = \sum_{j=1}^{n} v'_j e'_j,$$

it easily follows that these coordinates are linked through $v_i = \sum_{j=1}^{n} s_{ij}v'_j$, i.e.,

$$\begin{pmatrix} v_1 \\ \vdots \\ v_n \end{pmatrix} = \begin{pmatrix} s_{11} \ldots s_{1n} \\ \vdots \quad \ddots \quad \vdots \\ s_{n1} \ldots s_{nn} \end{pmatrix} \begin{pmatrix} v'_1 \\ \vdots \\ v'_n \end{pmatrix}.$$

The matrix (s_{ij}) is usually denoted by $\boldsymbol{S}_{B,B'}$ (or just $\boldsymbol{S}$, if no ambiguity arises) and referred to as the *substitution matrix* for B and B'. Its columns are exactly the coordinate vectors of the e'_j with respect to the basis B. If we denote by $\boldsymbol{v}$ resp. $\boldsymbol{v}'$ the coordinate vector of any $v \in V$ with respect to the basis B resp. B', then the previous relation may thus be rewritten as $\boldsymbol{v} = \boldsymbol{S}_{B,B'}\boldsymbol{v}'$.

It is easy to see that the matrix $\boldsymbol{S}_{B,B'}$ is invertible. Conversely, if $\boldsymbol{S} = (s_{ij})$ is an invertible matrix, then it yields for any basis $B = \{e_1 \ldots, e_n\}$ of V a new basis $B' = \{e'_1, \ldots, e'_n\}$ by putting $e'_j = \sum_{i=1}^{n} s_{ij}e_i$ for any $1 \leq j \leq n$, as one easily verifies.

2.3 Euclidean spaces

A symmetric bilinear form on the vector space V, i.e., a map

$$\langle -, - \rangle : V \times V \to \mathbb{R}$$

with the property that $\langle \alpha u + \beta v, w \rangle = \alpha \langle u, w \rangle + \beta \langle v, w \rangle$ and that $\langle u, v \rangle = \langle v, u \rangle$ for any $u, v, w \in V$ and any scalars α, β, is said to be a *scalar product*, if $\langle v, v \rangle$ is positive, for any $v \in V$ and if $\langle v, v \rangle = 0$ if, and only if, $v = 0$.

Let us assume for the rest of this section V to be a *Euclidean space*, i.e., to be endowed with a scalar product. We then define the norm of any $v \in V$ as

$$||v|| = \sqrt{\langle v, v \rangle}.$$

From the properties of the scalar product, it follows that the norm only takes positive values and that $||v|| = 0$ if, and only if, $v = 0$.

Vectors v and w are said to be *orthogonal* if $\langle v, w \rangle = 0$. A set $\{e_1, \ldots, e_n\}$ is said to be an *orthogonal basis* if it is a basis and if e_i and e_j are orthogonal for any $i \neq j$. If we also have $||e_i|| = 1$ for each i, then we speak of an *orthonormal basis*.

Proposition B.21. *Let $\boldsymbol{S}$ be a square matrix of dimension n. The following assertions are equivalent:*

1. $\boldsymbol{S}^{-1} = {}^t\boldsymbol{S}$ *or, equivalently,* ${}^t\boldsymbol{S}\boldsymbol{S} = \boldsymbol{I}_n$,
2. *the map*
$$f_{\boldsymbol{S}} : \mathbb{R}^n \to \mathbb{R}^n : \boldsymbol{x} \mapsto \boldsymbol{S}\boldsymbol{x}$$
has the property that $\langle f_{\boldsymbol{S}}(\boldsymbol{x}), f_{\boldsymbol{S}}(\boldsymbol{y}) \rangle = \langle \boldsymbol{x}, \boldsymbol{y} \rangle$ *for any* $\boldsymbol{x}, \boldsymbol{y} \in \mathbb{R}^n$,
3. *the columns* $\{\boldsymbol{s}_1, \ldots, \boldsymbol{s}_n\}$ *of* $\boldsymbol{S}$ *form an orthonormal basis for* $\mathbb{R}^n$,
4. *the rows* $\{\boldsymbol{t}_1, \ldots, \boldsymbol{t}_n\}$ *of* $\boldsymbol{S}$ *form an orthonormal basis for* $\mathbb{R}^n$,
5. $\boldsymbol{S} = \boldsymbol{S}_{B,B'}$ *for a pair of orthonormal bases B, B' of an n-dimensional Euclidean vector space.*

For any subset U of V, we denote by

$$U^{\perp} = \{v \in V; \forall u \in U, \langle u, v \rangle = 0\}$$

its so-called *orthogonal complement*. Note that this is always a subspace of V, even if U is not.

It is easy to see that:

Proposition B.22. *For any subspace U of V, we have:*

1. $V = U \oplus U^{\perp}$,

2. *if* $V = U \oplus U'$, *then* $U' = U^{\perp}$.

3 Linear maps

3.1 Definition and examples

A map $f : V \to W$ between vector spaces V and W is said to be *linear*, if it has the property that $f(v+v') = f(v) + f(v')$ and $f(\alpha v) = \alpha f(v)$, for any vectors $v, v' \in V$ and any scalar α. Clearly this is equivalent to $f(\sum_{i=1}^{n} \alpha_i v_i) = \sum_{i=1}^{n} \alpha_i f(v_i)$, for any vectors $v_i \in V$ and scalars α_i.

Example B.23.

1. The zero map $f : V \to W$, which assigns to any $v \in V$ the zero vector in W, is a linear map,

2. the identity $id_V : V \to V$ is a linear map,

3. for any $m \times n$ matrix $\boldsymbol{A}$, the map

$$f_{\boldsymbol{A}} : \mathbb{R}^n \to \mathbb{R}^m : \boldsymbol{x} \mapsto \boldsymbol{A}\boldsymbol{x}$$

is linear.

The *kernel* $\mathrm{Ker}(f)$ and the *image* $\mathrm{Im}(f)$ of any linear map $f : V \to W$ are defined by

$$\mathrm{Ker}(f) = \{v \in V; f(v) = o\}$$

and

$$\mathrm{Im}(f) = \{w \in W; \exists v \in V, f(v) = w\}.$$

Clearly, $\mathrm{Ker}(f)$ is a subspace of V and $\mathrm{Im}(f)$ a subspace of W. The *rank* $\mathrm{rk}(f)$ of f is defined to be the dimension of $\mathrm{Im}(f)$.

These notions are related through:

Proposition B.24. *For any linear map $f : V \to W$ between finite-dimensional vector spaces, we have*

$$\dim(V) = \dim(\mathrm{Ker}(f)) + \mathrm{rk}(f).$$

Note also:

Lemma B.25. *If $f : V \to V$ is a linear map, then $f^2 = f$ if, and only if, $V = \mathrm{Ker}(f) \oplus \mathrm{Im}(f)$.*

3.2 Linear maps and matrices

Let $f : V \to W$ be a linear map with $\dim(V) = n$ and $\dim(W) = m$ and consider a basis $B = \{e_1, \ldots, e_n\}$ of V and a basis $C = \{f_1, \ldots, f_m\}$ of W. For each vector e_i in B, we may write

$$f(e_i) = \sum_{j=1}^{m} a_{ji} f_j.$$

We thus obtain a matrix $\boldsymbol{A} = (a_{ji})$, which we call the matrix of f with respect to B and C, and which completely determines f, once B and C are given. Indeed, for any $v \in V$ with coordinate vector $\boldsymbol{v}$ with respect to B, the coordinate vector of $f(v)$ with respect to C is exactly $\boldsymbol{A}\boldsymbol{v}$.

Note also:

Lemma B.26. *If $\boldsymbol{A}$ is the matrix associated to the linear map $f : V \to W$ (with respect to given bases B and C), then $\mathrm{rk}(f) = \mathrm{rk}(\boldsymbol{A})$.*

The next result describes how base change affects the matrix associated to a linear map:

Proposition B.27. *Consider a linear map $f : V \to W$, bases B and B' for V, bases C and C' for W and let $\boldsymbol{A}$ (resp. $\boldsymbol{A}'$) be the matrix associated to f with respect to B and C (resp. with respect to B' and C'). Then*

$$\boldsymbol{A}' = \boldsymbol{S}_{C',C} \boldsymbol{A} \boldsymbol{S}_{B,B'},$$

where $\boldsymbol{S}_{B,B'}$ is the substitution matrix for B and B' and $\boldsymbol{S}_{C',C} = \boldsymbol{S}_{C,C'}^{-1}$ is the substitution matrix for C' and C.

3.3 Orthogonal projections

Assume $V = V_1 \oplus V_2$ for subspaces V_1 and V_2 of V. Then any vector $v \in V$ may be written in a unique way as $v = v_1 + v_2$, with $v_1 \in V_1$ and $v_2 \in V_2$. The linear map

$$p : V \to V : v \mapsto v_1$$

is said to be the *projection* on V_1 along V_2.

The next result gives an alternative description of projections:

Proposition B.28. *For any linear map $p : V \to V$, the following assertions are equivalent:*

1. *there exist subspaces V_1 and V_2 of V such that p is the projection on V_1 along V_2,*
2. *the map p is idempotent ($p^2 = p$).*

Let us now assume V to be a Euclidean vector space. In this case, any subspace U of V has a unique orthogonal complement $U^\perp$, hence determines the projection on U along $U^\perp$. We denote this map by p_U and call it the *orthogonal projection* of V on U.

Orthogonal projections may also be described as follows:

Proposition B.29. *Let V be a Euclidean vector space. For any linear map $p : V \to V$ the following assertions are equivalent:*

1. *there exists a subspace U of V such that $p = p_U$,*
2. *the matrix associated to p with respect to any orthogonal basis of V is idempotent and symmetric.*

Note also:

Proposition B.30. *Let $p : V \to V$ be a linear map. If U is a subspace of V with the property that $p(u) = u$ for any $u \in U$ and $p(u) = o$ for any $u \in U^\perp$, then $p = p_U$.*

From this it follows:

Corollary B.31. *For any matrix $\boldsymbol{A}$, the product $\boldsymbol{A}\boldsymbol{A}^\dagger$ is the orthogonal projection on* $\mathrm{Im}(\boldsymbol{A})$.

Indeed, it suffices to note that $\boldsymbol{A}\boldsymbol{A}^\dagger\boldsymbol{x} = \boldsymbol{x}$ for any $\boldsymbol{x} \in \mathrm{Im}(\boldsymbol{A})$ and $\boldsymbol{A}\boldsymbol{A}^\dagger\boldsymbol{x} = \boldsymbol{0}$ for any $\boldsymbol{x} \in \mathrm{Im}(\boldsymbol{A}^\perp)$, and to apply the previous result to $f_{\boldsymbol{A}\boldsymbol{A}^\dagger}$.

4 Diagonalization

As pointed out before, to any matrix $\boldsymbol{A} \in M_n(\mathbb{R})$, we may associate the linear map

$$f_{\boldsymbol{A}} : \mathbb{R}^n \longrightarrow \mathbb{R}^n : \boldsymbol{x} \mapsto \boldsymbol{A}\boldsymbol{x},$$

with respect to the canonical basis of $\mathbb{R}^n$. The matrix of f with respect to another basis B of $\mathbb{R}^n$ is of the form $\boldsymbol{S}^{-1}\boldsymbol{A}\boldsymbol{S}$ (where $\boldsymbol{S}$ is the corresponding substitution matrix). The matrices $\boldsymbol{A}$ and $\boldsymbol{S}^{-1}\boldsymbol{A}\boldsymbol{S}$ are said to be *similar*.

In this section, we will take a look at the question whether any matrix $\boldsymbol{A}$ is similar to a diagonal matrix or, equivalently, whether there exists a basis B of $\mathbb{R}^n$ such that the matrix associated to $f_{\boldsymbol{A}}$ with respect to B is a diagonal matrix.

4.1 Eigenvalues and eigenvectors

Let us fix a vector space V and a linear map $f : V \to V$. A scalar λ is an *eigenvalue* of f if $f(v) = \lambda v$ for some nonzero vector $v \in V$, which is then said to be an *eigenvector* of f associated to the eigenvalue λ. We call the set

$$V_\lambda = \{v \in V; f(v) = \lambda v\},$$

which consists of the zero vector and all eigenvectors associated to λ, the *eigenspace* associated to λ.

One easily verifies the following result:

Proposition B.32. *Let $\boldsymbol{A}$ be the matrix of the linear map $f : V \to V$ with respect to any basis of the n-dimensional vector space V, and let id_V denote the identity function on V. For any $\lambda \in \mathbb{R}$, we then have:*

1. $V_\lambda = \mathrm{Ker}(f - \lambda id_V)$,

2. $\dim(V_\lambda) = n - \mathrm{rk}(\boldsymbol{A} - \lambda \boldsymbol{I}_n)$,

3. λ *is an eigenvalue of* f *if, and only if,* $\det(\boldsymbol{A} - \lambda \boldsymbol{I}_n) = 0$.

As we saw in the previous result, a scalar λ is an eigenvalue of f if, and only if, $\det(\boldsymbol{A} - \lambda \boldsymbol{I}_n) = 0$. Since $\det(\boldsymbol{A} - \lambda \boldsymbol{I}_n)$ is a polynomial in λ of degree n, the so-called *characteristic polynomial* of f, and since the zeros of this polynomial are exactly the eigenvalues of f, it follows that f has at most n distinct eigenvalues.

Let us also point out that the characteristic polynomial of f does not depend on the chosen basis, i.e., similar matrices have the same characteristic polynomial. Indeed, if $\boldsymbol{B} = \boldsymbol{S}^{-1}\boldsymbol{A}\boldsymbol{S}$, for some invertible matrix $\boldsymbol{S}$, then

$$\det(\boldsymbol{B}-\lambda\boldsymbol{I}_n) = \det(\boldsymbol{S}^{-1}\boldsymbol{A}\boldsymbol{S}-\lambda\boldsymbol{I}_n) = \det(\boldsymbol{S}^{-1})\det(\boldsymbol{A}-\lambda\boldsymbol{I}_n)\det(\boldsymbol{S}) = \det(\boldsymbol{A}-\lambda\boldsymbol{I}_n).$$

Note B.33. Since we work over the real numbers, the number of eigenvalues of a matrix may be smaller than the dimension of this matrix (in view of the occurrence of complex zeroes). For example, the matrix

$$\boldsymbol{A} = \begin{pmatrix} 0 & 1 \\ -1 & 0 \end{pmatrix}$$

has no (real!) eigenvalues, since its characteristic polynomial is

$$\det(\boldsymbol{A} - \lambda \boldsymbol{I}_2) = \lambda^2 + 1.$$

On the other hand, eigenvalues, viewed as zeroes of the characteristic polynomial, may have multiplicity higher than 1. For example, the matrix

$$\boldsymbol{A} = \begin{pmatrix} 1 & 1 \\ -1 & -1 \end{pmatrix}$$

has characteristic polynomial

$$\det(\boldsymbol{A} - \lambda \boldsymbol{I}_2) = \lambda^2$$

and this polynomial has a single zero ($\lambda = 0$) with multiplicity 2.

Suppose that f has r distinct eigenvalues $\lambda_1, \ldots, \lambda_r$. The *algebraic multiplicity* α_i of λ_i is the multiplicity of λ_i as zero of the characteristic polynomial of f, while the *geometric multiplicity* of λ_i is defined to be

$$d_i = \dim(V_{\lambda_i}) = n - \mathrm{rk}(\boldsymbol{A} - \lambda_i \boldsymbol{I}_n).$$

It is easy to see that for each $1 \leq i \leq r$, we always have $1 \leq d_i \leq \alpha_i$.

4.2 Diagonalizable matrices

A square matrix $\boldsymbol{A}$ is said to be *diagonalizable* if it is similar to a diagonal matrix. Alternatively, we call a linear map $f : V \to V$ *diagonalizable* if there exists a basis for V with respect to which the corresponding matrix is diagonal.

Let us first point out the following result:

Proposition B.34. *A linear map $f : V \to V$ is diagonalizable if, and only if, V possesses a basis consisting of eigenvectors of f.*

Note also:

Lemma B.35. *The eigenvectors associated to different eigenvalues are linearly independent.*

The next result completely answers the question whether a given linear map or matrix is diagonalizable:

Proposition B.36. *A linear map $f : V \to V$ with r eigenvalues $\lambda_1, \ldots, \lambda_r$ is diagonalizable if, and only if,*

1. $\alpha_1 + \cdots + \alpha_r = n$,

2. $d_i = \alpha_i$ *for every* $1 \leq i \leq r$.

Using the previous results, one now easily proves:

Corollary B.37. *Any square matrix of dimension n with n different eigenvalues is diagonalizable.*

Let us now assume V to be a Euclidean vector space. A linear map $f : V \to V$ is the said to be *symmetric* if $\langle f(u), v\rangle = \langle u, f(v)\rangle$ for all $u, v \in V$. It is clear that this is equivalent to asserting that the matrix associated to f with respect to any othonormal basis of V be symmetric.

The reason for introducing this notion stems from:

Proposition B.38. *Let V be a Euclidean vector space. If $f : V \to V$ is a symmetric linear map, then the characteristic polynomial of f only has real zeroes, i.e., all eigenvalues of f are real.*

We may now conclude with the following fundamental result:

Theorem B.39. (Spectral Theorem) *Any symmetric (real) matrix is diagonalizable.*

Corollary B.40. *For any symmetric (real) matrix $\boldsymbol{A}$ we may find an orthogonal matrix $\boldsymbol{S}$ and a diagonal matrix $\boldsymbol{D}$ such that ${}^t\boldsymbol{SAS} = \boldsymbol{D}$.*

Bibliography

[1] D. H. Ackley. *A Connectionnist Machine for Genetic Hill-climbing.* Kluwer, Boston, 1987.

[2] L. Altenberg. Fitness distance correlation analysis: an instructive counterexample. In T. Bäck, editor, *Proceedings of the 7th International Conference on Genetic Algorithms*, pages 57–64. Morgan Kaufmann, San Francisco, 1997.

[3] T. Bäck, D. B. Fogel, and T. Michalewicz. *Evolutionary Computation 1: Basic Algorithms and Operators.* Institute of Physics Publishing, Boston and Philadelphia, 2000.

[4] T. Bäck, D. B. Fogel, and T. Michalewicz. *Evolutionary Computation 2: Advanced Algorithms and Operators.* Institute of Physics Publishing, Boston and Philadelphia, 2000.

[5] D. L. Battle and M. D. Vose. Isomorphisms of genetic algorithms. In G. J. E. Rawlins, editor, *Foundations of Genetic Algorithms*, pages 242–251. Morgan Kaufmann, San Francisco, 1991.

[6] K. Beauchamp. *Walsh Functions and their Applications.* Academic Press, London, 1975.

[7] A. Ben-Israel and T. N. E. Greville. *Generalized Inverses.* John Wiley, New York, 1971.

[8] A. D. Bethke. Genetic algorithms as function optimizers. PhD thesis, Dissertation Abstracts International, 41(9), University of Michigan Microfilms No.8106101, 1981.

[9] N. Bourbaki. *Eléments de Mathématique. Algèbre, chapitres 1 à 3.* Hermann, Paris, 1970.

[10] C. Darwin. *The origin of species.* John Murray, London, 1859.

[11] Y. Davidor. Epistasis variance: a viewpoint on representations, GA hardness and deception. *Complex Systems*, 4:369–383, 1990.

[12] Y. Davidor. Epistasis variance: a viewpoint on GA-hardness. In G. J. E. Rawlins, editor, *Foundations of Genetic Algorithms*, pages 23–35. Morgan Kaufmann, San Francisco, 1991.

[13] L. Davis. *Handbook of Genetic Algorithms.* Van Nostrand Reinhold, New York, 1991.

[14] K. Deb and D. E. Goldberg. Analyzing deception in trap functions. In L. D. Whitley, editor, *Foundations of Genetic Algorithms 2*, pages 93–108. Morgan Kaufmann, San Francisco, 1993.

[15] D. Dubois and H. Prade. *Fuzzy Sets and Systems: Theory and Applications.* Academic Press, New York, 1980.

[16] L. J. Eshelman and J. D. Schaffer. Crossover's niche. In S. Forrest, editor, *Proceedings of the 5th International Conference on Genetic Algorithms*, pages 9–14. Morgan Kaufmann, San Francisco, 1993.

[17] P. Field. *A Multary Theory for Genetic Algorithms: Unifying Binary and Nonbinary Problem Representations.* PhD thesis, University of London, London, 1996.

[18] S. Forrest and M. Mitchell. Relative building-block fitness and the building-block hypothesis. In L. D. Whitley, editor, *Foundations of Genetic Algorithms 2*, pages 109–126. Morgan Kaufmann, San Francisco, 1993.

[19] S. Forrest and M. Mitchell. What makes a problem hard for a genetic algorithm? Some anomalous results and their explanation. *Machine Learning,* 13:285–319, 1993.

[20] M. R. Garey and D. S. Johnson. *Computers and Intractability – A Guide to the Theory of NP-Completeness.* W. H. Freeman, San Francisco, 1979.

[21] J. Garnier and L. Kallel. How to detect all maxima of a function. In L. Kallel, B. Naudts, and A. Rogers, editors, *Theoretical Aspects of Evolutionary Computing,* Natural Computing, pages 343–370. Springer-Verlag, Berlin Heidelberg New York, 2000.

[22] H. Geiringer. On the probability theory of linkage in mendelian heredity. *Annals of Math. Stat.*, 15:25–57, 1944.

[23] D. E. Goldberg. Simple genetic algorithms and the minimal deceptive problem. In L. Davis, editor, *Genetic Algorithms and Simulated Annealing,* pages 74–88. Morgan Kaufmann Publishers, San Francisco, 1987.

[24] D. E. Goldberg. Genetic algorithms and Walsh functions: Part I: a gentle introduction. *Complex Systems,* 3:129–152, 1989.

[25] D. E. Goldberg. Genetic algorithms and Walsh functions: Part II: deception and its analysis. *Complex Systems,* 3:153–171, 1989.

[26] D. E. Goldberg. *Genetic Algorithms in Search, Optimization and Machine Learning.* Addison–Wesley, Reading, MA, 1989.

[27] D. E. Goldberg. *The Design of Innovation.* Kluwer Academic Publishers, Boston Dordrecht London, 2002.

[28] D. E. Goldberg, B. Korb, and K. Deb. Messy genetic algorithms: Motivations, analysis and first results. *Complex Systems,* 3:493–530, 1989.

[29] J. J. Grefenstette. Deception considered harmful. In L. D. Whitley, editor, *Foundations of Genetic Algorithms 2,* pages 75–92. Morgan Kaufmann, San Francisco, 1993.

[30] G. Harik, E. Cantú-Paz, D. E. Goldberg, and B. L. Miller. The gambler's ruin problem, genetic algorithms, and the sizing of populations. *Evolutionary Computation*, 7(3):231–253, 1999.

[31] R. Heckendorn. Polynomial time summary statistics for two general classes of functions. In D. Withley, et al., editors, *Proceedings of the Genetic and Evolutionary Computation Conference 2000*, pages 919–926. Morgan Kaufmann, San Francisco, 2000.

[32] R. Heckendorn, S. Rana, and D. Whitley. Polynomial time summary statistics for a generalization of MAXSAT. In W. Banzhaf et al., editors, *Proceedings of the Genetic and Evolutionary Computation Conference*, pages 281–288. Morgan Kaufmann, San Francisco, 1999.

[33] R. B. Heckendorn and D. Whitley. Predicting epistasis from mathematical models. *Evolutionary Computation*, 7(1):69–101, 1999.

[34] J. H. Holland. *Adaptation in Natural and Artificial Systems*, 2nd edition. MIT Press, Cambridge, MA, 1992.

[35] M. T. Iglesias, C. Vidal and A. Verschoren. A Global Approach to Schemata. In: *International Conference on Intelligent Technologies in Human-Related Sciences (ITHURS'96)*, vol I, pages 147–152. University of León Printing Service, Léon, 1996.

[36] M. T. Iglesias. *Algoritmos Genéticos Generalizados: Variaciones sobre un Tema.* PhD thesis, Servicio de Publicacións da Univerdidade da Coruña, Monografía No 62. University of La Coruña, 1997.

[37] M. T. Iglesias, C. Vidal, D. Suys and A. Verschoren. Multary Epistasis. *Bull. Soc. Math. Belg. Simon Stevin*, 8:1–21, 2001.

[38] M. T. Iglesias, C. Vidal, D. Suys and A. Verschoren. Epistasis and Unitation. *Computers and AI,* 18(5):467–483, 1999.

[39] M. T. Iglesias, C. Vidal and A. Verschoren. Computing Epistasis through Walsh Transforms. In *Quinto Encuentro de Álgebra Computacional y Aplicaciones (EACA'99),* pages 309-318. Tenerife, 1999.

[40] M. T. Iglesias, C. Vidal and A. Verschoren. Template Functions and their Epistasis. In *Proceedings of MS'2000 International Conference on Modelling and Simulation,* pages 539-546. Las Palmas de Gran Canaria, 2000.

[41] M. T. Iglesias, C. Vidal, D. Suys and A. Verschoren, *Generalized Walsh Transforms and Epistasis.* Submitted to Computers and Artificial Intelligence.

[42] M. T. Iglesias, C. Vidal and A. Verschoren. Computing Epistasis of Template Functions through Walsh Transforms. Submitted to Bull. Soc. Math. Belg. Simon Stevin.

[43] E. Ising. Beitrag zur Theorie des Ferromagnetismus. *Z. Physik*, 31:235, 1924.

[44] T. Jansen and I. Wegener. Real Royal Road functions – where crossover provably is essential. In Lee Spector et al., editor, *Proceedings of the Genetic and Evolutionary Computation Conference 2001*, pages 1034–1041. Morgan Kaufmann, San Francisco, 2001.

[45] T. Jones. *Evolutionary Algorithms, Fitness Landscapes and Search.* PhD thesis, The University of New Mexico, 1995.

[46] T. Jones and S. Forrest. Fitness distance correlation as a measure of problem difficulty for genetic algorithms. In L. J. Eshelman, editor, *Proceedings of the 6th International Conference on Genetic Algorithms*, pages 184–192. Morgan Kaufmann, San Francisco, 1995.

[47] T. Jones and S. Forrest. Fitness distance correlation as a measure of problem difficulty for genetic algorithms. In L. J. Eshelman, editor, *Proceedings of the 6th International Conference on Genetic Algorithms,* pages 184–192. Morgan Kaufmann, San Francisco, 1995.

[48] L. Kallel, B. Naudts, and M. Schoenauer. On functions with a fixed fitness–distance relation. In *Proceedings of the 1999 Congress on Evolutionary Computation*, volume 3, pages 1910–1916. IEEE Press, 1999.

[49] S. A. Kauffman. Adaptation on rugged fitness landscapes. In *Lectures in the Sciences of Complexity*, volume I of *SFI studies*, pages 619–712. Addison–Wesley, Reading, MA, 1989.

[50] S. Kirkpatrick, C. D. Gelatt and M. P. Vecchi. *Optimization by Simulated Annealing.* Science, 220:671–680, 1983.

[51] J. R. Koza, *Genetic Programming.* The MIT Press, Cambridge, MA, 1992.

[52] G. E. Liepins and M. D. Vose. Representational issues in genetic optimization. *J. Expt. Theor. Artif. Intell.*, 2:101–115, 1990.

[53] G. E. Liepins and M. D. Vose, Deceptiveness and genetic algorithm dynamics. In G. J. E. Rawlins, editor, *Foundations of Genetic Algorithms*, pages 36–52. Morgan Kaufmann, San Francisco, 1991.

[54] E. MacIntyre, P. Prosser, B. Smith, and T. Walsh. Random constraint satisfaction: theory meets practice. In *Proceedings of Fourth International Conference on Principles and Practice of Constraint Programming*, volume 1520 of *LNCS*, pages 325–339. Springer-Verlag, Berlin Heidelberg New York, 1998.

[55] B. Manderick, M. de Weger, and P. Spiessens. The genetic algorithm and the structure of the fitness landscape. In R. K. Belew and L. B. Booker, editors, *Proceedings of the 4th International Conference on Genetic Algorithms*, pages 143–150. Morgan Kaufmann, San Francisco, 1991.

[56] M. Manela and J. A. Campbell. Harmonic analysis, epistasis and genetic algorithms. In R. Manner and B. Manderick, editors, *Proceedings of the 2nd Conference on Parallel Problem Solving from Nature*, pages 57–64. North Holland, 1992.

[57] A. J. Mason, Partition coefficients, static deception and deceptive problems for non-binary alphabets. In R. K. Belew and L. B. Booker, editors, *Proceedings of the 4th International Conference on Genetic Algorithms,* pages 210–214. Morgan Kaufmann, San Francisco, 1991.

[58] N. Metropolis, A. Rosenbluth, M. Rosenbluth, A. Teller, and E. Teller. Equations of state calculations by fast computing machines. *Journal of Chemical Physics*, 21:1087–1091, 1953.

[59] Z. Michalewicz. *Genetic Algorithms+Data Structures=Evolution Programs,* 3rd edition. Springer-Verlag, Berlin Heidelberg New York, 1996.

[60] M. Mitchell, S. Forrest, and J. H. Holland. The Royal Road for genetic algorithms: Fitness landscapes and GA performance. In F. J. Varela and P. Bourgine, editors, *Proceedings of the First European Conference on Artificial Life-93*, pages 245–254. MIT Press/Bradford Books, Cambridge, MA, 1993.

[61] M. Mitchell and J. H. Holland. When will a genetic algorithm outperform hillclimbing? In S. Forrest, editor, *Proceedings of the 5th International Conference on Genetic Algorithms*, page 647. Morgan Kaufmann, San Francisco, 1993.

[62] M. Mitchell. *An introduction to genetic algorithms.* MIT Press, Cambridge, MA, 1996.

[63] H. Mühlenbein and T. Mahnig. FDA – a scalable evolutionary algorithm for the optimization of additively decomposed functions. *Evolutionary Computation*, 7(4):353–376, 1999.

[64] H. Mühlenbein and T. Mahnig. Evoluationary algorithms: from recombination to search distributions. In L. Kallel, B. Naudts, and A. Rogers, editors, *Theoretical Aspects of Evolutionary Computing,* Natural Computing, pages 135–173. Springer-Verlag, Berlin Heidelberg New York, 2000.

[65] H. Mühlenbein and D. Schlierkamp-Voosen. Analysis of selection, mutation and recombination in genetic algorithms. In W. Banzhaf and F. H. Eekman,

editors, *Evolution and biocomputation*, volume 899 of *LNCS*, pages 188–214. Springer-Verlag, Berlin Heidelberg New York, 1995.

[66] B. Naudts. *Measuring GA-Hardness.* PhD thesis, University of Antwerp, Belgium, 1998.

[67] B. Naudts and L. Kallel. A comparison of predictive measures of problem difficulty in evolutionary algorithms. *IEEE Transactions on Evolutionary Computing*, 4(1):1–16, 2000.

[68] B. Naudts and L. Schoofs. GA performance distributions and randomly generated binary constraint satisfaction problems. *J. Theoretical Computer Science*, 287(1):167–185, 2002.

[69] B. Naudts, D. Suys and A. Verschoren. Generalized Royal Road Functions and their Epistasis. *Computers and Artificial Intelligence,* 19:317–334, 2000.

[70] B. Naudts and J. Naudts. The effect of spin-flip symmetry on the performance of the simple GA. In A. E. Eiben et al., editors, *Proceedings of the 5th Conference on Parallel Problem Solving from Nature*, volume 1498 of *LNCS*, pages 67–76. Springer-Verlag, Berlin Heidelberg New York, 1998.

[71] C. H. Papadimitriou. *Computational Complexity.* Addison Wesley, Reading, MA, 1993.

[72] R. Penrose. A Generalized Inverse for Matrices. *Proc. Cambridge Philos. Soc.*, 51:406–413, 1955.

[73] A. Prügel-Bennett and A. Rogers. Modelling GA dynamics. In L. Kallel, B. Naudts, and A. Rogers, editors, *Theoretical Aspects of Evolutionary Computing*, Natural Computing, pages 59–86. Springer-Verlag, Berlin Heidelberg New York, 2000.

[74] L. M. Rattray and J. L. Shapiro. The dynamics of a genetic algorithm for a simple learning problem. *Journal of Physics A*, 29:7451–7473, 1996.

[75] G. J. E. Rawlins, editor. *Foundations of Genetic Algorithms.* Morgan Kaufmann, San Francisco, 1991.

[76] C. Reeves. Predictive Measures for Problem Difficulty. In *Proceedings of the 1999 Congress on Evolutionary Computation*, pages 736–743. IEEE Press, 1999.

[77] C. R. Reeves. Experiments with tuneable fitness landscapes. In M. Schoenauer et al., editors, *Proceedings of the 6th Conference on Parallel Problem Solving from Nature*, volume 1917, pages 139–148. Springer-Verlag, Berlin Heidelberg New York, 2000.

[78] C. R. Reeves. Direct statistical estimation of GA landscape properties. In W. Martin and W. Spears, editors, *Foundations of Genetic Algorithms 6*, pages 91–107. Morgan Kaufmann, San Francisco, 2001.

[79] C. Reeves and C. Wright. An experimental design perspective on genetic algorithms. In L. D. Whitley and M. D. Vose, editors, *Foundations of Genetic Algorithms 3*, pages 7–22. Morgan kaufmann, San Francisco, 1995.

[80] S. Rochet, G. Venturini, M. Slimane and E. E. Kharoubi. A critical and empirical study of epistasis measures for predicting GA performances: a summary. In J. K. Hao et al., editors, *Artificial Evolution 97*, volume 1363 of *LNCS*, pages 275–286. Springer-Verlag, Berlin Heidelberg New York, 1998.

[81] A. Rogers and A. Prügel-Bennett. The dynamics of a genetic algorithm on a model hard optimization problem. *Complex Systems*, 11(6):437–64, 2000.

[82] G. Rudolph. Convergence analysis of the canonical genetic algorithm. *IEEE Transactions on Neural Networks*, 5(1):96–101, 1994.

[83] I. Satake. *Linear Algebra.* Marcel Dekker, New York, 1975.

[84] L. Schoofs. *An empirical study of heuristically solving constraint satisfaction problems with stochastic evolutionary algorithms.* PhD thesis, Department of Mathematics and Computer Sciece, University of Antwerp, Belgium, 2002.

[85] J. L. Shapiro. Statistical mechanics theory of genetic algorithms. In L. Kallel, B. Naudts, and A. Rogers, editors, *Theoretical Aspects of Evolutionary Computing*, Natural Computing, pages 87–108. Springer-Verlag, Berlin Heidelberg New York, 2000.

[86] J. L. Shapiro and A. Prügel-Bennett. Genetic algorithm dynamics in a two-well potential. In R. K. Belew and M. D. Vose, editors, *Foundations of Genetic Algorithms 4*, pages 101–116. Morgan Kaufmann, San Francisco, 1997.

[87] D. Sherrington and S. Kirkpatrick. Solvable model of a spin-glass. *Phys. Rev. Lett.*, 35:1792–1796, 1975.

[88] P. F. Stadler. Landscapes and their correlation functions. *J. Math. Chem.*, 20:1–45, 1996.

[89] C. R. Stephens. Effect of mutation and recombination on the genotype-phenotype map. In W. Banzhaf et al., editors, *Proceedings of the Genetic and Evolutionary Computation Conference*, pages 1382–1389. Morgan Kaufmann, San Francisco, 1999.

[90] C. R. Stephens. Some exact results from a coarse grained formulation of genetic dynamics. In Lee Spector et al., editors, *Proceedings of the Genetic and Evolutionary Computation Conference 2001*, pages 631–638. Morgan Kaufmann, San Francisco, 2001.

[91] C. R. Stephens and H. Waelbroeck. Schemata evolution and building blocks. *Evolutionary Computing*, 7(2):109–124, 1999.

[92] M. M. Stickberger. *Genetics.* Collier-MacMillan, London, 1968

[93] G. Strang. *Linear Algebra and its Applications*, 3rd edition. Saunders, Philadelphia, 1988.

[94] D. Suys and A. Verschoren. Extreme Epistasis. In *International Conference on Intelligent Technologies in Human-Related Sciences (ITHURS'96),* vol II, pages 251–258. University of León Publishing Service, Léon, 1996.

[95] D. Suys. *A Mathematical Approach to Epistasis.* PhD thesis, Department of Mathematics and Computer Sciece, University of Antwerp, Belgium, 1998.

[96] E. Tsang. *Foundations of Constraint Satisfaction.* Academic Press, New York, 1993.

[97] H. Van Hove. *Representational Issues in Genetic Algorithms.* PhD thesis, Department of Mathematics and Computer Sciece, University of Antwerp, Belgium, 1995.

[98] H. Van Hove and A. Verschoren. On Epistasis. *Computers and Artificial Intelligence,* 14(3):271–277, 1994.

[99] H. Van Hove and A. Verschoren. What is Epistasis? *Bull. Soc. Math. Belg. Simon Stevin,* 5:69–77, 1998.

[100] C. Van Hoyweghen, B. Naudts and D. E. Goldberg. Spin-flip symmetry and synchronization. *Evolutionary Computation,* 10(4):317–344, 2002.

[101] P. J. van Laarhoven and E. H. L. Aarts. *Simulated Annealing: Theory and Applications.* Kluwer Academic Press, Dordrecht, The Netherlands, 1987.

[102] J. H. Van Lint. *Introduction to Coding Theory.* Springer-Verlag, Berlin Heidelberg New York, 1982.

[103] A. Verschoren and H. Van Hove. A Fuzzy Schema Theorem. *Fuzzy sets and systems,* 94(1):93–99, 1998.

[104] M. D. Vose. Generalizing the notion of schema in genetic algorithms. *Artificial Intelligence,* 50:385–396, 1991.

[105] M. D. Vose. *The Simple Genetic Algorithm: Foundations and Theory.* Complex Adaptive Systems, MIT Press/Bradford Books, Cambridge, MA, 1998.

[106] G. R. Walsh. *Methods of Optimization.* John Wiley, New York, 1975.

[107] R. Watson, G. S. Hornby, and J. B. Pollack. Modeling building block interdependency. In A. E. Eiben et al., editors, *Proceedings of the 5th Conference on Parallel Problem Solving from Nature*, pages 97–106. Springer-Verlag, Berlin Heidelberg New York, 1998.

[108] L. D. Whitley. Fundamental principles of deception in genetic search. In G. J. E. Rawlins, editor, *Foundations of Genetic Algorithms*, pages 221–241. Morgan Kaufmann, San Francisco, 1991.

[109] C. Williams and T. Hogg. Exploiting the deep structure of constraint problems. *Artificial Intelligence*, 70:73–117, 1994.

[110] A. H. Wright, J. E. Rowe, R. Poli, and C. R. Stephens. A fixed-point analysis of a gene-pool GA with mutation. In W.B. Langdon et al., editors, *Proceedings of the Genetic and Evolutionary Computation Conference 2002*, pages 642–649. Morgan Kaufmann, San Francisco, 2002.

Index

MATHEMATICAL MODELLING:
Theory and Applications

---

1. M. Křížek and P. Neittaanmäki: *Mathematical and Numerical Modelling in Electrical Engineering.* Theory and Applications. 1996
ISBN 0-7923-4249-6
2. M.A. van Wyk and W.-H. Steeb: *Chaos in Electronics.* 1997
ISBN 0-7923-4576-2
3. A. Halanay and J. Samuel: *Differential Equations, Discrete Systems and Control.* Economic Models. 1997 ISBN 0-7923-4675-0
4. N. Meskens and M. Roubens (eds.): *Advances in Decision Analysis.* 1999
ISBN 0-7923-5563-6
5. R.J.M.M. Does, K.C.B. Roes and A. Trip: *Statistical Process Control in Industry.* Implementation and Assurance of SPC. 1999
ISBN 0-7923-5570-9
6. J. Caldwell and Y.M. Ram: *Mathematical Modelling.* Concepts and Case Studies. 1999 ISBN 0-7923-5820-1
7. 1. R. Haber and L. Keviczky: *Nonlinear System Identification - Input-Output Modeling Approach.* Volume 1: Nonlinear System Parameter Identification. 1999 ISBN 0-7923-5856-2; ISBN 0-7923-5858-9 Set
   2. R. Haber and L. Keviczky: *Nonlinear System Identification - Input-Output Modeling Approach.* Volume 2: Nonlinear System Structure Identification. 1999 ISBN 0-7923-5857-0; ISBN 0-7923-5858-9 Set
8. M.C. Bustos, F. Concha, R. Bürger and E.M. Tory: *Sedimentation and Thickening.* Phenomenological Foundation and Mathematical Theory. 1999
ISBN 0-7923-5960-7
9. A.P. Wierzbicki, M. Makowski and J. Wessels (eds.): *Model-Based Decision Support Methodology with Environmental Applications.* 2000
ISBN 0-7923-6327-2
10. C. Rocşoreanu, A. Georgescu and N. Giurgiţeanu: *The FitzHugh-Nagumo Model.* Bifurcation and Dynamics. 2000 ISBN 0-7923-6427-9
11. S. Aniţa: *Analysis and Control of Age-Dependent Population Dynamics.* 2000
ISBN 0-7923-6639-5

MATHEMATICAL MODELLING: Theory and Applications

12. S. Dominich: *Mathematical Foundations of Informal Retrieval.* 2001 ISBN 0-7923-6861-4

13. H.A.K. Mastebroek and J.E. Vos (eds.): *Plausible Neural Networks for Biological Modelling.* 2001 ISBN 0-7923-7192-5

14. A.K. Gupta and T. Varga: *An Introduction to Actuarial Mathematics.* 2002 ISBN 1-4020-0460-5

15. H. Sedaghat: *Nonlinear Difference Equations.* Theory with Applications to Social Science Models. 2003 ISBN 1-4020-1116-4

16. A. Slavova: *Cellular Neural Networks: Dynamics and Modelling.* 2003 ISBN 1-4020-1192-X

17. J.L. Bueso, J.Gómez-Torrecillas and A. Verschoren: *Algorithmic Methods in Non-Commutative Algebra.* Applications to Quantum Groups. 2003 ISBN 1-4020-1402-3

18. A. Swishchuk and J. Wu: *Evolution of Biological Systems in Random Media: Limit Theorems and Stability.* 2003 ISBN 1-4020-1554-2

19. K. van Montfort, J. Oud and A. Satorra (eds.): *Recent Developments on Structural Equation Models.* Theory and Applications. 2004 ISBN 1-4020-1957-2

20. M. Iglesias, B. Naudts, A. Verschoren and C. Vidal: *Foundations of Generic Optimization.* Volume 1: A Combinatorial Approach to Epistasis. 2005 ISBN 1-4020-3666-3